INTRODUCTION TO LIQUID STATE PHYSICS

INTRODUCTION TO LIQUID STATE PHYSICS

N H March
Oxford University, UK

M P Tosi
Scuola Normale Superiore, Italy

Published by

World Scientific Publishing Co. Pte. Ltd.

P O Box 128, Farrer Road, Singapore 912805

USA office: Suite 1B, 1060 Main Street, River Edge, NJ 07661

UK office: 57 Shelton Street, Covent Garden, London WC2H 9HE

British Library Cataloguing-in-Publication Data
A catalogue record for this book is available from the British Library.

INTRODUCTION TO LIQUID STATE PHYSICS

ISBN 981-02-4639-0
ISBN 981-02-4652-8 (pbk)

Printed in Singapore by World Scientific Printers (S) Pte Ltd

Preface

Some time ago, the authors collaborated on a book entitled "Atomic Dynamics in Liquids", which has subsequently been reprinted by Dover. This book, it is fair to say, was motivated by advanced lecture courses the two authors had presented at a variety of venues, notable among these being the Abdus Salam International Centre for Theoretical Physics in Trieste.

Subsequently, because of our mutual interests in charged fluids, we followed up the above Volume (Dover, 1991) with "Coulomb Liquids". This naturally had a narrower range of coverage: dominantly classical ionic melts and liquid metals, where the valence electrons are fully quantal.

Trends in the subject of the "Liquid State" since these two books were published have impressed us to the extent that both of us judged that the time was ripe for a more general book on this area. Thus, the subject has become important for workers in a wide range of disciplines. Two that came to our minds were the glassy (amorphous) state, concerning materials which have large technological relevance, and the need to understand, fully quantitatively eventually, the phenomena of turbulence.

It is also true that the massive increase in computational power available now in science and technology has had a big impact on the development of the ability to calculate a whole variety of liquid state properties. However, we were also conscious that the output of massive computer simulations is often analogous to, say, that of radiation scattering experiments. No one doubts, in the latter area, the importance of attempting to interrelate many experimental facts by some simple theoretical ideas. In turn, such ideas often spring from somewhat oversimplified models (e.g. hard spheres or one-component plasma models, depending on whether the force law between building blocks has a very well defined region of excluded volume, or whether the forces are long-range).

This is the background against which the present Volume has come into being. It has, for reasons expounded above, much greater breadth than our two earlier Volumes. We feel as an Introduction to the liquid state, including importantly a variety of chemically bonded liquids, it should be useful to students, both at advanced undergraduate and research levels, in a variety of disciplines: physicists, chemists, chemical and mechanical engineers, and also to workers in the important interfaces between chemistry, biology and medicine, as well as in environmental technology.

Because of breadth, we have also covered areas in which our own personal contributions have been minimal (unlike our earlier Volumes). Therefore our indebtedness to other workers is greater in the present case. Especially, we acknowledge that we have at times drawn heavily on existing books, e.g. Faber's on hydrodynamics and turbulence, and various accounts on polymers and liquid crystals. We trust, without being more specific (and inevitably then more tedious) such workers will accept our grateful thanks.

Should our book prove useful, we hope that readers who feel that there are places where we should do better will write to us and we shall do our utmost to respond constructively in the future.

Finally, we thank staff of World Scientific for their patience and understanding.

Oxford and Pisa
N. H. March
M. P. Tosi

Contents

Chapter 1

Qualitative Description of Liquid Properties

Everyday experience testifies to the classification of three different phases of matter: solid, liquid and gas. Solids are rigid, and when studied by X-ray diffraction give rise to sharp Bragg reflections. This is the hall-mark of crystallinity: an ordered array of the building blocks, be they atoms or groups of atoms (e.g. C_{60} in crystalline fullerites). In such a state of long-range order the neighbours of every molecule are arranged in a regular pattern so that, even when two molecules are separated by tens of intermediate ones, their distance apart is fixed.

A very characteristic property of liquids and gases is that they flow under a shear stress, however small that may be. As will be discussed in Chap. 4, when X-rays are scattered from a dense liquid, such as argon near its triple point, there are no longer sharp Bragg reflections, demonstrating that there is no long-range order among the atoms or molecules.

Though this distinction between crystals and fluids is clear, glasses and amorphous solids yield quite similar X-ray diffraction patterns to those of dense liquids. These solid states may be viewed as liquids with their disorder frozen in. We shall address the (finer) distinctions between liquid and glassy states in Chap. 10.

Though we mentioned above the three phases of matter — solid, liquid and vapour, the qualitative distinction drawn from diffraction experiments between crystal and liquid does not carry through to distinguish liquid and vapour. We elaborate on this below, but historically it should be noted that van der Waals was well aware of the continuity of liquid and gaseous phases. Below a critical temperature, two fluid phases are able to coexist in equilibrium. The denser phase is liquid and the less dense is vapour. Above the critical temperature coexistence of the two fluid phases is no longer observed, but a single fluid phase exists. Thus it is possible to pass continuously from vapour at low temperature to liquid at low temperature by (i) heating above the critical temperature and (ii) compressing and cooling. The difference between liquid and vapour phases is then essentially a difference of density: this means that intermolecular interactions play a much greater role in the denser liquid phase than in the dilute vapour.

Reference was made already to liquid argon. Rare gas atoms, including argon, are spherical and chemically saturated, and in a fluid assembly of such atoms the only disorder that is possible is that connected with translational motions. When the building blocks of the condensed phase are complex molecular units which are not spherical, there is the possibility of rotational disorder. This may, indeed, already occur in a crystal: the translational order being retained while rotational disorder exists — this is the case of plastic crystals. But in other systems rotational order can persist over a temperature range after the translational order is lost. One example then is that of liquid crystals: this area will be covered together with that of polymeric fluids, both classes of materials being very important for technology.

Before going on to discuss other challenges that understanding the physics of the liquid state poses, let us elaborate on the three phases available to matter: solid, liquid and gas.

1.1 Three Phases of Matter: pVT Behaviour of Pure Materials

Figure 1.1, taken from Blinder[1] illustrates a section of the surface representing the equation of state

$$F(p, V, T) = 0\,, \tag{1.1}$$

for a fixed mass of a chosen substance in the space defined by the variables pressure p, volume V and temperature T. This plot actually embodies, in a qualitative way, a whole body of empirically observed facts on the thermal and mechanical properties of matter.

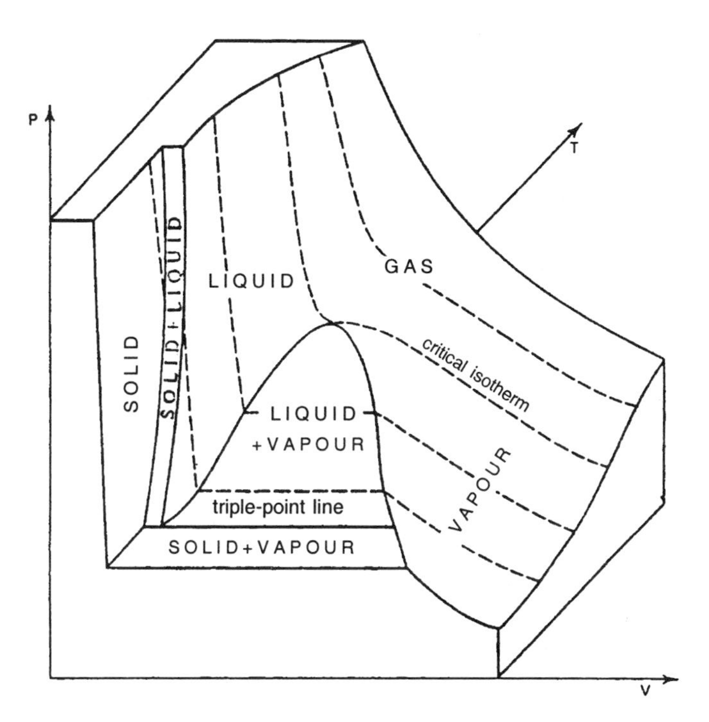

Fig. 1.1. A section of the surface in the (p, V, T) space representing the equation of state for a pure substance, with isotherms shown as dashed curves. The regions are named from the physical state of the material. (Schematic; redrawn from Blinder, Ref. 1.)

The surface shown in Fig. 1.1 is divided into various regions separated by solid lines and classified by the state of matter there. Within the regimes labelled "solid" and "liquid-gas-vapour", the system exists as a single homogeneous phase. In the other three regions, there co-exist two distinct phases in equilibrium. When one of the solid lines in Fig. 1.1 is crossed in some physicochemical process, a phase transition takes place. For instance, on crossing the boundaries which delimit the solid phase one enters into the regimes of solid–liquid coexistence (above the triple-point line) or of solid–vapour coexistence (below the triple-point line). After further boundary crossings the system ends into the homogeneous liquid or vapour phases.

In such different phases the substance under consideration has distinctly different properties: discontinuous changes in physical properties occur across a phase transition.

1.1.1 *Critical isotherm*

The dashed curves in Fig. 1.1 represent isotherms, i.e. lines of constant temperature. Geometrically, these curves are the intersections of the p, V, T surface with planes corresponding to T = constant. The tangent to the liquid + vapour regime is termed the critical isotherm. The critical point is the point of tangency, the uppermost point of the liquid + vapour region.

As a matter of terminology, the region above the critical isotherm is marked as "gas", whereas the contiguous regimes below the isotherm are denoted by "liquid" and "vapour", on either side of the two-phase region. There is, it must be stressed, no physical discontinuity observed in going across the critical isotherm: say from liquid to gas or from gas to vapour.

1.1.2 *Triple point*

The boundary marked triple-point line in Fig. 1.1 derives from the fact that the three states of matter can coexist in equilibrium at the so-called "triple point". For instance, the triple point of the ice-water-aqueous vapour system lies at $T = 0°\text{C}$ (or 273 K) and 0.006 bar (1 bar $= 10^5$ Pa). By comparison, the critical point of the water-vapour coexistence curve lies at $T = 374°\text{C}$ and $p = 221$ bar. The densities of the two phases differ by a factor of about 2×10^5 at the triple point, whereas at the critical point they are equal and correspond to a specific volume of about 5.7 times the intrinsic volume of an H_2O molecule.

In fact, it is seen from Fig. 1.1 that, inside the boundaries of the liquid-vapour, solid-vapour and solid–liquid sections of the pVT surface, the isotherms are parallel to the axis labelled V. This implies that the volume V is not a single-valued function of pressure p and temperature T within the two-phase regions. In these regions, the further variable determining the volume is the relative proportion of the two phases.

For instance, in the liquid-vapour coexistence region each value V of the volume (per mole, say) is given by

$$V = c_l V_l + c_v V_v \,, \tag{1.2}$$

where V_l and V_v are the molar volumes of the two phases and c_l and c_v are their concentrations (with $c_l + c_v = 1$). In this situation liquid droplets are formed in which the molecules pack more tightly, while the remainder of the molecules are dispersed thus keeping the pressure constant.

1.1.3 *Phase diagram of a pure material (e.g. argon)*

A schematic projection of the whole pVT surface on to the (p, T) plane is shown in Fig. 1.2. This figure essentially summarises the phase diagram of a pure material. The two-phase regions, where the isotherms as discussed above are parallel to the V axis, project on to curves in the (p, T) plane. Specifically, the liquid-vapour region projects on to the vaporisation curve, the solid-vapour region on to the sublimation curve and the solid–liquid region on to the curve labelled fusion. In the (p, T) plane, the triple-point line projects on to a single point.

The vaporisation curve in Fig. 1.2 depicts the dependence on temperature of the saturated vapour pressure, which is the pressure of the vapour in equilibrium with the liquid at a given temperature. Equilibrium between liquid and vapour implies that in a given period of time the number of molecules which from the vapour would be condensing on the liquid surface after hitting it balances the number of molecules escaping from the liquid into the vapour. An increase in vapour pressure increases the frequency of collisions with the liquid surface and favours condensation against evaporation.

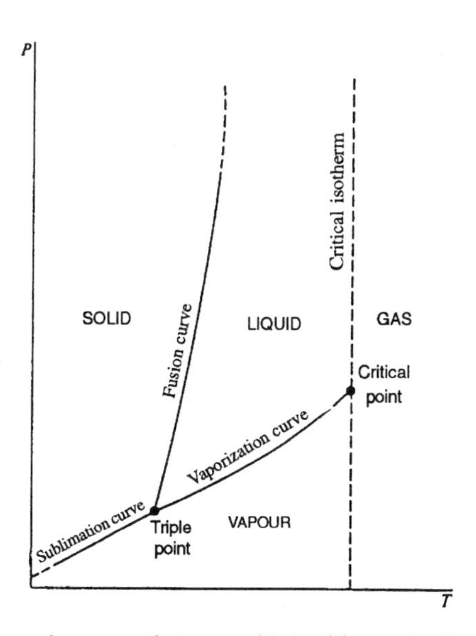

Fig. 1.2. Phase diagram of a pure substance, obtained by projecting the pVT surface onto the (p, T) plane. (Schematic; redrawn fram Blinder, Ref. 1.)

In analogous fashion, the sublimation curve gives the equilibrium vapour pressure above the solid. The sublimation curve terminates at the triple point and the vaporisation curve ends in the critical point. The slopes of both curves are always positive, since vapour pressure increases with temperature. Evaporation absorbs energy but the system as a whole is more disordered when molecules are randomly dispersed: these two effects balance each other at equilibrium, as on the two curves representing the saturated vapour pressure against temperature.

The slope of the fusion curve is normally positive. Increased pressure favours the solid, which is, in most cases, denser than the liquid. A quite crucial exception here is water, for which the fusion curve has a negative slope, ice being less dense than water near freezing. The functional forms of three-phase equilibrium lines can be obtained by thermodynamic arguments, resulting in the Clapeyron and Clausius–Clapeyron equations (see Chap. 3).

1.1.4 *Phase change from gas to liquid*

Consider now compressing water vapour, whose properties can be represented as a function of its thermodynamic state in the (p, T) plane. When the temperature exceeds the critical value $T_c \approx 647$ K, the effect of increasing the pressure leads to no sharp transformation. Provided the final pressure of the compressed vapour exceeds the critical value $p_c \approx 221$ bar, then we can verify that, on decreasing the temperature at constant pressure, it is possible to pass in a continuous fashion from the gaseous to the liquid state. Releasing the pressure will now bring us back to the vaporisation curve from above.

On the contrary, below the critical temperature T_c as we increase the pressure there is a definite discontinuity on reaching the vaporisation curve from below. Here the molecules do not become closer in a continuous way if the volume is made smaller, but as already described small isolated droplets are formed inside which the molecules pack tightly. These droplets grow until the state of the homogeneous liquid is reached. The abrupt change in volume that occurs at condensation is substantial at lower pressures but decreases as the pressure approaches the critical pressure p_c. At the critical point the densities of the liquid and the gas become equal.

In fact, because of fluctuations small droplets of higher-than-average density are already continuously forming and dissolving in the gas phase. If the critical point is approached from the gas phase by lowering the temperature at constant volume, such density fluctuations will continue to grow in size and live longer: in this situation the gas will start scattering light strongly as the size of the fluctuating regions becomes comparable to or larger than the light wavelength. This phenomenon is known as critical opalescence (see Chap. 3): the gas near criticality comes to look "milky" in visible light (milk being a suspension of droplets of fat whose diameter is of order 1 μm, comparable to the wavelength of sunlight at about 0.5 μm).

In the two-phase (liquid + vapour) region, on the other hand, liquid droplets will be larger than in the gas away from criticality and will persist for longer times. Some of them, most often nucleated on dust particles, grow to be very large by absorbing other droplets and further vapour molecules and become visible by scattering light of wavelength comparable to their size. This is not critical opalescence, but the effect is similar. As an example, clouds in the sky are made from water droplets or ice particles with a characteristic size in the range from 1 to 10 μm and are seen through their scattering of sunlight.

The conclusion is that the liquid, since it can be reached continuously from the gas where on average the molecules are homogeneously distributed in a random fashion, must have a disordered structure. The difference between the gas and the liquid arises from the change in density — passing from low to high density. However, as we shall see in Chap. 4 the liquid acquires short-range order from its high density inducing tight molecular packing.

1.1.5 *A liquid open to the atmosphere*

A liquid exposed to the open air must also be in equilibrium with its vapour at the value of the saturated vapour pressure corresponding to the prevailing temperature. The atmosphere does not affect the equilibrium state and, if the molecules escaping from the liquid surface can freely migrate away, the liquid will in time evaporate completely. Evaporation is delayed if the diffusive motions of the escaping molecules are in some way slowed down, so that a layer enriched in vapour content may form above the liquid surface. A layer of water in an open vessel with a depth of a few centimeters will be stable for a few hours, thanks to the humidity (i.e. water vapour) which is already present in air.

Boiling occurs when the saturated vapour pressure becomes equal to the atmospheric pressure above the liquid. Take again water exposed to air as an example: the total pressure is $P = 1$ atm and the partial pressure of air on the water surface is $P - p$, where p is the vapour pressure. Therefore, the system can be in equilibrium only as long as the water vapour pressure is below 1 atm, this value being attained at 100°C. When water is heated in open air, its temperature increases continuously till it reaches 100°C and remains constant thereafter while the water boils away into vapour.

1.2 Melting and Lindemann's Law

The phase diagram illustrated in Fig. 1.2 shows the crystalline solid as coexisting with the vapour on the sublimation curve and with the liquid on the fusion curve. Across these lines there is a fundamental discontinuity, since they separate the ordered state from the disordered states. It is impossible to go continuously through this essential structural change.

Of course, a static model for crystalline order is not completely correct. The atoms are always in thermal motion and in a crystal each of them is executing small-amplitude vibrations around a fixed point belonging to an ordered lattice. It is the average atomic positions which are ordered over long distances.

From this viewpoint, it is a very remarkable aspect of fusion that the melting point is so sharp for a crystalline solid. The cohesive forces maintain crystalline order in spite of the atomic vibrations, up to a temperature where the amplitude of the vibrations becomes so large that the solid melts. In 1910 Lindemann proposed that one could estimate the melting temperature by assuming that melting occurred when the amplitude of vibrational motions in the hot crystal exceeds a "critical" fraction of the atomic spacing.

Let x_c be that vibration amplitude which leads to melting in the above viewpoint. If f denotes the force constant between one atom and its neighbour, then for harmonic motion the mean total energy during vibration will be $\langle E \rangle = \frac{1}{2} f x_c^2$. Since melting points are generally high, we assume use of the principle of equipartition of energy and equate $\langle E \rangle$ to $k_B T$ (a linear harmonic oscillator having two degrees of freedom, each contributing $\frac{1}{2} k_B T$). Hence

$$\langle E \rangle = k_B T_m = \frac{1}{2} f x_c^2 , \tag{1.3}$$

where T_m is the melting temperature.

But for a simple cubic structure, $f = Ya$ where a is the atomic spacing and Y the Young modulus of the solid. Hence it follows that $k_B T_m = Y a x_c^2 / 2$.

Since Lindemann's assumption was that melting occurs when x_c is some fixed fraction of the atomic spacing a, write $x_c = \beta a$ where $0 < \beta < 1$. Then $T_m = Y\beta^2 a^3/2k_B$ and assuming cubic structure ($a^3 = \Omega = M/\rho$, where Ω is the molecular volume, ρ is the density of the solid and M is the molar mass) we get

$$T_m = \left(\frac{YM}{2\rho R}\right)\beta^2 , \tag{1.4}$$

with $R = 8.31$ J/(mol K). Assuming $\beta^2 = 1/50$, Table 1.1 shows the melting point T_m (K) for some solids.

Table 1.1. Melting point from Lindemann criterion.

Solid	$Y \times 10^{-10}$ $(\mathrm{N \cdot m^{-2}})$	$\rho \times 10^{-3}$ $(\mathrm{kg \cdot m^{-3}})$	M (kg)	T_m (K) criterion	observed
Lead	1.6	11.3	0.207	400	600
Silver	8.3	10.5	0.108	1100	1270
Iron	21.2	7.9	0.056	1800	1800
Tungsten	36.0	19.3	0.184	4200	3650
Sodium chloride	4.0	2.16	0.057	1200	1070
Quartz	7.0	2.6	0.060	1900	2000

1.3 Molecular Thermal Movements in the Liquid Phase: Brownian Motion

As a result of thermal agitation, no static molecular model is adequate to describe a dense liquid. However, it is plain that since in a dense liquid the molecules are packed rather closely together, their motions cannot be as free as they are in a dilute gas of the same substance.

Thermal motions are indeed easy to visualise in a dilute gas. The atoms fly around with a kinetic energy which on average is proportional to temperature, with speeds of the order of hundreds of meters per second at ordinary temperatures. During these motions the atoms collide with each other and with the walls of the container, exerting a pressure. A collision deflects a colliding atom from a straight path, so that the trajectory becomes a complicated pattern of broken lines. Such zig-zag trajectories result in a random distribution of the atomic positions.

In a tightly packed liquid, on the other hand, many of the atoms are so confined by their neighbours that each of them can only vibrate as if inside a cage. Almost as soon as the atom moves away from the centre of the cage, collisions from its neighbours reverse its velocity and send it back. Motions of this sort have frequencies of order 10^{12}–10^{13} Hz, similar to those of the vibrational motions in a solid or of the internal vibrations in a molecule, and may last on average for time intervals of the order of a few picoseconds before being damped out.

However, the "cage" is not a rigid one but is made of other atoms, which are going through their own thermal motions. If it so happens that its neighbours move in some appropriate concerted way, the "central atom" may succeed in exiting from the cage and start on a diffusive type of motion which will ultimately bring it far away from its initial position. A picture emerges in which each atom in the dense liquid is hopping along a zig-zag trajectory made of discrete microscopic jumps interspersed with oscillations in discrete sites of residence. The hopping frequencies may be of the order of 10^{10}–10^{12} Hz.

Macroscopic evidence for diffusive motions in a liquid is provided by the fact that two miscible liquids will slowly mix together even if no stirring is done. More directly, the diffusive motions of finely divided particles suspended in a liquid can be observed under an optical microscope, as was first done in 1827 by the English botanist R. Brown. In this so-called "Brownian motion" particles with typical dimensions of 1 μm or less are seen to dance around in an

Fig. 1.3. Schematic plane projection of Brownian motion for three particles suspended in a liquid.

irregular manner under the effect of random collisions by the molecules of the medium. Again a zig-zag trajectory is observed for the suspended mesoscopic particle (see Fig. 1.3).

In trying to visualize how such a trajectory arises, one should recall an argument first given by Stokes: a macroscopic particle floating in a fluid and experiencing collisions from its molecules only feels viscous friction against its direction of motion, since the collisions compensate each other in the other directions. The particle must be large compared with the mean free molecular path, so that it feels the buffeting by the molecules as if the fluid were a continuum. As the size of the diffusing particle decreases, the probability increases for an unbalanced collisional event which may deflect it into a particular direction. Further collisions produce viscous friction as the particle proceeds in that direction towards the next deflection. This picture may be brought down to the microscopic level to describe the hopping motion of an atom in a liquid under the effect of collisions against its partner atoms, over time scales longer than that of the localised vibrations inside the cage of first neighbours.

In 1908 Langevin described the diffusive motions of a mesoscopic particle in a fluid by partitioning the forces that it feels into the sum of a viscous force and of a random collisional force. This yields the equation of motion

$$m\ddot{x} = -f\dot{x} + F_{\mathrm{r}}\,, \tag{1.5}$$

where m is the particle mass, x is the component of its displacement in any one of the three spatial directions, f is a friction coefficient and F_{r} is the random force. We multiply Eq. (1.5) by x and average over a large number of collisions. This averages away the random collisional term, since positive and negative values of F_{r} are equally probable in the long run. We get

$$m\langle \ddot{x}x\rangle = -f\langle \dot{x}x\rangle\,, \tag{1.6}$$

where the brackets denote the average.

We assume that the mean square kinetic energy of the particle equals the thermal energy: $\frac{1}{2}m\langle \dot{x}^2\rangle = \frac{1}{2}k_{\mathrm{B}}T$ (here $k_{\mathrm{B}} = 1.38 \times 10^{-23}$ J/K is Boltzmann's constant). From the identity $\ddot{x}x = (d(\dot{x}x)/dt) - \dot{x}^2$ we have $md\langle \dot{x}x\rangle/dt + f\langle \dot{x}x\rangle = k_{\mathrm{B}}T$. This integrates to

$$\langle \dot{x}x\rangle = \left(\frac{k_{\mathrm{B}}T}{f}\right) + Ae^{-ft/m}\,, \tag{1.7}$$

A being an integration constant. The exponential term decays rapidly to zero, so that over long time intervals we have $\langle \dot{x}x \rangle = k_B T/f$ or equivalently a mean square displacement of the diffusing particle which increases linearly with time:

$$\lim_{t\to\infty} \langle x^2 \rangle = \left(\frac{2k_B T}{f} \right) t\,. \tag{1.8}$$

This behaviour is characteristic of diffusive motions and indeed Eq. (1.8) can be used to relate the diffusion coefficient D to the viscous coefficient f as

$$D = \frac{k_B T}{f}\,. \tag{1.9}$$

Diffusion in liquids will be the subject of Chap. 5 in this book. We shall see there that Eq. (1.9) yields at once the Stokes–Einstein relation between diffusion coefficient, particle size and shear viscosity, the latter transport coefficient being also determined by collisions. Here we add the remark that, while Brownian motion results from spontaneous fluctuations, the atomic diffusion coefficient D that these determine can be shown to be proportional to the particle mobility μ, according to the Nernst–Einstein relation

$$D = k_B T \mu \tag{1.10}$$

(see Chap. 5). The mobility μ is accessible to measurements of driven transport under a constant external field, the most common case being that of charged particles in an electric field.

1.4 Qualitative Considerations Continued: Flow Properties of Dense Liquids

We have already emphasised that, while solids exhibit resistance to shear, a liquid flows under an arbitrarily small shear stress. This may be termed a "collective" or "cooperative" property on the macroscopic scale. A sophisticated question is what distinguishes between equilibrium and non-equilibrium liquids.[2] Let us begin to discuss the flow properties of liquids at an elementary, though basic level (cf. the book by Tabor[3]; see also the book by Faber[4]).

1.4.1 *Ideal liquids and Bernoulli's equation*

As with gases, one can postulate the existence of ideal liquids in which internal forces play an unimportant part: leading to neglect, at first of surface tension and viscosity. Naturally no real liquids can have such properties: they would not be liquids if that were the case. But if the inertial forces dominate, they can often behave as ideal liquids in flow. This is the circumstance in which one is led to an important flow equation, due to Bernoulli.

Let us consider then the flow of such an idealized liquid and try to follow the trajectory of some particle in it. If the liquid moves in a continuous steady state, one can draw a line such that the tangent at any point gives the direction of flow of the particle. Such lines are termed streamlines. These are smooth continuous lines throughout the fluid and cannot intersect. No liquid particles can flow across from one streamline to another.

Consider Fig. 1.4, in which AB represents an imaginary tube in the liquid bounded by streamlines. At A the liquid is at a height h_1 with a flow velocity v_1, the pressure is p_1 and the cross-sectional area of the tube is a_1. At B the corresponding quantities are h_2, v_2, p_2 and a_2. Consider now the energy balance during a short time interval dt. The pressure drives through the tube a volume $dV = a_1 v_1 dt$ of liquid at A and an equal amount $dV = a_2 v_2 dt$ of liquid leaves at B. The sum of the pressure work $p_1 dV$, the kinetic energy $\frac{1}{2}\rho v_1^2 dV$

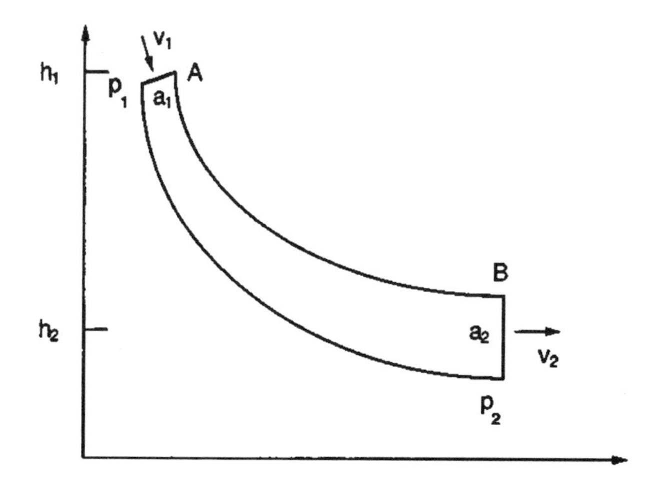

Fig. 1.4. Streamlines of flow in a gravitational field.

and the gravitational potential energy $\rho g h_1 dV$ at A must be equal to that at B, since we assume steady motion and an ideal fluid. The result is

$$p_1 + \frac{1}{2}\rho v_1^2 + \rho g h_1 = p_2 + \frac{1}{2}\rho v_2^2 + \rho g h_2 \,. \tag{1.11}$$

This is Bernoulli's equation of flow for an ideal liquid. It is just a statement of the law of energy conservation as applied to a fluid in which mechanical and thermal energy are uncoupled.

In deriving Eq. (1.11) we have assumed that the mass density ρ of the fluid does not vary with position ("incompressible flow"). This is not permissible in handling flows where large pressure differences arise (e.g. flows through narrow channels or the circulation of the atmosphere over a large range of height) nor in treating the propagation of sound waves (see Chap. 6). Excluding such situations, Eq. (1.11) becomes applicable when the speed of flow v is small compared with the speed of sound c (the ratio v/c is known as the Mach number).

As an example of application of Bernoulli's law let us consider the outflow of liquid driven by gravity through a small hole near the bottom of a container. We can set in Eq. (1.11) $p_1 \cong p_2$ and $v_1 \cong 0$. The outflow velocity thus is

$$v_2^2 = 2g(h_1 - h_2) \,. \tag{1.12}$$

The flow of such an ideal liquid does not involve any particular molecular model. Density is the only property involved, and we shall only discuss it briefly. Thus, it may be noted that in general an increase in flow velocity is accompanied by a drop in pressure. This accounts, for instance, for aerodynamic lift: because the top surface of an aerofoil is larger, the air velocity in flight is higher and the pressure lower than over the bottom surface.

Similarly, forcing water flowing inside a horizontal channel to go through a narrow constriction at sufficiently high velocity may lead to a negative pressure inside the constriction. The critical velocity can be estimated from $\rho(v_2^2 - v_1^2)/2 = p_1 - p_2$ by taking $p_1 = 1$ atm and $v_1 \ll v_2$ ahead of the constriction and $p_2 = 0$ at the constriction: this yields $v_2 \approx 10\ \mathrm{m \cdot s^{-1}}$. At negative p_2 the liquid, being prevented from evaporation by the walls of the channel, cavitates through the formation and growth of vapour-filled bubbles either at the walls or in its interior.

1.4.2 *Flow in real liquids: Introduction of viscosity*

Real liquids experience a viscous resistance to flow. If the velocity gradient between two neighbouring planes is dv_x/dz the force F_{xz} per unit area to overcome viscous resistance is

$$F_{xz} = \eta \frac{dv_x}{dz}, \tag{1.13}$$

where η is defined as the shear viscosity. More precisely, F_{xz} is the shear stress needed to maintain the velocity gradient in a steady state. Fluids which obey this relation are known as Newtonian fluids (see Chaps. 6 and 11). The dimensions of η are $[\eta] = L^{-1}MT^{-1}$.

The viscosity of gases is explained in terms of transfer of molecular momentum across the flow. In a dense liquid the simple molecular model developed by Eyring[5] explains viscous resistance by considering that the molecules are so close together that considerable energy must be expended in dragging one molecular layer over its neighbour. Here we describe it in words and shall discuss viscosity more quantitatively in Chap. 6.

Consider an instantaneous snapshot of the liquid. There is short-range order (SRO) and we may represent a small region of two neighbouring molecular planes by an array similar to that of a solid. Dragging a molecule in the top plane, say, from its equilibrium position to a neighbouring equilibrium position requires overcoming the attractions from its neighbours: there is a potential energy barrier and energy must be provided to bring the molecule to the top of the barrier. When no stress is applied, thermal fluctuations will induce such jumps but, in the absence of a preferred direction, no net flow can occur. An applied shear stress instead performs work on the molecules and favours sliding in one direction against all others. The work dissipated in this process is restored as heat, which is released when the molecule slides down from the top of the barrier into its new equilibrium position.

1.4.3 *Poiseuille's formula: Viscous flow through a tube*

As an important example of flows which are dominated by viscosity rather than by inertia as in Bernoulli's equation, let us consider an incompressible viscous fluid flowing through a cylindrical pipe of length l and radius a. The experiment allows liquid to enter one end at pressure p_1 and to leave the other end at pressure p_2. We ask how large must be the longitudinal pressure drop to force through a volume Q of fluid per unit time.

If the flow is uniform, the streamlines are all parallel to the axis of the pipe. There is no slip of liquid at the solid boundary and the velocity profile is parabolic across the pipe diameter. The rate of flow is then

$$Q = \int_0^a 2\pi v_z r dr = \frac{\pi a^4}{8\eta}\left(\frac{p_1 - p_2}{l}\right) . \tag{1.14}$$

Poiseuille first used this formula to determine the viscosity of blood in horses' arteries. It was independently derived by Hagen, a German hydraulic engineer.

1.4.4 *Turbulence and Reynolds number*

If the flow is steady and stable the work done is expended solely in overcoming viscosity and appears as heat. However, if the velocity of flow is too high or if there are some other unfavourable circumstances, vortices may develop and some of the work goes in providing their kinetic energy. Vortices often assemble on a solid boundary in a boundary layer.

The condition for turbulent flow was first established by Reynolds. Consider a liquid of mass density ρ and viscosity η flowing with velocity v along a channel of lateral dimension a. There is a critical velocity v_c above which orderly streamlined flow gives way to turbulent motion. The Reynolds number *Re* is defined by purely dimensional analysis as

$$Re = \frac{\rho a v}{\eta} . \tag{1.15}$$

It is dimensionless and the transition from streamlined to turbulent motion occurs when its value is around 1000 to 2000 (see Chap. 12 below).

Turbulence is essentially a condition of instability and the force involved in flow is not necessarily a direct indication of whether or not turbulence has occurred. The force is determined primarily by viscous or inertial factors. A simple example of these two conditions is afforded by considering the steady movement of a solid sphere of radius a through a fluid. The resistance to motion can be derived from arguments of dimensional analysis. One solution for the resistive force is

$$F \propto a\eta v . \tag{1.16}$$

This is Stokes' law. Further analysis proves that the proportionality constant equals 6π, so that the friction coefficient f in Eq. (1.9) equals $6\pi a\eta$. Another solution is

$$F \propto \rho a^2 v^2 . \tag{1.17}$$

This does not involve viscosity but kinetic energy, the force being determined by the momentum of the incoming fluid.

The laminar flow pattern in a pipe to which Poiseuille's law refers becomes turbulent when primary eddies get out of control before viscosity may quench them and start generating strings of further eddies. Experiments show that above the critical Reynolds number the pressure gradient needed to drive a fluid through a pipe increases more rapidly than linearly with the rate of flow Q. Hydraulic engineers, who need to transfer fluid losing as little pressure head as possible, may achieve values of the critical Reynolds number as high as 10^5 by taking special care in pipe construction and lay-out.

1.5 Rigidity of Liquids

We mentioned in Sec. 1.3 that the molecules in a dense liquid jump around with frequencies typically in the range 10^{10}–10^{12} Hz. If the rate of shear is sufficiently great there may not be time for the molecules to progress to a neighbouring site, as assumed in our qualitative discussion of shear viscosity in Sec. 1.4.2. In this case the liquid will not show viscous flow, but will exhibit a finite elastic rigidity.

For simple liquids the rates of shear for this to occur in macroscopic flow are enormous, but visco-elastic effects may arise in collective motions at a microscopic level and have been observed in water by experimental techniques of inelastic photon scattering (see Chap. 6). For liquids with large bulky molecules visco-elastic behaviour may instead arise in macroscopic flow at moderate rates of shear. There is a similar behaviour in polymers (see Chap. 11) such as polyethylene, nylon and perspex. The flow of polymeric chains which occurs at slow rates of loading resembles the viscous flow of liquids. This may be impossible at high rates of strain and the material will then deform elastically.

Another example of this type of behaviour is the material silicone putty. This material will bounce with a very high resilience, as there is not enough time for the molecules to flow. Again if the material is in the form of a rod and is pulled rapidly it will first stretch and then snap in a brittle manner. However, at very low rates of deformation it will flow in a viscous fashion.

1.6 Surface Properties

In this section, basic concepts of surface properties of liquids will be introduced. This will suffice to introduce the reader to the account of the liquid-vapour interface given in Chap. 13.

1.6.1 *Surface free energy and surface tension*

It is a well known fact that liquids have a tendency to draw up into drops. Small drops form in spherical geometry and because a sphere is that geometrical form which has the smallest ratio of surface area to volume, one may conclude that the surface of a liquid has higher energy than that of the bulk fluid. It is correct to say, then, that a liquid is always endeavouring to achieve its lowest energy configuration by reducing its surface area. The free energy excess per unit surface area is termed the free surface energy γ and is conveniently measured in $\mathrm{J \cdot m^{-2}}$. For many simple liquids γ has values in the range 10–100 $\mathrm{mJ \cdot m^{-2}}$.

For a liquid having area A, the free surface energy is therefore γA. If one increases the surface area by stretching, then the work done is given by

$$\frac{d(\gamma A)}{dA} = \gamma + A\frac{d\gamma}{dA}\,. \tag{1.18}$$

In the case of a liquid, however much the surface is stretched, the initial configuration of the surface is regained in a very short interval of time. Thus, in contrast to a solid, the equilibrium structure remains unchanged so that $d\gamma/dA = 0$ and γ gives the work done per unit area of extension. Numerous experiments have shown that very thin films of water have the same surface energy as two surfaces separated by bulk water, down to a thickness of only 2 nm. This implies that the molecular forces responsible for the surface energy have short range.

The free surface energy is equivalent to a line tension acting in all directions parallel to the surface. Consider a liquid surface of width L and length X (see Fig. 1.5). Suppose at the edge AB one applies a force F parallel to the surface and normal to AB so that the length of the surface is extended by an amount x. The work done in increasing the surface area is $\gamma L x$ and this must be equal

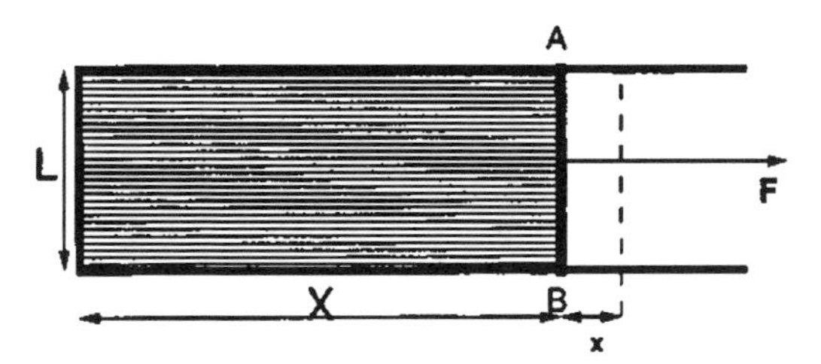

Fig. 1.5. Illustrating the equivalence of surface energy per unit area and surface tension per unit length.

to the external work Fx. Hence

$$\gamma = \frac{F}{L}. \tag{1.19}$$

Therefore the free surface energy is equivalent to a surface tension γ (in $\mathrm{N \cdot m^{-1}}$).

The free surface energy of a liquid lends itself to a rather simple molecular interpretation. Whereas molecules in the bulk are subject to attraction from isotropically distributed neighbours, molecules at the surface are pulled towards the bulk liquid. There is practically no attraction from the side of the vapour, except on the approach to the liquid-vapour critical point where the two phases are taking equal densities. An increase in surface area can only be achieved by pulling molecules up to the surface from the bulk against this one-sided attraction. The bond breaking in this process is practically confined to the last few liquid layers.

Without going into more detail it is evident that the surface energy is of the order of one-half the energy required to break all bonds per molecular layer. If L is the latent heat of vaporisation per mole, M is the molecular mass and ρ the mass density, one estimates

$$\gamma \approx 0.3 \left(\frac{L}{N_\mathrm{A}}\right) \left(\frac{N_\mathrm{A}\rho}{M}\right)^{2/3} \tag{1.20}$$

as an approximate relation, N_A being Avogadro's number. This gives (i) for liquid argon $\gamma = 14\ \mathrm{mJ \cdot m^{-2}}$ compared with the observed value of 13, (ii) for nitrogen 11 compared with 10.5, (iii) for benzene 110 compared with 40 and (iv) for mercury 630 compared with 600 $\mathrm{mJ \cdot m^{-2}}$.

1.6.2 *Surface energy versus surface free energy*

Thermodynamics considers also another quantity, the surface energy h. When a surface is increased in area, apart from the work γ there is also a "latent heat" term. Heat must be applied to keep the temperature constant. The magnitude is $-Td\gamma/dT$. Hence the surface energy is

$$h = \gamma - T\frac{d\gamma}{dT}. \tag{1.21}$$

Equation (1.20) does not distinguish between these two concepts: it is really a model to estimate γ at $T = 0$. On the other hand the difference between h and γ is often significant: as an example, for water $\gamma = 72$ and $h = 118\ \mathrm{mJ \cdot m^{-2}}$.

If fine drops of water are permitted to coalesce so as to destroy their surface area, the increase in temperature is determined by h, not by γ. This provides a method of determining areas of fine particles. They are first equilibrated with water vapour so that their surface is fully covered with a condensed film of water only a few tens of a nanometer thick. They are then immersed in water inside a calorimeter. If the heat given out is ΔQ, the surface area is $\Delta Q/h$.

1.6.3 *Contact angle*

Liquids with low surface tension readily wet most solids, giving a contact angle of zero, while those with high surface tension often show a finite contact angle. In molecular terms, if the cohesion between the molecules of the liquid is greater than the adhesion between liquid and solid, then the liquid will not wet the solid and will exhibit a finite contact angle.

Contact equilibrium is depicted in Fig. 1.6. The free surface energy of the liquid is γ_{L}; the solid–liquid interface has a free surface energy γ_{SL} and the exposed portion of the solid adjacent to the liquid where vapour has been absorbed has a free surface energy γ_{SV}. If one considers a virtual process which expands the area of wetting by 1 m^2, then the virtual work done on the liquid surface ($\gamma_{\mathrm{L}} \cos\theta$) and on the solid–liquid interface (γ_{SL}) must be balanced by the virtual work released at the solid-vapour interface (γ_{SV}):

$$\gamma_{\mathrm{L}} \cos\theta + \gamma_{\mathrm{SL}} = \gamma_{\mathrm{SV}}\,. \tag{1.22}$$

This equation, going back to Young (1805) and Dupré (1869), determines the contact angle θ.

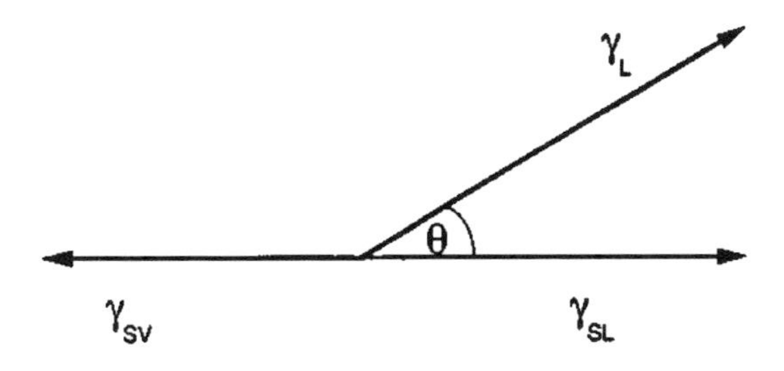

Fig. 1.6. Illustrating the contact angle equilibrium.

Table 1.2. Interfacial free energies of mica ($\mathrm{mJ \cdot m^{-2}}$).

Fluid	γ_{SV}	γ_{SL}	$\gamma_{SV} - \gamma_{SL}$	γ_L
Water	183	107	76	73
Hexane	271	255	16	18

Equation (1.22) also corresponds to balancing the horizontal components of the surface tension forces. One may then raise the question as to what has happened to the vertical component $\gamma_L \sin\theta$. The answer is that this force is very small and usually has negligible effect on the solid. However, if the solid has a thin, flexible sheet structure as in mica, the vertical forces may be sufficient to visibly distort the solid surface.

As an example, Eq. (1.22) has been studied experimentally for mica, by measuring the force needed to open a crack in this "perfectly brittle" solid first exposed to water vapour and then immersed in liquid water (see Tabor[3]). These experiments measure γ_{SV} and γ_{SL}. Since water wets mica completely ($\theta = 0$), one should have $\gamma_L = \gamma_{SV} - \gamma_{SL}$. This is seen in Table 1.2 to be the case for water and also for hexane, another liquid which wets mica.

1.6.4 *Capillarity*

Some of the general properties of capillarity will now be discussed (see the book of Rowlinson and Widom[6] for an advanced account). The starting point is a simple relation for the pressure difference across a curved liquid-vapour interface,

$$p_1 - p_2 = \gamma \frac{(R_1 + R_2)}{(R_1 R_2)}, \tag{1.23}$$

where R_1 and R_2 are the principal radii of curvature of the interface, p_1 is the pressure on the vapour side and p_2 is that on the liquid side. The pressure on the concave side is greater than on the convex side of the liquid surface.

The relation (1.23) becomes clear in the following two simple examples. For a spherical drop of radius R, upon expanding its radius by dR and using the principle of virtual work, the work done in stretching the surface is $(8\pi\gamma R dR)$ and is balanced by the work $(4\pi p R^2 dR)$ released by the internal pressure. Therefore, the excess pressure inside the drop is

$$p = \frac{2\gamma}{R}. \tag{1.24}$$

The second example is that of a soap bubble. Here two surfaces are involved and the excess pressure inside the bubble is related to the surface tension γ_S of the soap solution by

$$p = \frac{4\gamma_s}{R}. \tag{1.25}$$

We can apply Eq. (1.23) to discuss capillary rise. A uniform tube of small radius R is held vertically and lowered into a liquid of density ρ_L (see Fig. 1.7). The liquid rises to a height h wetting the capillary. For zero contact angle the meniscus is a hemisphere of radius R and the pressure on the liquid side is lowered by $2\gamma/R$. Therefore, the liquid is drawn up the tube to a height h

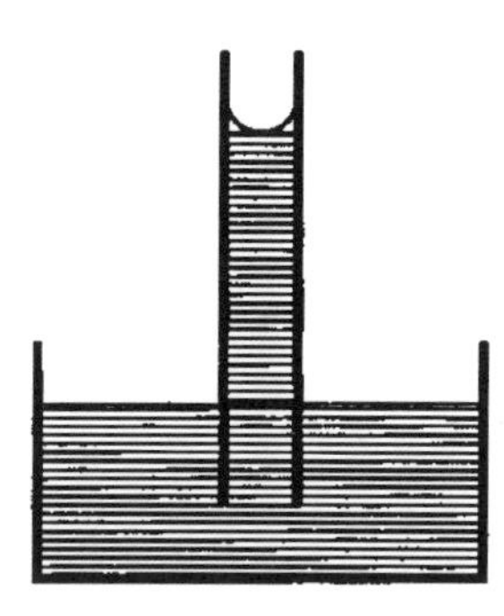

Fig. 1.7. Capillary rise in terms of pressure defect.

determined by balancing the weight of the liquid column against the pressure deficit:

$$\rho_L g h = \frac{2\gamma}{R}. \tag{1.26}$$

From the above argument, the capillary rise depends only on what happens at the meniscus, in agreement with observations. If the liquid ascent is large compared with the diameter of the meniscus, then the meniscus is hemispherical in shape since the height of the rise is practically constant over its entire width. If the contact angle is finite, the radius of curvature becomes $R/\cos\theta$ and the pressure defect is $2\gamma\cos\theta/R$. As the capillary is made wider the height of the rise becomes small compared with the capillary diameter. The detailed shape of the meniscus can then be calculated by equating the pressure defect given in Eq. (1.23) at any given point of the liquid surface to the value of $\rho_L g h$ at that point, h being the local height of the liquid above the bulk level far away from the walls of the container.

Referring again to Fig. 1.7 in the case of a narrow capillary, we see that the vapour pressure above the meniscus is reduced by the amount $\rho_V g h$, where ρ_V is the mean density of the vapour. We may write

$$\rho_V g h = \frac{2\gamma}{R}\frac{\rho_V}{\rho_L}. \tag{1.27}$$

More generally, the vapour pressure over a concave meniscus surface is less than that over a flat liquid surface by the amount shown in Eq. (1.27). The vapour pressure over a convex meniscus is likewise greater than over a flat surface by the same amount.

1.6.5 *Energy for capillary rise*

By rising to a height h in a narrow tube of radius R the liquid gains a gravitational energy equal to $\frac{1}{2}\pi R^2 h^2 \rho_L g$, corresponding to raising a mass $\pi R^2 h \rho_L$ through a height of $\frac{1}{2}h$. We ask the question as to what is the source of this energy. The answer is that it comes from the wetting of the walls of the tube by the liquid.

Let us consider again a completely wetting liquid. If the liquid advances along the tube so as to cover 1 m^2 of the surface, one loses 1 m^2 of γ_{SV} and gains 1 m^2 of γ_{SL}. There is no change in the area of the liquid meniscus. The energy given up by the system in this wetting process is then $\gamma_{SV} - \gamma_{SL}$, which from

Eq. (1.22) is equal to γ_L. Therefore, the energy released in a rise of h is given by $2\pi Rh\gamma_L$, or $\pi R^2h^2\rho_L g$ from Eq. (1.26). This implies that if the liquid were non-viscous it would rise to height $2h$ and oscillate between 0 and $2h$ with its mean position at h. In practice viscosity dissipates the excess energy very rapidly.

Notice a further implication of the above result. If the liquid is completely wetting, then the capillary rise does not depend on the material of which the capillary is made. It depends only on the surface tension of the liquid-vapour interface.

1.7 Water and Ice Revisited

We have often appealed in this chapter to liquid argon and to water as examples of liquid systems. The specific non-spherical shape of the H_2O molecule and the specific interactions that it gives rise to have a number of novel and important consequences.

In the H_2O molecule the oxygen atom binds two hydrogens by electron pairing, thereby imparting an electric dipole moment to the molecule, and arranges its further four valence electrons in two lone-pair bonds. The four bonds point towards the vertices of an almost perfect tetrahedron and each lone pair can interact with an electron-deficient hydrogen atom belonging to a neighbouring H_2O molecule. By virtue of this so-called hydrogen bond, each H_2O molecule in the dense liquid or solid phases is tetrahedrally coordinated by four other H_2O molecules and structural correlations can build up in space as a consequence of this constraint. Of course, this type of bonding is responsible for much of what happens in the field of biology.

From X-ray diffraction experiments (see Chap. 4 for an introduction to this technique of structure determination) it is possible to determine how the density of molecules in water builds up with increasing distance from a central "average" water molecule. This is shown in the top part of Fig. 1.8, from the work of Narten *et al.*[7] As in all liquids, there is a region of "excluded volume" (i.e. of essentially zero local density) within a molecular diameter from the central molecule: the electronic energy rises very sharply from both the exclusion principle and the Coulomb repulsions as the valence electrons of two chemically saturated molecules are squeezed close together, so that each molecule looks at very short range like a hard wall. Further out the liquid density profile in Fig. 1.8 is seen to rise into a shell of first neighbours and then to oscillate

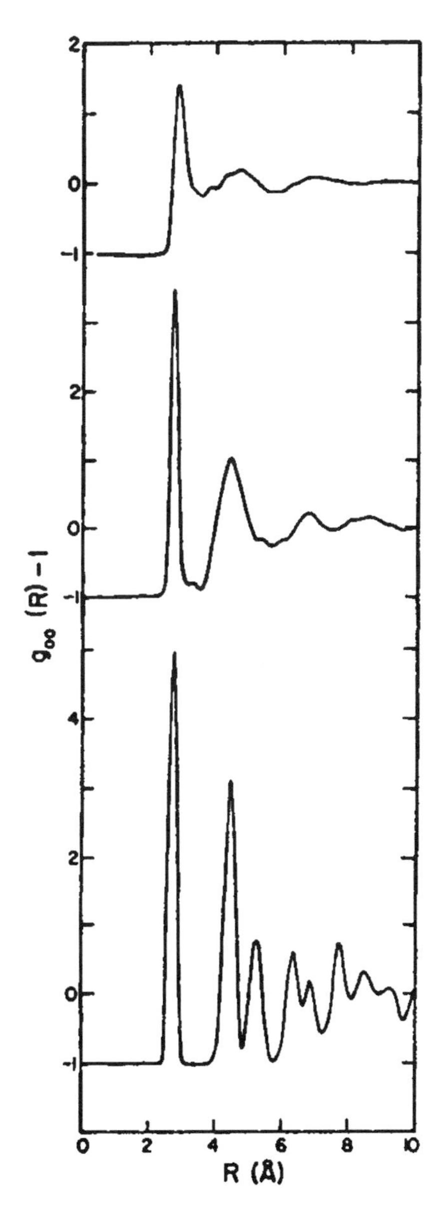

Fig. 1.8. Oxygen–oxygen correlations determined by X-ray scattering from liquid H_2O (top), amorphous solid H_2O (middle) and polycrystalline ice (bottom). (Redrawn from Narten *et al.*, Ref. 7.)

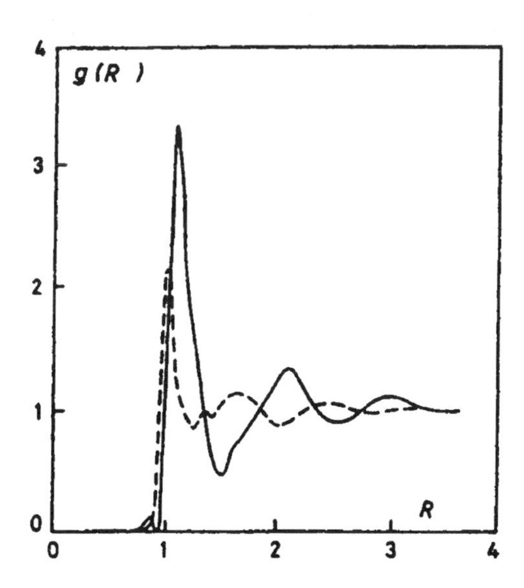

Fig. 1.9. Radial distribution functions for oxygen atoms in water (dashed line) and for liquid argon (solid line). Both liquids are near their respective freezing points and the distance $R = r/\sigma$ is scaled by the van der Waals diameter σ (2.82 Å in water and 3.4 Å in argon). (Redrawn from Franks, Ref. 9.)

and decay to the homogeneous liquid density at a few intermolecular distances further out. Figure 1.8 also contrasts the liquid structure with that of amorphous solid water (middle part) and with that of polycrystalline ice (bottom). In fact X-rays are almost blind to hydrogen atoms, so that Fig. 1.8 essentially shows us the distribution of oxygen atoms in these states of aggregation of water. Special techniques in neutron diffraction have allowed detailed mapping of the spatial arrangement of the two atomic species in water and also around foreign cations and anions in electrolyte solutions.[8] Finally, in Fig. 1.9 (from the booklet by Franks[9]) the structure of water is contrasted with that of liquid argon near freezing. The average number of first neighbours is estimated from these data to be about 4.4 in water and about 10 in liquid argon, the latter being, however, very sensitive to temperature changes.

As water crystallises into ice at atmospheric pressure, the local tetrahedral arrangement of the oxygens is frozen into a periodic lattice consisting of layers of rippled hexagons (see Fig. 1.10, taken from the book of Petrenko and Whitworth[10]). This microscopic hexagonal arrangement of the H_2O molecules in the ordered crystalline phase is beautifully revealed at the macroscopic level

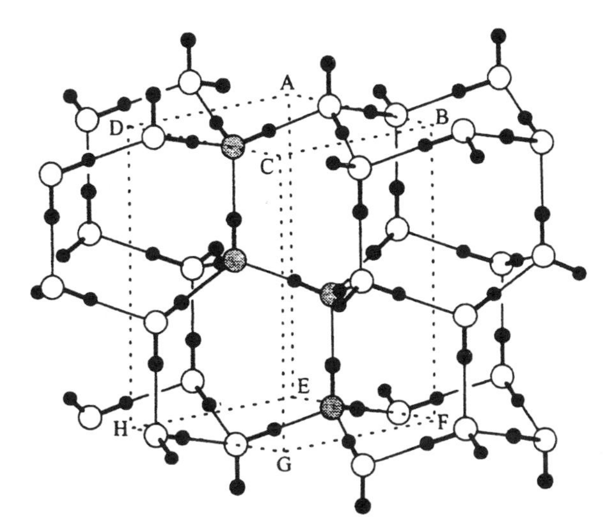

Fig. 1.10. Crystal structure of hexagonal ice, with the molecules frozen in a particular hydrogen-bond configuration and showing in gray the four oxygen atoms belonging to the unit cell of the average structure (shown in dashes). (Redrawn from Petrenko and Whitworth, Ref. 10.)

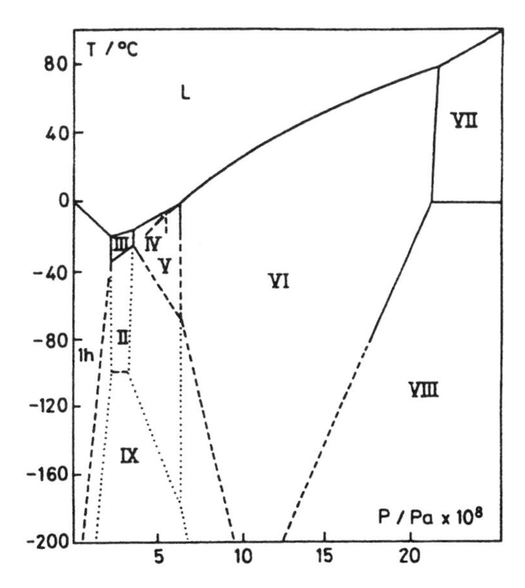

Fig. 1.11. Phase diagram of the ice-water system. (Redrawn from Franks, Ref. 9.)

in the sixfold symmetry of snowflakes. There evidently is a lot of empty space in the very open structure of ice and this has two main consequences: (i) melting is accompanied by a partial structural collapse, leading to the observed increase in density; and (ii) the crystalline structure of ice is very sensitive to pressure and some ten polymorphs are known — each of them being stable in a limited range of pressure and temperature and all showing fourfold coordination of the oxygen atoms. The phase diagram of H_2O, showing the liquid and various solid polymorphs, is reported in Fig. 1.11 (taken from the booklet by Franks[8]).

In crystalline ice both the translational and rotational symmetries of the liquid water phase are broken. Systems with intermediate symmetries also occur in nature: in particular we have already mentioned liquid crystals, in which the full symmetry of the liquid phase is broken in steps through a series of mesophases. It is also worth stressing that the dynamical behaviours of the crystalline and liquid phase (and of an amorphous state if attainable) are crucially dependent on the time scale. Having mentioned in Sec. 1.5 the visco-elastic behaviour shown by water under a high-frequency probe, we may conclude this Chapter by recalling that ice flows over very long time scales, as is evidenced by the behaviour of glaciers. Such flow properties of crystalline materials are determined by the presence of line and plane defects in the real crystal.

Chapter 2

Excluded Volume, Free Volume and Hard Sphere Packing

In the first chapter, a bird's eye view has been attempted of the liquid state of matter, in largely qualitative terms. While the concept of short-range order in liquids has already been introduced, in the present chapter we press this in the simplest model with at least a measure of realism, namely an assembly of hard spheres.

2.1 Excluded Volume and Packing Problems

There is a sense in which van der Waals was already appealing to a model of this kind when he wrote his equation of state,

$$\left(p + \frac{aN^2}{V^2}\right)(V - Nb) = RT\,. \tag{2.1}$$

For in Eq. (2.1), the term $V - Nb$ acknowledges the idea of some "excluded volume", whereas the "correction" aN^2/V^2 to the pressure in the perfect gas equation of state is taking some gross account of attractive forces between atoms. Although Eq. (2.1) has severe quantitative limitations, it already embodies some essential ideas of the gross structure of a liquid.

Since the pioneering studies of Bernal[11] and Scott[12] on static packing models of hard spheres, to be discussed below, assemblies of spherical particles interacting *via* a hard sphere potential (i.e. experiencing only impulsive forces at contact) have been studied quantitatively in order to model dense liquids[13] and also glasses,[14] as will be discussed in a later chapter. It can be asserted by now, with confidence, that notwithstanding the simplicity of such hard sphere interactions, lacking both attractions and directionality, hard sphere assemblies exhibit a number of the properties of real dense liquids. Just as the hard sphere fluid provides a useful reference system for understanding the geometric consequences of interatomic interactions and some features of liquid structure for at least monatomic systems (see Chap. 4), it also is the simplest system which is known to exhibit a fluid–solid transition.

The excluded volume property of the hard sphere model can be stated as follows: with fixed diameter σ there is an upper limit to the number of molecules that can be contained in a fixed volume V. Put another way, for a given number N of hard spheres, there is a lower limit V_c to the volume V needed to accommodate them. Finally, there is a lower limit v_c, which may depend on N and on the shape of the container for small N, to the volume v per molecule.

In the limit as N tends to infinity, certain exact bounds on v_c are available for static hard sphere models. Though we are mainly interested in three dimensions, among two-dimensional assemblies of hard disks the hexagonal close-packed arrangement is the densest possible,[15] with a lower limit given by $3^{1/2}\sigma^2/2$ for the area per disk. For three dimensions[16] the densest possible packing is likely that of the face-centred-cubic and the hexagonal close-packings, each of which has $v_c = \sigma^3/2^{1/2}$, but to our knowledge there is no rigorous proof. The best lower bound for v_c in the three-dimensional case, again due to Rogers,[15] is some 5% smaller. It is found in practice that when hard spheres are packed as tightly as possible in an irregular three-dimensional static arrangement, the largest value that can be attained for the ratio of the volume of the spheres to the total volume is about 0.64. This is 14% less than the value 0.7405 for the ordered crystalline arrangements mentioned above.

2.2 Accessible Configuration Space

Salsburg and Wood[16] have constructed different close-packed arrangements of equal numbers of hard disks inside equal rectangular containers. For instance, they compare two arrangements of twelve hard disks of diameter $43^{1/2}/24$ in a rectangular container of sides $3^{1/2}/2$ and 1, the first arrangement being part of the regular hexagonal lattice while the second is a close-packed arrangement of coordination number 4. Here, the term "close-packed" is used without further qualification to describe an arrangement of molecules in which no molecule can be moved while the others are held fixed. Thus, in a two-dimensional system each molecule must touch at least three others, while for three dimensions it must touch at least four others.

Salsburg and Wood point out that it is quite likely that the above two arrangements are not accessible from each other, even for somewhat larger values of the total available area. The issue emerges of how much of configuration space is accessible from a given *static* arrangement of the molecules, as opposed to the freedom that a molecular assembly at high temperature has to explore configuration space through collective rearrangements.

Similarly, and for our purposes more importantly, the randomly packed arrangements studied by Bernal[11] (see immediately below for some details of such experiments) and by Scott[12] appear to be stable against shaking and may (for finite N and fixed V and σ) be inaccessible from the closest-packed arrangements.

2.3 Experiments on Random Packing Models

Therefore, to begin the study of the liquid state without making any direct appeal to either gaseous or solid phases, it is natural to elaborate on the studies of Bernal and Scott (see also Tabor's book[3] and the review by Finney[17]). The limitations of these static models will become apparent in the sequel, when we shall turn to dynamic studies of hard sphere assemblies.

Bernal from the outset recognised not only the disorder inherent in the liquid structure, but also its relatively close-packed character. Prompted by this fact, his approach was extremely simple and practicable, and went as follows. Starting with a large number of plasticine spheres, which were chalked to prevent sticking, Bernal placed them in a football bladder, removing the air to avoid bubbles. Then he squeezed them together until they filled the whole of the available space. On examining the aggregate it was found that the spheres had been changed into polyhedra of various irregular shapes. The most common number of faces was thirteen and the most common number of sides to a face was five. The former number is an upper estimate for the first-neighbour coordination number, because the centres of some of the neighbours were more than the average distance apart. Associated with the high incidence of five-fold faces, there are five-fold rings of neighbouring atoms. This is a feature which for any crystal lattice is excluded by translational symmetry.

In a later study Bernal emphasised that the packing of the molecules which occurs in condensed states is mainly determined by the repulsive part of the interatomic forces: the molecules behave in this respect very much like impenetrable spheres. He therefore considered the random packing, not of deformable plasticine spheres as in his earlier study, but that of a large number of hard steel balls. The number of contacts and near contacts could be studied by pouring black paint on the assembly. The model led to a coordination number varying between four and eleven. The average radial distribution of pairs of hard spheres, as a function of the distance between them, has the same qualitative features as that measured in experiments of X-ray diffraction on real liquids (see Fig. 2.1 and Chap. 4): around each sphere there is a region of excluded volume and a shell of first neighbours, followed by progressively more poorly defined shells of further neighbours. This is a good picture of short-range order for monatomic liquids in thermodynamic states close to freezing.

The Bernal model tries to provide an instantaneous picture of the structure of a liquid (this is known as the random-close-packing model of a dense disordered system). If we add to this the thermal motions we can imagine that the molecules are continuously shuffling around and occasionally squeezing through their neighbours. The radial distribution functions in Fig. 2.1

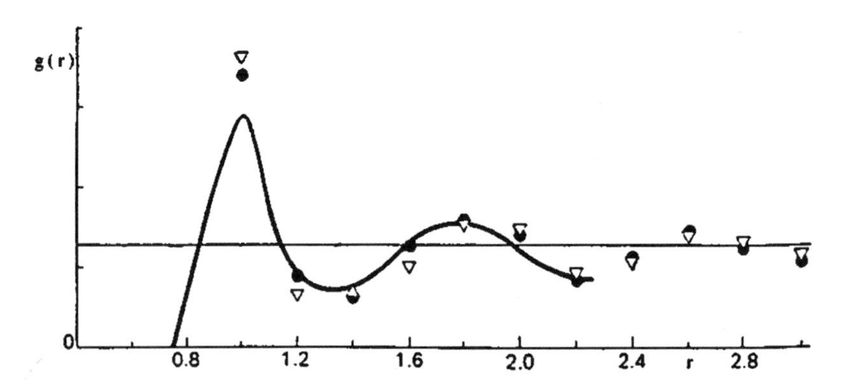

Fig. 2.1. Radial distribution function $g(r)$ as a function of the reduced distance r/σ for the random packing model of Bernal (•) and Scott (∇), compared with a schematic plot of diffraction data on an argon-like liquid. For the hard sphere fluid one has $g(r) = 0$ for $r/\sigma < 1$. (Redrawn from Tabor, Ref. 3.)

essentially represent averages over various regions of the sample — a static disordered sample in the Bernal model and a dynamic one in the diffraction experiments. The marks of thermal motions in the measured radial distribution function are evidently rather subtle and can only be extracted from it through quantitative analysis, especially on the detailed shape of the main first-neighbour peak and of the valley following it. Through this valley the exchange of atoms takes place between the first-neighbour shell and rest of the fluid.

As Bernal had already pointed out, it is the decrease in the coordination number with increasing temperature, rather than an essentially uniform expansion of the interatomic distances, which leads to the rather large thermal expansion of liquids. He made the proposal that the transition from the liquid to the gas phase occurs when the coordination number falls to an average value of three or four. This implies that the density at the liquid–vapour critical point may be between one-quarter and one-third of the density of the liquid near freezing. This prediction is in reasonable agreement with experiment for simple fluids.

2.4 Origins of Method of Molecular Dynamics

Bernal's models for liquids have been criticized for being static. The method of molecular dynamics enables one to construct models in which the molecules are continuously on the move. There are whole books devoted to this method (see e.g. Allen and Tyndesley[18]), which has had and will continue to have a strong impact on our quantitative understanding of the liquid state. The basic concept is simple, though elaborate computations are involved. The approach is associated with names like B. Alder and A. Rahman, though many others were involved from an early stage. The very first studies by this technique were concerned with the dynamics of assemblies of oscillators and with the thermodynamics of a classical gas of charged particles.

A surprising result coming from early computer studies[13,19] was that, with increasing density, a fluid of equisized hard spheres exhibits a first-order freezing transition to form a close-packed crystal (see Sec. 1.2 on Lindemann's law of melting). It must follow then that since the internal energy of such a system of hard spheres is purely kinetic, the above transition must be entropically driven.[20] The volume change on melting from these dynamical studies is only 5%, against the value of 14% predicted from the static hard sphere models (see Sec. 2.1). We have already noticed that a dynamical model has much greater freedom to wander in configuration space and thereby to go from one arrangement of high structural packing to another.

However, to return to the basic concept, in such early studies molecules were treated as hard, smooth and perfectly elastic spheres obeying the laws of classical mechanics. Later such refinements were embodied in the method as representing the atoms as attractive square-wells having a rigid core or more "realistic" interactions described by smooth pair potentials consisting of a repulsive core and an attractive tail (see Sec. 2.7). Even more recent developments have been the ability to simulate systems of particles obeying the laws of quantum mechanics and to treat fluids as assemblies of ionic inner cores and valence electrons (see Chap. 14).

The idea of the method in its early developments was to simulate the behaviour of a limited number of molecules placed inside a given volume by numerically solving Newton's equations on a computer. The particles are started off with random velocities and periodic boundary conditions are usually imposed, i.e. a particle which leaves the box across one face re-enters with the same velocity through the opposite face. In this way the number of molecules in the box and the total energy are kept constant. After relatively few collisional times the velocity distribution equilibrates to a Maxwellian form. The mean kinetic energy measures the temperature of the sample.

The behaviour of the system of hard spheres was found to depend on the kinetic energy and on the packing of the particles (see Fig. 2.2). At one extreme the particles oscillate about a set of equilibrium lattice positions: this corresponds to the solid state [Fig. 2.2(a)]. In the other limit the particles can move through the box quite freely, as in a dilute fluid state. In an intermediate dense-fluid regime the particles oscillate within first-neighbour cages about a disordered set of positions, but are also able to exchange places and to slowly diffuse through the sample [Fig. 2.2(b)]. Of course diffusion also occurs in crystals (often *via* vacancies), but the rate of diffusion is much lower than in the fluid. The results reported in Fig. 2.2 refer to 32 particles and map out a view of the trajectories as seen from one face of the simulation box.

The results reproduced in Fig. 2.2 are quite striking, even though they refer to an assembly of particles without attractive interactions. From the point of view of packing, the structure of a dense monatomic fluid should not be too sensitive to the details of the interactions (see the work of Bernal and its discussion given above). However, without attractions the system has no

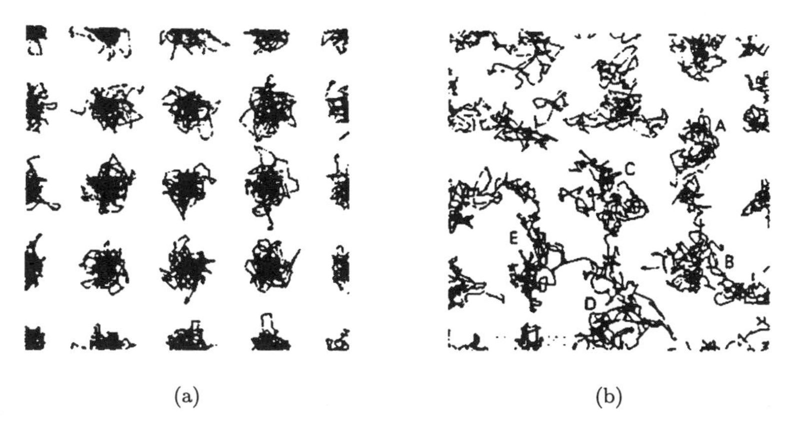

Fig. 2.2. End face of simulation box for 32 hard spheres: Solid-like behaviour for particle trajectories in (a) is contrasted with fluid-like behaviour in (b). Notice particles swapping places in (b): A and B; C, D and E. (Redrawn from Tabor, Ref. 3.)

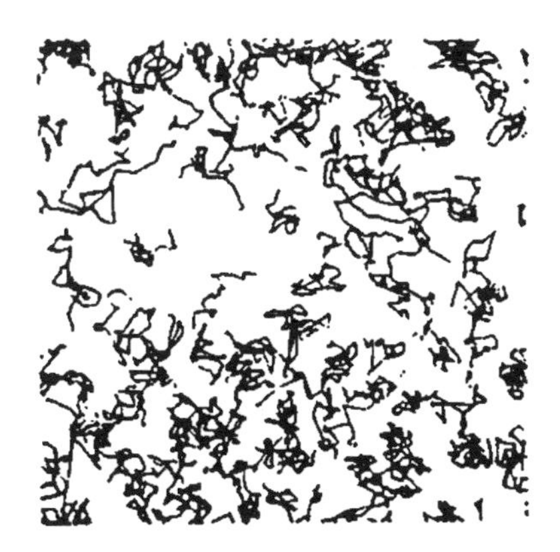

Fig. 2.3. End face of simulation box for 108 particles interacting *via* a hard-core repulsion and a square-well attraction. (Redrawn from Tabor, Ref. 3.)

cohesion and external pressure has to be applied to keep it in a condensed state. Perhaps more seriously, in the absence of attractive interactions there is no condensation from vapour to liquid (i.e. a single fluid phase exists). It is therefore appropriate to report in Fig. 2.3 the behaviour shown by a system of 108 spherical particles possessing a hard elastic core and a square-well attractive field. This seems to depict a liquid–vapour transition region.

2.5 Free-Volume Approximation

Before proceeding to a discussion of the entropically driven freezing transition in the system of hard spheres, it will be useful to introduce the so-called free volume approximation to the equation of state of hard spheres. The limiting configuration is taken to be that of a regular close-packed lattice. The discussion below follows closely those of Kirkwood and Wood.[21]

The volume of the fluid is divided up into N cells $\Delta_i (i = 1, 2 \cdots N)$, one for each molecule. These cells are the Voronoi polyhedra, constructed by drawing planes to bisect at right angles the lines that join adjacent molecules. By such a construction the cells pack so as to completely fill the volume. The approximation of single occupancy of the cells Δ_i is implicit and is closely related to the assumption of disconnected regions of configuration space (see Sec. 2.2). The free volume for each sphere essentially is the volume within which it can move without requiring changes in the positions of other spheres.

These assumptions, when developed quantitatively, lead to an equation of state having the form in d dimensions:

$$\frac{pV}{Nk_{\mathrm{B}}T} = 1 + \left[\left(\frac{V}{V_0}\right)^{1/d} - 1\right]^{-1} \tag{2.2}$$

(see also Salsburg and Wood[16]).

The essential approximations of the free volume theory, in summary, are (i) single occupancy and (ii) the assumption that the molecules may be treated as moving independently within their own cells. Each molecule is thus confined within its own cell, instead of being able to ultimately wander over the whole volume of the fluid.

Such confinement is quite appropriate for a crystalline solid and can be usefully invoked in the evaluation of the equation of state of the dense fluid. However, it evidently leads to an underestimate of the entropy of the fluid state. Further study indicates that this entropy deficit becomes increasingly severe in a fluid as its density decreases.

2.6 Free Volume and Entropically Driven Freezing Transition

The freezing transition of an assembly of hard spheres arises from a competition between configurational entropy and the entropy associated with the amount of local free volume available to the spheres. As for instance Eldridge *et al.*[22] note in the context of their work on binary hard sphere mixtures, at high concentrations efficient ordered packings, which can provide greater free volume than "amorphous" packings, are favoured thermodynamically.

As already mentioned in Sec. 2.1, it was demonstrated by Bernal that the maximum packing or volume fraction of an amorphous system of hard spheres is about 0.64. At such a density, the spheres are totally constrained by their neighbours and have no free volume for local motions. But in a crystal, at volume fraction 0.64, the spheres have considerable free volume, the concentration of a fully compressed close-packed crystal of such hard spheres being 0.74. Thus, from the above discussion one might anticipate that hard spheres freeze at a concentration smaller than 0.64. This is in agreement with the results of Hoover and Ree,[19] whose computer studies revealed a first-order freezing transition, the maximum (freezing) volume fraction of an equilibrium hard sphere fluid being 0.494 and the minimum (melting) value of the crystal being 0.545. Evidently, even in the solid phase near melting the thermal motions of the spheres keep them well apart from one another.

The liquid–solid phase transition may be detected in these computer experiments not only by extracting the coordinates of the particles and verifying that crystalline order has set in, but also by examining the mean pressure exerted by the hard spheres on the walls of the container. Since the internal energy U of the hard sphere system is entirely kinetic, it does not depend upon the volume at constant temperature. From the thermodynamic identities $(\partial U/\partial V)_T = -p + T(\partial S/\partial V)_T$ and $(\partial S/\partial V)_T = (\partial p/\partial T)_V$ (see Eq. (3.20)) we then find

$$\frac{p}{T} = \left(\frac{\partial p}{\partial T}\right)_V , \tag{2.3}$$

i.e. p/T is a function of the volume only. Therefore all isotherms may be made to coincide in a single curve by plotting the dimensionless quantity $pV_{\mathrm{cp}}/Nk_{\mathrm{B}}T$ against the reduced volume V/V_{cp}, where V_{cp} is the volume that the spheres would occupy if they were close packed.

Such a scaled plot for the solid and fluid isotherms of the hard sphere model is reported in Fig. 2.4.[23] There is some uncertainty on the position of the horizontal tie-line which joins the solid and liquid branches. The melting pressure p_{m} is

$$p_{\mathrm{m}} \cong \frac{5.8Nk_{\mathrm{B}}T}{V_{\mathrm{cp}}} . \tag{2.4}$$

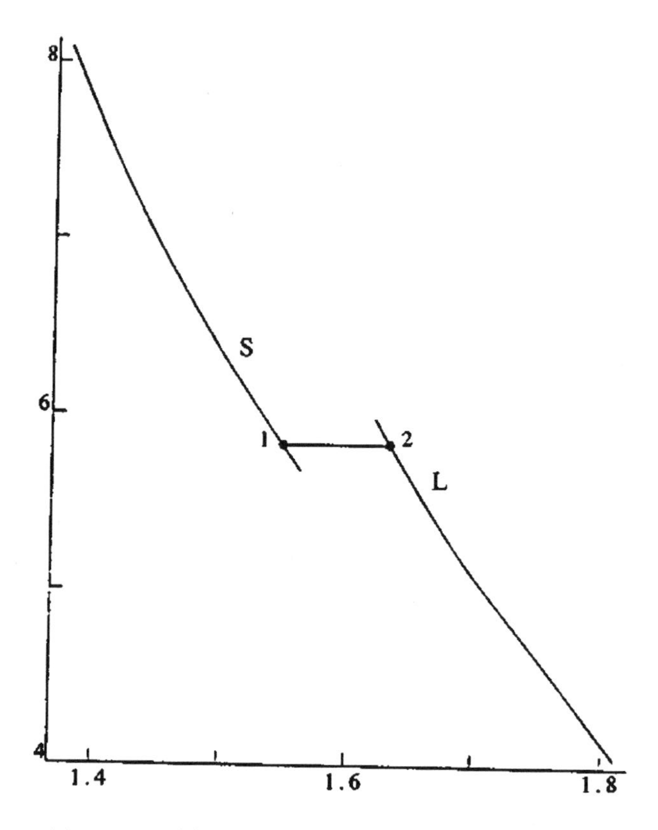

Fig. 2.4. Solid (S) and fluid (L) branches of the isotherms for the hard sphere system. The plot shows $pV_{\rm cp}/Nk_{\rm B}T$ against $V/V_{\rm cp}$, with $V_{\rm cp}$ the volume at close packing (Redrawn from Faber, Ref. 23.)

From the equality of the Gibbs free energies of the two phases at coexistence (see Sec. 3.4), we have $p_{\rm m}\Delta V = T\Delta S$. Hence, the change of entropy on melting is

$$\Delta S = p_{\rm m}\frac{\Delta V}{V} \cong 5.8Nk_{\rm B}\frac{\Delta V}{V} \cong 0.5Nk_{\rm B}\,. \tag{2.5}$$

These results define the curve of fusion for the hard sphere system: on this curve the pressure increases linearly with temperature and the entropy change on melting is constant.

The isotherms of Alder and Wainwright for the fluid phase of the hard sphere system are analytically represented with high accuracy by the formula proposed by Carnahan and Starling.[24] The hard sphere pressure p_{hs} is written as

$$\frac{p_{\mathrm{hs}}V}{Nk_{\mathrm{B}}T} = \frac{(1+\eta+\eta^2-\eta^3)}{(1-\eta)^3}, \tag{2.6}$$

where

$$\eta = \frac{\pi N\sigma^3}{6V} \tag{2.7}$$

is the packing fraction. Much use has been made of the expression (2.6) in treating simple liquids. One example will be illustrated in the next section.

A more accurate analytic expression for the equation of state of the hard sphere fluid has been proposed by Hall[25] through fits of the simulation data by means of Padé approximants. He also used the same method to obtain an approximate formula for the equation of state of the hard sphere solid, on which simulation data have been available from the work of Alder and coworkers.[26] More recent studies of crystalline phases of the hard sphere system include the evaluation of its radial distribution function in the face-centred cubic and hexagonal close-packed structures over a very broad range of density[27] and studies of thermodynamic properties and density distributions of crystals of hard hyperspheres by an extension of the free-volume approach, with special focus on the body-centred cubic crystal.[28]

2.7 Building on Hard Sphere Equation of State: the Model of Longuet–Higgins and Widom

A point pressed by workers such as Hoover and Ross in early studies is that a hard sphere model can be viewed as a limiting case of a repulsive pair potential decaying with increasing separation r between an atomic pair as an inverse power r^{-n}, in the limit $n \to \infty$. For argon, for example, the Lennard–Jones (LJ) potential

$$\phi_{\mathrm{LJ}}(r) = \frac{A}{r^{12}} - \frac{C_6}{r^6} \tag{2.8}$$

has evidently $n = 12$ for the repulsive interactions and includes the van der Waals interactions to account for an attractive tail, with a constant conventionally written as C_6 in accord with quantum chemical notation. Clearly, C_6

is to be put to zero in a purely repulsive limit such as hard spheres, whereas n is then in this model allowed to tend to infinity. The discussion of thermodynamic properties, to be given in Chap. 3, starts from such inverse power potentials, which are a broad enough class to embrace the hard sphere model on which this chapter is focused.

Here we refer to the work of Longuet–Higgins and Widom,[29] who have suggested an elegant way by which the effects of attractions may be included in the equation of state through a simple modification of the hard sphere result. Their main attention was on simple monatomic liquids at high density, specifically on liquid argon near the triple point.

In essence, Longuet–Higgins and Widom show that in this regime the equation of state is well represented by a modified form of the van der Waals Eq. (2.1). Let us rewrite the van der Waals equation in the form

$$p = \frac{\rho k_B T}{1 - b\rho} - a\rho^2 , \tag{2.9}$$

where $\rho = N/V$ is the particle number density. The first term on the RHS is designed to take account of the "finite size" of the atoms, i.e. it is the analogue of a hard sphere term, while the second term takes account of the attractive forces. Longuet–Higgins and Widom suggest that the potential energy of each atom in argon is proportional to the density of its neighbours, i.e. to ρ. In present terms we may thus want to write

$$p = p_{hs}(\rho, T) - a\rho^2 . \tag{2.10}$$

The function $p_{hs}(\rho, T)$ may be taken from studies of the hard sphere system and the quantity a from a model of the interatomic attractions.[30]

Longuet–Higgins and Widom show that this equation of state gives a quantitative account of the properties of argon near the triple point. To show this, we reproduce in Table 2.1 some results that they obtained, together with the corresponding experimental results. The dimensionless quantities shown in Table 2.1 are: first the ratio of the liquid and solid volumes at the triple point;

Table 2.1. Properties of argon at triple point.

	V_L/V_S	p	$\Delta S/Nk_B$	E_c
Theory	1.19	−5.9	1.64	−8.6
Experiment	1.11	−5.88	1.69	−8.53

second the pressure; third the entropy of melting ΔS, in units of Nk_B; and fourth the cohesive energy E_c of the liquid.

Some second derivatives of the free energy are given less satisfactorily by the model of Longuet–Higgins and Widom, and in particular there is no configurational contribution to the specific heat. Nevertheless, their idea has been the germ for some modern approaches making use of the hard sphere model as a reference system for liquid structure.

2.8 Hard-Particle Fluid Equation of State Using Nearest-Neighbour Correlations

With the above introduction to the hard sphere model, we shall go on to note a number of approximate analytical results for such a model. Thus, a more recent form of its equation of state will be presented in this section. In the chapters on mass, momentum and energy transport, it will be shown that diffusion, viscosity and thermal conductivity are elegantly related using such a model. However, it must occasion no surprise that while such results are very valuable for obtaining order-of-magnitude estimates and for gaining physical and chemical insight, deviations from the hard sphere predictions are often quantitatively serious and sometimes are even qualitative.

An approach for determining the free energy of classical fluids developed by Edgal[31] has been employed by Edgal and Huber[32] in an approximate form for the hard sphere fluid. These workers formulate the equation of state of this fluid in terms of a nonlinear differential equation with a single unknown parameter.

Edgal and Huber write the hard sphere equation of state as

$$Z = [1 - k\eta\varepsilon(\eta)]^{-1}\left[1 + k\eta^2\frac{d\varepsilon(\eta)}{d\eta}\right], \tag{2.11}$$

where $Z \equiv p/\rho k_B T$ and in one, two and three dimensions k is 1, $(2/\pi)3^{1/2}$ and $(3/\pi)2^{1/2}$. The packing fraction η is defined as $\eta = \rho v_0$, with $\rho = N/V$ and v_0 the hard-core volume of a particle. In obtaining Eq. (2.11), it has been assumed that the partition function may be written in terms of a reduced volume $\tilde{V}$ as $(\Omega\tilde{V})^N/N!$. The quantity Ω has the dimensions of an inverse volume, resulting in the partition function being dimensionless. $\tilde{V}$ is then expressed as $\tilde{V} = V - k\varepsilon(\eta)Nv_0$. The parameter ε is a dimensionless quantity which can be expected to approach unity as the packing fraction tends to its maximum value $\eta_{\max} = 1/k$.

To complete Eq. (2.11), one needs a way to determine the function $\varepsilon(\eta)$. Edgal and Huber write for this function a differential equation in the form

$$1 + k\eta^2 \frac{d\varepsilon(\eta)}{d\eta} = \exp[-4\eta(1-k\eta)^{-m(\eta)}] \exp(Z-1) \,. \tag{2.12}$$

The equation of state of the hard-particle fluid is therefore characterized by the function $m(\eta)$. To date, this function is only known precisely for small η through the expansion

$$m(\eta) = 2.407 - 0.9468\eta + \cdots \,, \tag{2.13}$$

where known exact results for the third and fourth virial coefficients have been used to obtain the numerical values shown.

Finally Edgal and Huber use the above results to construct a differential equation for the compressibility ratio Z, which is characterised by the function $m(\eta)$. In particular, the high-density asymptotic solution for Z is

$$Z \approx 4\eta(1-k\eta)^{-m(\eta)} = 4\eta\left(1 - \frac{\eta}{\eta_{\max}}\right)^{-m(\eta)} \,. \tag{2.14}$$

The merit of writing the compressibility ratio in this way is that $m(\eta)$ is a relatively slowly varying function of η. Examining their Fig. 1, and taking account of computer simulation data suggests values of $m \approx 1.84$ and $m \approx 1.14$ corresponding to simulation data for the low and the high density branch.

2.9 Free Volume Revisited in Hard Sphere Fluid

The fact that the properties of the hard sphere fluid stem from strictly entropic terms, that is from purely geometrical considerations, underlies the ongoing interest in this model. This is reflected in considerable emphasis directed to gaining understanding of the statistical geometry of dense sphere packings (see Sastry *et al.*[33] for detailed references; see also Gonzales and Gonzales[34] and Speedy[35,36]).

In the study of Sastry *et al.* a method is presented for the efficient calculation of free volumes and corresponding surface areas in the hard sphere assembly. Their method is then used to evaluate the free-volume distribution of the fluid in a range of densities near to freezing.

From the distribution of free volumes, the equation of state can be obtained by a geometric analysis (for the reader with some acquaintance with computer simulation, this allows the calculation of the pressure in Monte Carlo studies). Moreover, Sastry *et al.* obtain the cavity-volume distribution indirectly from the free-volume distributions in a density range where direct measurement is inadequate. They also point out that direct measurement of the first moment of the cavity-volume distribution makes it possible to calculate the chemical potential near freezing.

More generally, quantities that describe the void space (volume available for the insertion of an additional hard sphere), the free volume (volume within which a given hard sphere can move without requiring changes in the positions of other spheres) and the corresponding surface areas are quite directly related to thermodynamic properties.

2.9.1 *Statistical geometry of high-density fluid*

Speedy and Reiss[37] have shown that the free volume and cavity volume distribution functions are related — their arguments are given below following the account of Sastry *et al.*[33]

They define $p(v)dv$ as the probability that a cavity has volume between v and $v+dv$, while $p(v_\mathrm{f})dv_\mathrm{f}$ is the probability that the free volume of a sphere lies between v_f and $v_\mathrm{f}+dv_\mathrm{f}$. Probability densities analogous to these can be defined for the cavity surface $p_\mathrm{s}(s)$ and the free surface $f_\mathrm{s}(s_\mathrm{f})$. For a given configuration of spheres, the union volume of the cavities represents the available space V_0. The surface area S_0 available comprises the surface area of the individual cavities. The average cavity volume and surface area may be expressed as

$$\langle v\rangle = \frac{\langle V_0\rangle}{N_\mathrm{c}} = \int_0^\infty xp(x)dx \qquad (2.15)$$

and

$$\langle s\rangle = \frac{\langle S_0\rangle}{N_\mathrm{c}} = \int_0^\infty yp_s(y)dy\,. \qquad (2.16)$$

N_c being the number of cavities in the system, averaged over all realisations of the particles.

Speedy[38] has demonstrated that the equation of state of a hard sphere fluid at equilibrium can be written in terms of the statistical geometry of the cavities as

$$\frac{p}{\rho k_B T} = 1 + \frac{\sigma}{2D}\frac{\langle s\rangle}{\langle v\rangle}\rho^2 \tag{2.17}$$

where D is the dimensionality of the system. He also proved the valuable relationship

$$\frac{\langle s_f\rangle}{\langle v_f\rangle} = \frac{\langle s\rangle}{\langle v\rangle} = \frac{\langle S_0\rangle}{\langle V_0\rangle} \tag{2.18}$$

and hence the pressure can be found from free-volume information alone:

$$\frac{p}{\rho k_B T} = 1 + \frac{\sigma}{2D}\frac{\langle s_f\rangle}{\langle v_f\rangle}. \tag{2.19}$$

It is worthy of note in the present context that Eq. (2.19) was suggested long ago by Hoover *et al.*[39] during their considerations on the dynamics of a light particle in a classical system.

2.9.2 *Chemical potential in terms of statistical geometry*

The chemical potential μ of the hard sphere system can also be directly connected to its statistical geometry (see e.g. Sastry *et al.*[33]):

$$\mu = k_B T \ln\left(\frac{\lambda^d N}{\langle V_0\rangle}\right) = k_B T \ln\left(\frac{\lambda^d N\langle 1/v_f\rangle}{N_c}\right) \tag{2.20}$$

where N is the number of particles in the assembly while λ is the thermal de Broglie wavelength [$\lambda = (2\pi\hbar^2/mk_B T)^{1/2}$ with $\hbar = 1.0542 \times 10^{-34}$ J s being Planck's constant]. It is useful to separate μ into the sum of an ideal and an excess contribution:

$$\mu = \mu_{id} + \mu_{ex} = k_B T \ln\left(\frac{\lambda^d N}{V}\right) + k_B T \ln\left(\frac{V}{\langle V_0\rangle}\right). \tag{2.21}$$

The excess term embodies the reversible work required to form a cavity of radius σ.

While both analytical and numerical techniques have been used[40] to study the cavity volume and free volume distributions in two dimensions (hard disks),

Sastry *et al.* focus on their exact determination for the three-dimensional assembly. The excess chemical potential as calculated from the cavity volume statistics is compared in their work with earlier results using more conventional approaches. The pressure was also determined in the vicinity of the freezing transition. For further details the interested reader is referred to the original work.

2.10 Hard Particles in Low Dimensions

In bringing this chapter on a "gross" account of the short-range order in a liquid to its conclusion, we may summarise its main points as follows. The pioneering studies of Bernal and Scott on static assemblies of hard spheres did already evidence the gross features of the structure of a dense liquid such as argon near its triple point. By endowing the hard sphere model with thermal agitation through the techniques of computer simulation, Alder and Wainwright were able to show that it could, starting from the fluid phase, undergo a transition to a close-packed crystal with ample free volume for thermal vibrations, as envisaged by Lindemann for a solid near melting. There has been continued interest in the hard sphere model, not only as a reference system for real monatomic liquids but also in regard to its statistical geometry and to how this relates to thermodynamic properties. On the other hand, it should also be mentioned at this point that dense packing of spheres has little relevance to the numerous disordered systems in which strong directionality of the interatomic forces induces open network structures. While excluded-volume considerations still apply in the immediate neighbourhood of each molecule, short and intermediate-range order reflect rather specific aspects of the molecular interactions (see Figs. 1.8–1.10 and Chaps. 8 and 9).

Let us close this chapter, therefore, by (i) briefly recalling some properties of the equation of state for systems of hard particles in lower dimensionalities (disks and rods) and for systems of hard ellipses mimicking anisotropic mesophases; and (ii) by summaries of studies of the equation of state of hard-body fluids and of hard sphere fluids in narrow cylindrical pores. This account may provide the reader with the stimulus to continue this study on his own.

2.10.1 *Rods and disks*

The equation of state for impenetrable rods which are constrained to move on a line is

$$\frac{p}{\rho k_{\mathrm{B}}T} = (1 - \rho b)^{-1}, \tag{2.22}$$

where ρ is the linear number density of the rods and b is the length of one rod.[41] Making a virial expansion of the RHS of Eq. (2.22) in powers of the density, it is seen that the virial coefficient of order $n + 1$ is b^n.

According to Eq. (2.22), the pressure rises smoothly and monotonically to become infinite for the close-packed array. It follows that there is no phase change in dimensionality $d = 1$. This result has been generalised by van Hove[42] to all one-dimensional assemblies for which the molecules possess hard cores and attractive forces of finite range. A transition can, however, appear if the attractive forces have infinite range but are infinitely weak.[43]

Turning to the two-dimensional case, the equation of state for a system of hard disks is not known in exact closed form paralleling Eq. (2.22). However, Alder and Wainwright[44] carried out computer simulations of the equation of state of assemblies of hard disks, with densities increasing to quite near that of the close-packed ordered array. The isotherms thereby found have two branches: a disordered fluid state and an ordered solid phase whose structure is that of the close-packed hexagonal lattice. The phase behaviour of the two-dimensional system of hard disks is thus analogous to that of the three-dimensional system of hard spheres.

2.10.2 *Hard ellipses*

Vieillard–Baron[45] has simulated an assembly of 170 long hard ellipses (with an axis ratio equal to 6) as a two-dimensional model for a nematic liquid crystal (see Chap. 9). A close-packed configuration of such a system, corresponding to the minimal value A_0 for the area of the confining box, is shown in Fig. 2.5. The isotherms of the model are found to exhibit two phase transitions with decreasing areal density. A first transition occurs at an area $A = 1.15A_0$ to a nematic mesophase, in which the axes of the ellipses are still oriented whilst the centres of the ellipses show no longer any extended-range order. The orientational order is then lost at a density 1.5 times smaller.

As for the hard-disk system, the disorientation transition to the disordered fluid phase leads to a sudden increase in the configuration space available to

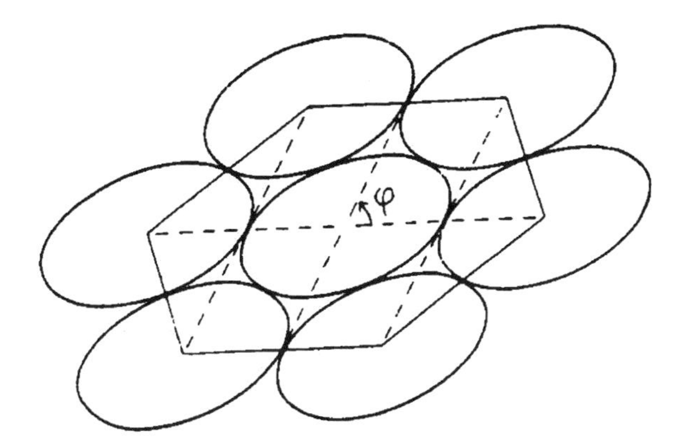

Fig. 2.5. A close-packed configuration for a hard-ellipse system. (Redrawn from Vieillard–Baron, Ref. 45.)

the system and hence in its entropy. However, since the disorientation affects only one degree of freedom per ellipse instead of two, the nematic and liquid branches of the isotherm are very close to each other. The observed entropy change $\Delta S/Nk_\mathrm{B}$ has a value between 0.05 and 0.12, much smaller than that for melting of hard disks which is 0.36.

2.11 Equation of State of Hard-Body Fluids

We have referred earlier to approximate forms of the equation of state for the hard sphere fluid. Here we record an extension to fluids in which the basic building block is not spherical.

Kolafa and Nezbeda[46] have considered a fluid of hard tetrahedra and raised the question as to its possible relevance as a model for the structure of water. Without going into details, these authors adopt as an equation of state valid for large non-sphericities the expression

$$\frac{p}{\rho k_\mathrm{B}T} = \frac{1 + (3\alpha - 2)\eta + (\alpha^2 + \alpha - 1)\eta^2 - \alpha(5\alpha - 4)\eta^3}{(1-\eta)^3} . \tag{2.23}$$

As usual η represents the packing fraction. The non-sphericity parameter α has the value $\alpha = 2.2346$ for a fluid of hard tetrahedra.

Kolafa and Nezbeda compare the analytic form (2.23) with Monte Carlo results available in the literature on a fluid of regular hard tetrahedra and find it to be good up to moderately large packing ($\eta < 0.3$). Beyond the equation of state results, they conclude that nowithstanding the fact that there are no attractive forces in their model, the tetrahedral shape of the hard basic units leads at high densities to a structure which resembles that of water.

2.12 Hard Sphere Fluid in Narrow Cylindrical Pores

Fluids in restricted geometries have been studied over a long period (see e.g. Warnock *et al.*,[47] Klafter and Drake[48] and Henderson[49]) and are often found to behave very differently from bulk fluids. Here we record results from a study by Mon and Percus[50] of the pore radius dependence of hard sphere fluids in very narrow cylindrical pores with hard walls over a substantial range of pressure and density.

These workers deal with sixty hard spheres of diameter σ in cylindrical pores of radius R. The length of the cylinder is denoted by L and the two ends obey periodic boundary conditions. To investigate the effect of such restricted geometry on the equation of state, Mon and Percus consider a constant-pressure Monte Carlo (MC) simulation (see Allen and Tildesley[18]). The radius of the pore remains constant but the length of the cylinder is permitted to fluctuate with Boltzmann weight of $\exp(-pV/k_{\mathrm{B}}T)$, where p is the pressure while V denotes the cylinder volume.

Figure 2.6 shows the constant-pressure MC results for the density corresponding to three values of the pore radius. The important point to note from this figure is the non-monotonic dependence on the pore radius with a minimum at $R = \sigma$ and an increase towards the bulk limit with increasing R. This non-monotonic behaviour persists over a very wide range of pressure of density, both near and much below the dense solid. To illustrate this further, Fig. 2.7 shows a plot of density against pore radius for a pressure $p = 0.5k_{\mathrm{B}}T/\sigma^3$. Two system sizes are shown and no significant differences appear.

Mon and Percus explain the above behaviour as follows. For pore radius less than σ, the fluid density is maximum at the smallest allowable radius $\sigma/2$. On increasing the radius, the density initially will fall since the pore radius is still too small to accommodate two or more hard spheres across the diameter. It is only above a pore radius of σ that much more dense packing configurations are allowed, which then increase the fluid density. Mon and Percus also observe an inflexion point near the onset of configurations with three hard spheres across

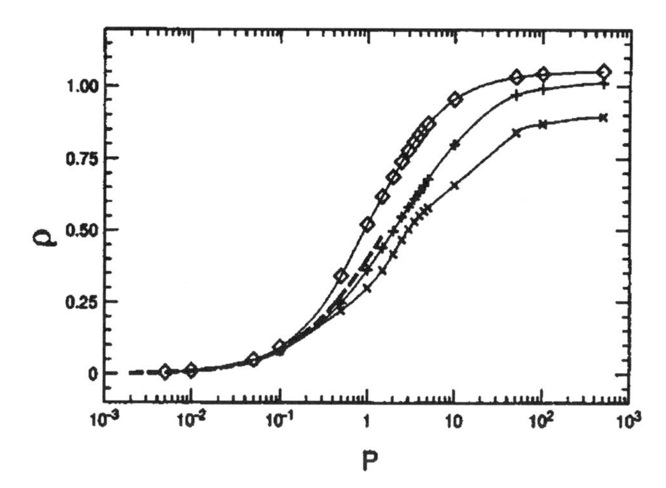

Fig. 2.6. Monte Carlo results for density against pressure in a hard sphere fluid inside narrow cylindrical pores. For ◇, x and + the pore radius is 0.55, 1.0 and 1.75 times the hard sphere diameter σ. Density is in units of number of particles per σ^3, βp in units of σ^3. The estimated sampling errors are smaller than the size of the plotting symbols and the solid lines are only to guide the eye. The dashed line gives the Carnahan–Starling equation of state for the bulk hard sphere fluid. (Redrawn from Mon and Percus, Ref. 50.)

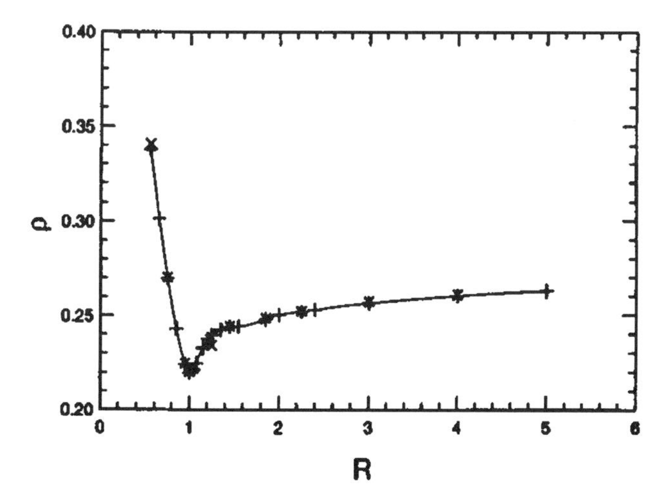

Fig. 2.7. Monte Carlo results for the density (in units of number of particles per σ^3) against pore radius R (in units of σ) at constant pressure $\beta p = 0.50/\sigma^3$. The two symbols refer to a system of 60(+) and 120(x) particles. (Redrawn from Mon and Percus, Ref. 50.)

the pore. This occurs at a radius of $[(1/2) + (1/\sqrt{3})]\sigma$. The density then approaches the bulk fluid density smoothly for larger radius.

The interested reader should consult the study of Mon and Percus who discuss another way to understand the origin of the non-monotonic dependence by using a low-pressure expansion for the mean density.

Chapter 3

Thermodynamics, Equipartition of Energy and Some Scaling Properties

In this chapter, some thermodynamics plus a little relevant statistical mechanics will be summarised. After an introduction to the thermodynamic functions which are needed to describe the fluid state of matter, we shall discuss specific heats and compressibilities, which play an important role in theories of especially classical liquids, and shall emphasise their role in governing fluctuations around the equilibrium state. Thermodynamics relevant to the melting transition will then be surveyed, again quite briefly.

In the latter part of the chapter we shall pause on the evaluation of thermodynamic functions in statistical mechanics. Specific attention will be given to the principle of equipartition of energy and to thermodynamic properties of the hard sphere fluid. Finally, scaling properties of thermodynamic quantities for a special, but important class of repulsive potentials, namely inverse power potentials $\phi(r) \propto r^{-n}$ with $n \geq 1$ will be presented. As already mentioned in Chap. 2, the limit $n \to \infty$ recovers the hard sphere system.

3.1 Thermodynamic Functions for a Fluid

We recall in this section some basic facts of thermodynamics for a (one-component) fluid in contact with a thermal bath and subject to pressure forces only.[51] Considering the entropy $S(U, N, V)$ of the fluid as a function of its internal energy U, its number N of particles and its volume V, the definitions of absolute temperature T, chemical potential μ, and pressure p follow by imposing conditions of thermal equilibrium through maximisation of the total entropy. The notion of entropy as a measure of the number of accessible microscopic states is then to be used to obtain the rules of statistical mechanics for the evaluation of the thermodynamic functions.[52]

In particular the definition of inverse temperature,

$$\frac{1}{T} = \left(\frac{\partial S}{\partial U}\right)_{N,V} \tag{3.1}$$

follows from imposing equilibrium against the net exchange of energy between the fluid and the thermal bath. Of course, thermal contact in isothermal conditions allows spontaneous exchange of energy, so that small fluctuations of energy occur in the fluid at thermal equilibrium. By the same argument it also follows that, if initially the fluid and the thermal bath are not at the same temperature, then there is a net flow of energy from the hotter to the cooler system.

Let us next consider the more general case in which both thermal contact and diffusive contact are established. Diffusive contact means that molecules can freely move from one system to the other across their permeable boundary. The equilibrium against net transfer of matter is determined by the condition of constancy of chemical potential, the latter being defined by

$$\mu = -T\left(\frac{\partial S}{\partial N}\right)_{U,V} . \tag{3.2}$$

Thus, two bodies that can exchange energy and particles are in mutual equilibrium when their temperatures and their chemical potentials are equal. Again, this situation allows spontaneous exchanges of particles between the two bodies and small fluctuations in the number of particles. Away from diffusive equilibrium, a net flow of matter passes from the body of high chemical potential to that of low chemical potential.

Finally, the notion of pressure is introduced by considering the mechanical work done on the fluid through a quasi-static displacement of its boundaries and by equating it to the change in its internal energy:

$$p = -\left(\frac{\partial U}{\partial V}\right)_{S,N} . \tag{3.3}$$

Such a quasi-static process is reversible and the entropy remains constant in time, its rate of change being quadratic in the rate of displacement of the boundaries. Small fluctuations in pressure on a boundary surface will occur and unbalance of pressure across a boundary surface will act as a net mechanical force driving its displacement.

3.1.1 *Thermodynamic identity and the first principle of thermodynamics*

The expression (3.3) for the pressure may also be written in the form[a]

$$p = T\left(\frac{\partial S}{\partial V}\right)_{U,N} . \tag{3.4}$$

This expression is valid only for reversible changes.

It then follows from Eqs. (3.1), (3.2) and (3.4) that during an infinitesimal thermodynamic process the change in the entropy function $S(U, N, V)$ is given by $dS = (dU - \mu dN + pdV)/T$, or equivalently that the change in the internal energy function $U(S, N, V)$ is

$$dU = TdS + \mu dN - pdV . \tag{3.5}$$

The thermodynamic identity in Eq. (3.5) is closely related to the first principle of thermodynamics, expressing conservation of energy in any thermodynamic process undergone by the fluid. The first principle states that the change dU of internal energy of the fluid in an infinitesimal process is the sum of the amount of heat $đQ$ exchanged with the fluid and of the work $đL$ done on it,

$$dU = đQ + đL \tag{3.6}$$

(we are using the symbol $đ$ to denote an infinitesimal change which depends on the path followed during the process). Energy added to the fluid by thermal contact with a reservoir is called heat, while energy added by all other means is called work.

In the case of a reversible process we can compare Eqs. (3.5) and (3.6) term-by-term and set $đQ = TdS$ and $đL = \mu dN - pdV$, the work $đL$ being the sum of a chemical term μdN and a mechanical term $-pdV$. Of course, chemical work is of special relevance in the context of electro-chemistry and for this we refer the reader to the book of Førland *et al.*[53]

[a]In deriving Eq. (3.4) from Eq. (3.3), we set $(\partial S/\partial V)_U + (\partial S/\partial U)_V(\partial U/\partial V)_S = (\partial S/\partial V)_S = 0$, whence $(\partial U/\partial V)_S = -(\partial S/\partial V)_U(\partial U/\partial S)_V = -T(\partial S/\partial V)_U$ (at constant number of particles).

On the other hand, in an irreversible process the inequality $TdS > đQ$ holds, expressing the fact that the entropy is increasing in such a process beyond and besides the quantity of heat that may be transferred to the fluid. We therefore reach the important inequality

$$đL_{\text{irrev}} > đL_{\text{rev}} = \mu dN - pdV\,. \tag{3.7}$$

This inequality shows that the work done in bringing the system from a given initial state to a given final state through an irreversible process is always greater than that done in a reversible process between the same two states. The inequality may arise from two main reasons: (i) irreversibility is due to friction and dissipation, and additional work is needed to compensate the accompanying loss of energy; or (ii) during the irreversible process no work is done on the system, whereas negative work would be done if the same final state were reached in a reversible manner from the same initial state. An example of case (ii) is the expansion of a gas into a larger volume through a hole opened in its container. No work is done on the gas in this process whereas, if the same final state were reached in a reversible manner, the work done would be $-pdV$ and this is negative since the gas has expanded ($dV > 0$).

3.1.2 *Helmholtz free energy and variational principle*

From the above summary on basic thermodynamic variables and on the first principle of thermodynamics we proceed to introduce the thermodynamic functions which are most useful for statistical mechanics. These are primarily the free energy functions.

Let us consider an infinitesimal reversible process carried out at constant T and N. The work done during this process becomes an exact differential, since the identity (3.5) gives $đL_{\text{rev}} = -pdV = d(U - TS)$. We therefore introduce the thermodynamic state function

$$F = U - TS\,, \tag{3.8}$$

which is known as the Helmholtz free energy. From Eq. (3.5) its differential is given by

$$dF = -SdT - pdV + \mu dN \tag{3.9}$$

i.e. the Helmholtz free energy is a function $F(T, V, N)$.

From Eq. (3.9) we find the following important results for the first derivatives of F:

$$S = -\left(\frac{\partial F}{\partial T}\right)_{V,N}, \tag{3.10}$$

$$p = -\left(\frac{\partial F}{\partial V}\right)_{T,N}, \tag{3.11}$$

$$\mu = \left(\frac{\partial F}{\partial N}\right)_{V,T}. \tag{3.12}$$

Thus, knowledge of the Helmholtz free energy enables entropy, pressure and chemical potential to be directly calculated. We also remark that the definition (3.12) of the chemical potential is more appealing to intuition than that given in Eq. (3.2): μ is seen to correspond to the average change in Helmholtz free energy associated with adding or taking away a particle at constant volume and temperature. This alternative meaning for the chemical potential explains the crucial role that this concept plays in phase equilibria and interfacial phenomena.

Let us now consider the spontaneous evolution of a fluid towards equilibrium in an irreversible process at constant T and N, no work being done on the system. Since $dF = đL_{\text{rev}} < đL_{\text{irrev}} = 0$, F must decrease in such process. The following variational principle thus holds: *in the equilibrium state F is at a minimum against changes of state which occur at constant T, N and V.* This principle affords a precise definition of the state of thermal equilibrium for a macroscopic body.

Finally, we note that Eq. (3.11) can be generalised into the Hellmann–Feynman theorem,

$$\left(\frac{\partial F}{\partial f}\right)_{V,N,T} = \left\langle \frac{\partial H}{\partial f} \right\rangle, \tag{3.13}$$

where f is any parameter describing a quasi-static external field and H is the Hamiltonian of the system, the symbol $\langle \cdots \rangle$ denoting the average value. For instance f could be an applied electric field $\mathbf{E}$ and in this case from Eq. (3.13) we can find the polarisation as $\mathbf{P} = -(\partial F/\partial \mathbf{P})_{V,N,T}$, since the coupling of the system to the electric field is $-\mathbf{P}\cdot\mathbf{E}$.

3.1.3 *Gibbs free energy*

The Gibbs free energy is defined as $G = F + pV$. From Eq. (3.9) its differential is

$$dG = -SdT + Vdp + \mu dN\,, \tag{3.14}$$

i.e. $G = G(T, p, N)$ and $\mu = (\partial G/\partial N)_{p,T}$. However, since T and p are intensive variables, the dependence of G on N is of the simple type $G = N * fn(T, p)$, yielding $\mu = (\partial G/\partial N)_{p,T} = G/N = fn(T, p)$. In a monatomic fluid the chemical potential is the Gibbs free energy per particle and is a function of T and p.

An immediate consequence is the Gibbs–Duhem relation, which follows from Eq. (3.14) by setting $G = N\mu$: the chemical potential is a function of temperature and pressure, and its differential is given by

$$d\mu = -\left(\frac{S}{N}\right) dT + \left(\frac{V}{N}\right) dp\,. \tag{3.15}$$

The quantities entering this relation are the entropy and the volume per particle.

The Gibbs free energy obeys a variational principle which is analogous to that holding for the Helmholtz free energy: *in the equilibrium state G is at a minimum against changes of state occurring at constant T, N and p.*

3.2 Specific Heats and Compressibilities

According to the variational principles on the Helmholtz (or Gibbs) free energy that we have met in Sec. 3.1, the equilibrium state is defined as a state of minimal free energy against changes occurring at constant T, N and V (or p). Let us consider an infinitesimal process which moves the system out of equilibrium by changing its internal energy, entropy and volume by amounts dU, dS and dV at constant T, N and p. We impose the condition that the Gibbs free energy increases in such a process,

$$dG = dU - TdS + pdV > 0\,, \tag{3.16}$$

and use the thermodynamic identity (3.5) to find that the first-order terms cancel away. Clearly, a condition of minimal free energy implies inequalities on its second derivatives.

These inequalities follow from Eq. (3.16) by expanding the change dU in the internal energy function $U(S, V, N)$ up to quadratic terms in dS and dV. One finds that the second-order change in free energy is always positive provided that two conditions are satisfied: (i) the heat capacity at constant volume

$$C_V \equiv T\left(\frac{\partial S}{\partial T}\right)_V = \left(\frac{\partial U}{\partial T}\right)_V \tag{3.17}$$

must be positive, and (ii) the isothermal compressibility

$$K_T \equiv -\left(\frac{1}{V}\right)\left(\frac{\partial V}{\partial P}\right)_T \tag{3.18}$$

must be positive. Here and in the following, except where explicitly noted, we are keeping constant the number of particles. Since from Eqs. (3.10) and (3.11) we have $C_V = -T(\partial^2 F/\partial T^2)_V$ and $1/K_T = V(\partial^2 F/\partial V^2)_T$, it is clear that the stability condition (3.16) determines the signs of the second derivatives of the free energy: the function $F(T, V, N)$ is everywhere concave in T and convex in V. We also see that the internal energy U must be a monotonically increasing function of temperature. Similarly, it can be shown that the function $G(T, p, N)$ is concave in T and p.

3.2.1 *Specific heat at constant pressure*

There is a well-known relation expressing the difference in specific heats $C_p - C_V$ in terms of other thermodynamic quantities (see for instance the book of Zemansky[51]):

$$C_p - C_V = -T\left(\frac{\partial V}{\partial p}\right)_T \left[\left(\frac{\partial p}{\partial T}\right)_V\right]^2 . \tag{3.19}$$

Here, the heat capacity at constant pressure is defined as $C_p = T(\partial S/\partial T)_p$. The further result[51]

$$\left(\frac{\partial U}{\partial V}\right)_T = T\left(\frac{\partial p}{\partial T}\right)_V - p \tag{3.20}$$

allows Eq. (3.19) to be rewritten in the form

$$C_p - C_V = \frac{VK_T}{T}\left[p + \left(\frac{\partial U}{\partial V}\right)_T\right]^2 . \tag{3.21}$$

This yields the well known results $C_p - C_V = Nk_B$ in the limit of the ideal classical gas, with equation of state $pV = Nk_BT$.

We have seen in Sec. 2.6 that the quantity $(\partial U/\partial V)_T$ in Eq. (3.20) vanishes for the fluid of hard spheres. Equation (3.21) then yields the result $C_p - C_V = VK_Tp^2/T$, which is exact for this model. However, for real dense liquids the term $(\partial U/\partial V)_T$ can be quantitatively important in determining the difference in specific heats from Eq. (3.21). We shall see in Chap. 4 on structural theories of liquids that $(\partial U/\partial V)_T$ can in an (assumed) pair potential model be calculated from a structural correlation function (see Eq. (4.14)).

3.2.2 *Specific heat properties of liquid metals near freezing*

We wish to note here for future reference some interesting empirical properties of the specific heats of liquid metals near freezing. Thus we have recorded in Table 3.1 the ratio $\gamma = C_p/C_V$ of the specific heats near the melting temperature T_m and the value of C_V/Nk_B.

As Table 3.1 shows, γ is not substantially greater than unity, while C_V is quite near to, but usually somewhat larger than $3Nk_B$. As we shall see in Sec. 3.6 below, the latter value is appropriate to a classical system in which the thermal agitation can be described in terms of a superposition of harmonic oscillatory motions. We shall take up the underlying physics of this "harmonic-like" behaviour of liquid metals near freezing in Chaps. 5 and 6. We shall also see that during thermal motions the coupling between fluctuations in particle density and in heat density becomes negligible as γ approaches unity.

The assumption $C_p = C_V$, which becomes valid in strictly harmonic assemblies, is a useful "zero-order" approximation for the liquid metals in Table 3.1. In contrast, for liquid argon near its triple point one has $\gamma = 2.2$ and no quasi-harmonic approximation can be useful.

Table 3.1. Specific heats of liquid metals near the melting temperature.

	Na	K	Rb	Zn	Cd	Ga	Tl	Sn	Pb	Bi
$\gamma = C_p/C_V$	1.12	1.11	1.15	1.25	1.23	1.08	1.21	1.11	1.20	1.15
C_V/Nk_B	3.4	3.5	3.4	3.1	3.1	3.2	3.0	3.0	2.9	3.1

3.2.3 *Compressibilities, both adiabatic and isothermal*

It is a further well known result of thermodynamics[51] that the ratio of specific heats equals the ratio of compressibilities, i.e.

$$\frac{C_p}{C_V} = \frac{K_T}{K_S}. \tag{3.22}$$

The adiabatic compressibilities $K_S = -(1/V)(\partial V/\partial P)_S$ is experimentally accessible through measurements of the speed of sound — such experiments will be referred to in Chaps. 6 and 7.

The isothermal compressibility K_T is related to liquid structure in an important way, as we shall see in the following chapter. This fact is related to the special role played by K_T in governing fluctuations in the volume occupied by a given number of particles (or alternatively the fluctuations in the number of particles contained in a given volume) in a fluid at equilibrium.

In view of this fact, we turn next to a brief reminder of fluctuation phenomena on the thermodynamic scale.

3.3 Fluctuation Phenomena

This section introduces what is an important aspect of any statistical viewpoint: namely the fluctuations that occur in any thermodynamic property in a system at thermal equilibrium. The concept of root-mean-square (RMS) fluctuation is equally applicable in all branches of statistics.

Two of the most striking examples of fluctuations have already been introduced in Chap. 1, in presenting critical opalescence and Brownian motion. As mentioned there, the optical phenomena which occur in a normally colourless fluid near the liquid–vapour critical point are very remarkable. When illuminated by a beam of light the fluid appears diffuse and shimmering and extremely white, but as the temperature is raised or lowered by even a fraction of a degree away from the critical point, the whiteness disappears and the vapour or liquid is seen to be colourless again. This behaviour provides direct evidence that droplets of liquid and bubbles of vapour are continuously forming and breaking up inside the fluid near criticality, on a size scale which is comparable with the light wavelength.

Similarly, the movement of a mesoscopic Brownian particle suspended in a liquid is determined through collisions by the mean momentum of the molecules within a small volume of the surrounding liquid, comparable in size with that of the particle itself. The non-smooth motion of the particle shows that this quantity fluctuates. The observation of Brownian motion tells us that, although in a stationary fluid the mean momentum crossing any plane in the fluid interior averages to zero over a sufficiently long time interval, its value departs from zero at any instant. Namely, fluctuations in this momentum about its zero average value must occur.

We focus below on the fluctuations that occur in the density and the temperature of a classical fluid around its state of thermal equilibrium, as these are governed by the compressibility and heat capacity parameters that we have introduced just above in Sec. 3.2.

3.3.1 *Fluctuations in a perfect gas*

Later in this section we shall show, by simple physical arguments, that if a volume, say V_0, of a fluid in equilibrium at given pressure and temperature is altered to volume V by either expansion or contraction occurring in a spontaneous fluctuation, then we have

$$\frac{\text{Probability of volume } V}{\text{Probability of volume } V_0} = \exp\left(-\frac{BV_0}{2k_\mathrm{B}T}\Delta^2\right), \tag{3.23}$$

where $B = 1/K_T$ is the isothermal bulk modulus and $\Delta = |V - V_0|/V_0$ is the fractional change of volume. This formula assumes that the deviation from the state of equilibrium is limited in relative magnitude. The inequality $B > 0$ (see Sec. 3.2) then ensures that the state of equilibrium is stable against such density fluctuations.

Accepting Eq. (3.23) for the moment, let us first apply it to a classical perfect gas, with equation of state $pV = Nk_\mathrm{B}T$. Then B is simply equal to the pressure. Working out the mean square volume fluctuation $\langle\Delta^2\rangle$ from Eq. (3.23) yields after a short calculation

$$\langle\Delta^2\rangle = \frac{k_\mathrm{B}T}{BV_0} \tag{3.24}$$

and therefore for the perfect gas

$$\langle \Delta^2 \rangle = \frac{k_B T}{p V_0} = \frac{1}{N} . \tag{3.25}$$

Namely, the relative RMS fluctuation in the volume occupied by a given number N of particles in a classical perfect gas vanishes in the thermodynamic limit like $1/\sqrt{N}$.

An alternative way to express this result is by calculating the number of molecules to be found in a given volume of a classical perfect gas. Suppose that number, on average, to be N. From the result (3.25) it is easily seen that the actual number will fluctuate in the range $N \pm \sqrt{N}$. Fluctuations can still be said to be "small", in the sense that the *relative* RMS fluctuation in the number of particles is equal to $1/\sqrt{N}$ (or, more generally, of order $1/\sqrt{N}$, as is immediately seen from the factor V_0 entering Eq. (3.24)).

3.3.2 *Effect of intermolecular forces*

Including intermolecular forces, the result for $\langle \Delta^2 \rangle$ in Eq. (3.24) is still applicable, but the bulk modulus B no longer equals the pressure p. In fact, the main conclusion from Eq. (3.24) is that the smaller the bulk modulus, the larger are the fluctuations. Returning to critical opalescence, the bulk modulus tends to zero at the critical point, which would seem to herald infinite fluctuations. This is not physical, but can soon be correctly dealt with. The essential point is that fluctuations in volume (or density) are long-ranged in the critical region and this is the reason light is strongly scattered by a fluid near its critical point.

The remaining point of this section is to sketch the derivation of Eq. (3.23) above. Since the volume is changed isothermally from V_0 to V, we have (now approximately if V is near to V_0) $B \cong -V_0(\partial p/\partial V)_T$ and the extra pressure is $p' = -B(V - V_0)/V_0$. The energy required to make a further volume increase dV in isothermal conditions is $-p'dV$, so that the total energy needed is related to the fractional volume change Δ by

$$\Delta U = \frac{B}{V_0} \int_{V_0}^{V} (V - V_0) dV = \frac{1}{2} B V_0 \Delta^2 . \tag{3.26}$$

We can now calculate the ratio between the probabilities of these two states having volume V_0 and V by using the Boltzmann factor $\exp(-\Delta U/k_B T)$. Hence the result (3.23) is obtained.

3.3.3 *Temperature fluctuations*

A similar treatment can be given for the isochoric fluctuations of temperature, with the result that the probability of a fluctuation ΔT is proportional to $\exp[-C_V(\Delta T)^2/2k_BT^2]$. Namely, the "thermal stiffness" of a fluid is determined by its specific heat and the mean square fluctuation in temperature is given by

$$\langle(\Delta T)^2\rangle = \frac{k_B T^2}{C_V} \tag{3.27}$$

at constant volume. The relative RMS fluctuation is inversely proportional to $C_V^{1/2}$ and hence again vanishes like $1/\sqrt{N}$ as the number of particles increases.

We leave this topic and turn to a brief discussion of the thermodynamics of melting.

3.4 Clausius–Clapeyron Equation and Melting

Following Pippard,[54] we give below a derivation of the basic equation governing a first-order phase transition such as melting. We first note that along any equilibrium line separating two different phases (1 and 2, say) in the $p-T$ diagram, the Gibbs free energies (per mole, say) are equal, i.e.

$$G_1 = G_2\,. \tag{3.28}$$

Namely, the two phases coexist at the same pressure, temperature and chemical potential. This condition ensures that there is no drift in the boundary between the two phases and no net transfer of energy or matter across the interface. In what follows we choose the suffix 1 to label that phase which is stable on the low-temperature side of such an equilibrium line.

Due to Eq. (3.28), for small changes dp and dT of pressure and temperature which alter the state of the system to a neighbouring state still on the equilibrium line, the variations of G_1 and G_2 must be equal. From Eq. (3.14) we easily find

$$(S_2 - S_1)dT = (V_2 - V_1)dp\,. \tag{3.29}$$

We can use Eq. (3.29) as $dT \to 0$ to relate the slope dp/dT of the equilibrium line to the ratio of differences in specific entropy and volume of the two phases at equilibrium:

$$\frac{dp}{dT} = \frac{(S_2 - S_1)}{(V_2 - V_1)}. \tag{3.30}$$

This completes the derivation of the Clausius–Clapeyron equation. If one wishes, the RHS of Eq. (3.30) can be written in terms of the latent heat of the transition, $L = T(S_2 - S_1)$.

In Sec. 2.6 we have already used the relationship (3.30) in discussing the thermodynamic parameters of melting for the hard sphere system. More generally, Eq. (3.30) can be used to describe the dependence of the temperature T_m of melting on pressure, in the form

$$\frac{dT_m}{dp} = \frac{(V_L - V_S)}{S_m}, \tag{3.31}$$

where V_L and V_S are the molar volumes of the liquid and solid phase at coexistence and S_m is the molar entropy of melting. When the pressures used

Table 3.2. Thermodynamic parameters of melting for alkali halides.

Salt	S_m (e.u.)	ΔV_m (cm^3/mol)	$\Delta V_m/S_m$ (deg/bar)	$(dT/dp)_{obs}$ (deg/bar)
LiCl	5.6	5.88	0.025	0.0242
NaF	6.2	5.15	0.016	0.0161
NaCl	6.7	7.55	0.027	0.0238
NaBr	6.0	8.02	0.032	0.0287
NaI	5.6	8.58	0.037	0.0327
KCl	6.2	7.23	0.028	0.0265

Table 3.3. Parameters in the Simon melting equation.

Crystal	T_0 (K)	a (kbar)	c
^{4}He	2.046	0.05096	1.5602
^{3}He	3.252	0.11760	1.5178
Kr	115.745	2.376	1.6169
Xe	161.364	2.610	1.5892
In	429.76	35.800	2.30
Ni	1726.0	1020.0	2.2
Pt	2046.0	1020.0	2.0
Fe	1805.0	1070.0	1.76

are small the quantities on the RHS of Eq. (3.31) can usually be treated as independent of pressure: investigations of the melting curve at fairly low pressures have thus provided a general means of evaluating the volume change on melting from the more commonly known entropy change. An example for some alkali halides is reported in Table 3.2, from the book of Ubbelohde.[55] We shall discuss in Chap. 8 the empirical relationships that exist between volume and entropy changes on melting for a broad class of ionic systems, stressing their structural implications.

Empirical forms of the melting curve have also been proposed from studies over moderate pressure ranges. An example for crystals with simple structures is the empirical equation of melting due to Simon, which reads

$$p - p_0 = a\left[\left(\frac{T}{T_0}\right)^c - 1\right] . \tag{3.32}$$

Here, p_0 and T_0 are the pressure and temperature at the triple point, while a and c are empirical parameters characteristic of the material. For most systems p_0 in Eq. (3.32) can be neglected, and an illustrative tabulation of the other constants is given in Table 3.3.[56]

With the development of practicable means of applying very high pressures, as are allowed by the diamond anvil cell,[57] the study of solid–liquid equilibria has been expanding tremendously. Figure 3.1 shows experimental data on the melting curve of argon up to 717 K and 60 kbar, as obtained by an interferometric technique in a diamond anvil cell.[58] Examples of current foci of interest in the high-pressure area are (i) the emergence of electron delocalisation and metallic conduction in insulators as the overlap of the molecular orbits of neighbouring atoms is enhanced by the application of pressure,[59,60] and (ii) the equation of state for planetary materials (especially hydrogen) under pressure of astrophysical relevance[61] (see Sec. 14.6).

In Appendix 3.1 explicit examples are given of the use of the Clausius–Clapeyron equation (3.30) to characterise various kinds of phase transitions.

3.5 Free Energy from Partition Function

The evaluation of the Helmholtz free energy through the normalisation condition on the Gibbs' canonical ensemble is the basic law of equilibrium statistical mechanics.[52] $F(T, V, N)$ is written in terms of the partition function Q for N

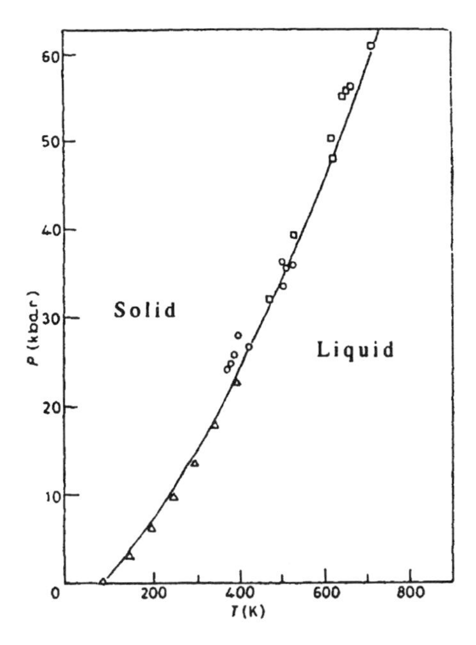

Fig. 3.1. Melting curve of argon at high pressure from smoothed piston-cylinder (Δ) and diamond-anvil-cell experiments ($\circ$ and $\square$), and from theory. (Redrawn from Zha *et al.*, Ref. 58.)

particles in an assembly in thermal equilibrium at temperature T as

$$F = -k_{\mathrm{B}} T \ln Q \,. \tag{3.33}$$

The partition function is in turn determined by the spectrum of energy levels E_n of the system under study according to

$$Q = \sum_{\text{all } n} g_{\mathrm{n}} e^{-E_n/k_{\mathrm{B}}T} \,, \tag{3.34}$$

where g_{n} is a degeneracy factor for each energy level.

As a particularly simple example we consider a set of N_{ho} identical harmonic oscillators, as is met in dealing with the vibrational motions in a gas of diatomic molecules or in the Einstein model for lattice dynamics. In this case the energy levels of each oscillator are

$$\varepsilon_n = \left(n + \frac{1}{2} \right) \hbar\omega \,, \tag{3.35}$$

where ω is the angular frequency of the oscillator and the index n can take all values $n = 0, 1, 2, \ldots$. The summation in Eq. (3.34) is readily carried out to yield the partition function

$$Q_{\text{ho}} = \frac{e^{-\hbar\omega/2k_{\text{B}}T}}{1 - e^{-\hbar\omega/k_{\text{B}}T}}, \tag{3.36}$$

for each oscillator. Hence, the partition function of the assembly of identical oscillators is

$$Q = (Q_{\text{ho}})^{N_{\text{ho}}} \tag{3.37}$$

and correspondingly, we find

$$F_{\text{ho}} = \frac{1}{2}N_{\text{ho}}\hbar\omega + N_{\text{ho}}k_{\text{B}}T\ln[1 - e^{-\hbar\omega/k_{\text{B}}T}]. \tag{3.38}$$

All other thermodynamic functions of this simple model can then be evaluated starting from Eqs. (3.10)–(3.12), if one knows the volume dependence (if any) of the vibrational frequency.

Until the final chapter of this book, however, we shall be dominantly concerned with classical liquids, and in the above example this means taking the limit such that the thermal energy $k_{\text{B}}T$ is very much greater than the level spacing $\hbar\omega$. Equation (3.38) reduces to

$$F_{\text{ho}} = N_{\text{ho}}k_{\text{B}}T\ln\left(\frac{\hbar\omega}{k_{\text{B}}T}\right), \tag{3.39}$$

whence

$$S_{\text{ho}} = -\left(\frac{\partial F_{\text{ho}}}{\partial T}\right)_V = N_{\text{ho}}k_{\text{B}}\left[1 - \ln\left(\frac{\hbar\omega}{k_{\text{B}}T}\right)\right] \tag{3.40}$$

and

$$U_{\text{ho}} = F_{\text{ho}} + TS_{\text{ho}} = N_{\text{ho}}k_{\text{B}}T. \tag{3.41}$$

This result will be utilised immediately below to motivate the statement of an important physical principle for classical liquids.

3.6 Principle of Equipartition of Energy

A crucial simplification in classical statistical mechanics over the quantal case is that the value of the average kinetic energy for a classical assembly in thermal equilibrium is independent of the nature of the interatomic interactions and is equal to $k_BT/2$ for each degree of freedom of (translational, rotational or vibrational) motion.

3.6.1 *Internal energy and other thermodynamic functions of a perfect gas*

Let us then consider the model of a monatomic perfect gas, in which we drop all interatomic interactions and omit electronic excitations arising only at very high temperatures. There are then no internal degrees of freedom and with each of the N atoms there are three translational degrees of freedom. Therefore, the above principle immediately allows the internal energy U (all kinetic energy in this example) to be written as

$$U = \frac{3}{2} N k_B T . \tag{3.42}$$

We next use the so-called Gibbs–Helmholtz relation,

$$U = F - T\left(\frac{\partial F}{\partial T}\right)_V = -T^2\left(\frac{\partial (F/T)}{\partial T}\right)_V , \tag{3.43}$$

which is obtained by combining Eqs. (3.8) and (3.10). The combination of Eqs. (3.42) and (3.43) then yields the first-order differential equation

$$\left(\frac{\partial (F/T)}{\partial T}\right)_V = -\frac{3}{2}\frac{N k_B}{T} , \tag{3.44}$$

which integrates to

$$F(T, V, N) = -N k_B T \left[\ln T^{3/2} + \mathrm{fn}(V/N)\right] . \tag{3.45}$$

The function of the specific volume V/N, that we have indicated by $\mathrm{fn}(V/N)$, enters Eq. (3.45) as a constant of integration.

If we now use in Eq. (3.11) the equation of state of the perfect gas, $pV = Nk_BT$, we easily find $\mathrm{fn}(V/N) = \ln(V/N) + \text{constant}$. The full result for the Helmholtz free energy, to be obtained from the calculation of the partition function for the case when the energy levels are determined by the translational kinetic energy of each atom, is (see e.g. Landau and Lifshitz[52])

$$F(T,V,N) = -Nk_{\mathrm{B}}T\left\{1+\ln\left[\frac{V}{N}\left(\frac{mk_{\mathrm{B}}T}{2\pi\hbar^2}\right)^{3/2}\right]\right\}. \tag{3.46}$$

In Eq. (3.46) we recognise the role of the thermal de Broglie wavelength $\lambda = (2\pi\hbar^2/mk_{\mathrm{B}}T)^{1/2}$, already introduced in Sec. 2.9, in giving a natural unit of length for the volume per particle V/N.

3.6.2 *Harmonic oscillator revisited*

In contrast to Eq. (3.42), the internal energy of a classical assembly of N_{ho} harmonic oscillators is given by Eq. (3.41). Since each oscillator corresponds to one degree of freedom for vibrational motion, the number of oscillators is related to the number of particles by $N_{\mathrm{ho}} = 3N$. We find from Eq. (3.41) that the vibrational internal energy in the classical regime is

$$U_{\mathrm{vib}} = 3Nk_{\mathrm{B}}T\,. \tag{3.47}$$

Of course, one-half of the vibrational internal energy in Eq. (3.47) arises from the kinetic degree of freedom, according to Eq. (3.42). The other half is associated with the potential energy of the oscillators. Since the Hamiltonian is in this case quadratic in both the momenta and the displacements, both sets of dynamical variables contribute the same amount to the internal energy, i.e. $3Nk_{\mathrm{B}}T/2$.

As a final remark, from Eq. (3.47) we find

$$C_V = \left(\frac{\partial U_{\mathrm{vib}}}{\partial T}\right)_V = 3Nk_{\mathrm{B}}\,, \tag{3.48}$$

for the specific heat of a set of classical harmonic oscillators. This was anticipated in Sec. 3.2.2.

3.7 Thermodynamic and Other Properties of Hard Sphere Fluid

In addition to the "ideal" term given in Eq. (3.46), the free energy of the hard sphere fluid contains an "excess" term due to the excluded volume from hard sphere packing. We write

$$F = F_{\text{id}} + F_{\text{ex}}\,, \tag{3.49}$$

where from Eq. (3.46)

$$F_{\text{id}} = -Nk_{\text{B}}T[1 - \ln(\rho\lambda^3)]\,. \tag{3.50}$$

Here, $\rho = N/V$ and λ is the thermal de Broglie wavelength.

The excess term and other thermodynamic properties of the hard sphere fluid can usually be derived with sufficient accuracy from the Carnahan–Starling equation of state (see Sec. 2.6). This reads

$$\frac{p_{\text{hs}}}{\rho k_{\text{B}}T} \equiv Z_{\text{hs}}(\eta) = \frac{(1 + \eta + \eta^2 - \eta^3)}{(1-\eta)^3} \tag{3.51}$$

with $\eta = \pi\rho\sigma^3/6$ the packing fraction (Eq. (2.6)). We therefore have

$$F_{\text{ex}} = Nk_{\text{B}}T \int_0^{\eta} d\eta' \frac{Z_{\text{hs}}(\eta') - 1}{\eta'} = Nk_{\text{B}}T \frac{\eta(4 - 3\eta)}{(1-\eta)^2}\,. \tag{3.52}$$

During the discussion of the structural properties of fluids in Chap. 4 it will be shown that within a pair-potentials model the pressure can be obtained from the interatomic force weighted by the radial distribution function $g(r)$, which gives the probability of finding two atoms in the fluid at a distance r from each other. However, in the hard sphere fluid interatomic forces arise only between particles at contact: one accordingly finds that the pressure is related to the value $g_{\text{hs}}(\sigma^+)$ of the radial distribution function at contact. Precisely,

$$Z_{\text{hs}}(\eta) = 1 + 4\eta g_{\text{hs}}(\sigma^+)\,, \tag{3.53}$$

(see Eq. (4.16) and the discussion following it). From Eqs. (3.51) and (3.53) one gets

$$g_{\text{hs}}(\sigma^+) = \frac{(1 - \eta/2)}{(1-\eta)^3}\,. \tag{3.54}$$

The contact value of $g(r)$ in the Bernal–Scott model was displayed in Fig. 2.1.

3.8 Scaling of Thermodynamic Properties for Inverse-Power Repulsive Potentials

From the very elementary examples above, let us turn to the thermodynamic properties of again a classical monatomic assembly, but this time with repulsive pair interactions described by a pair potential $\phi(r)$ having the form

$$\phi(r) = \varepsilon \left(\frac{\sigma}{r}\right)^n . \tag{3.55}$$

The merit of this reference system is that, for any inverse-power potential, thermodynamic properties are easy to calculate because only a single isotherm, isochore or isobar needs to be known: all others can then be determined, as discussed by Hoover and Ross.[62] This property follows from the partition function, which for a Hamiltonian H built from pair potentials is written explicitly in Appendix 3.2.

The essential feature to note for present purposes, which follows from the partition function constructed from pair-wise additive potentials of the form (3.55), is that the (now dimensionless) thermodynamic properties obtained by using Eq. (3.33) for F from the partition function depend only on the single density-temperature variable, denoted by x below:

$$x = \rho \left(\frac{\varepsilon}{k_\mathrm{B}T}\right)^{3/n} \tag{3.56}$$

with $\rho = N\sigma^3/V$.

3.8.1 *Consequence for melting transition*

Along any isotherm, isochore or isobar the melting transition is marked by discontinuities in $(\partial p/\partial V)_T$, $(\partial p/\partial T)_V$, $(\partial T/\partial V)_p$: all these derivatives being discontinuous at melting and again at freezing (see for example the isotherms of the hard sphere system in Fig. 2.4). The discontinuities, which signal the start and end of the phase transition, occur at two characteristic values of the density-temperature variable, say $x_\mathrm{S} = \rho_\mathrm{S}(\varepsilon/k_\mathrm{B}T)^{3/n}$ and $x_\mathrm{F} = \rho_\mathrm{F}(\varepsilon/k_\mathrm{B}T)^{3/n}$.

If we introduce the compressibility ratios Z_S and Z_F, with $Z \equiv p/\rho k_\mathrm{B}T$, for pure phase (solid and fluid) components, then at melting (with i = S or F)

$$\frac{p\sigma^3}{\varepsilon} = \rho_i^{(n+3)/3} Z_i x_i^{-n/3} , \tag{3.57}$$

$$\frac{p\sigma^3}{\varepsilon} = \left(\frac{k_B T}{\varepsilon}\right)^{(n+3)/n} Z_i x_i \tag{3.58}$$

and

$$\rho_i = \left(\frac{k_B T}{\varepsilon}\right)^{3/n} x_i \, . \tag{3.59}$$

Of course, these four constants (x_S, x_F, Z_S and Z_F) characterising the melting transition have to be determined by statistical calculations.

The empirical Simon expression for the melting curve under pressure in Eq. (3.32) is in fact exact for "soft-sphere" inverse-power repulsive potentials, with

$$c = 1 + \frac{3}{n} \, . \tag{3.60}$$

Of course, the situation is complicated once an attractive term is included in the interactions (see Chap. 4, especially Secs. 4.6 and 4.7).

Appendix 3.1 Analogues of the Clausius–Clapeyron Equation for Other Phase Transitions

In this appendix we first give as an example the extension of the Clausius–Clapeyron equation to a first-order transition in a magnetic fluid and then briefly refer to the discussion given by Pippard[54] for extensions to higher-order phase transitions.

A3.1.1 *A magnetic system*

According to the Hellmann–Feynman theorem (see Sec. 3.1.2), the thermodynamic identity for the Gibbs free energy of a magnetic fluid at given N is

$$dG = -SdT + Vdp - MdH \, , \tag{A3.1.1}$$

where M is the magnetic moment and H the magnetic field. All extensive properties in this equation are per mole and the free energy incorporates the field energy of the empty solenoid.

The analogues of the Clausius–Clapeyron equation follow immediately by the same argument as given in Sec. 3.4, except that here in imposing the equality $G_1 = G_2$ we have to deal with a transition surface instead of a transition line. By taking small changes dp, dT and dH on the transition surface we get $-S_1dT + V_1dp - M_1dH = -S_2dT + V_2dp - M_2dH$ and hence

$$\left(\frac{\partial H}{\partial T}\right)_p = -\frac{S_2 - S_1}{M_2 - M_1}, \tag{A3.1.2}$$

$$\left(\frac{\partial H}{\partial p}\right)_T = \frac{V_2 - V_1}{M_2 - M_1} \tag{A3.1.3}$$

and

$$\left(\frac{\partial p}{\partial T}\right)_H = \frac{S_2 - S_1}{V_2 - V_1}. \tag{A3.1.4}$$

The last equation is identical to the standard Clausius–Clapeyron equation. The relations holding for a phase transition in an electrically polarised system are also evident from the above.

A3.1.2 *Higher-order phase transitions*

For the equilibrium between a superconductor (s) and the normal (n) fluid at the critical field H_c, the Meissner effect gives $M_n \ll M_s = -V_sH_c/4\pi$. Equation (A3.1.2) becomes

$$\left(\frac{\partial H_c}{\partial T}\right)_p = -4\pi\frac{(S_n - S_s)}{V_sH_c}, \tag{A3.1.5}$$

showing that the transition becomes of the second order at the critical temperature T_c where both H_c and $(S_n - S_s)$ vanish. Here we follow Pippard[54] in adopting Ehrenfest's classification of phase transitions: in a second-order transition first derivatives of the free energy (such as the entropy) are continuous while second derivatives (such as the specific heat) are discontinuous.

Thus, in a second-order transition we take S and V as continuous, but allow for discontinuities in the specific heat C_p and in the volume expansion coefficient β. We need to consider in this case the second-order terms in the equality $G_1 = G_2$ on the transition line:

$$\frac{dp}{dT} = -\left[\left(\frac{\partial S_2}{\partial T}\right)_p - \left(\frac{\partial S_1}{\partial T}\right)_p\right] \Big/ \left[\left(\frac{\partial S_2}{\partial p}\right)_T - \left(\frac{\partial S_1}{\partial p}\right)_T\right]$$

$$= (VT)^{-1}\frac{(C_{p2} - C_{p1})}{(\beta_2 - \beta_1)}\,. \tag{A3.1.6}$$

This relation, which replaces the standard Clausius–Clapeyron equation, may also be rewritten in terms of the compressibility difference between the two phases (see Eq. (3.19)).

Pippard[54] goes on to discuss the equilibrium equations holding for phase transitions of still higher order and the appropriateness of the Ehrenfest classification. We must, however, refer the interested reader to the illuminating discussion given in his book.

Appendix 3.2 Partition Function, Phase Space and Configurational Integral for Inverse Power Repulsive Potentials

The canonical partition function Q can be formally written as the trace of $\exp(-H/k_BT)$, with H the appropriate Hamiltonian. For a classical monatomic fluid this implies an integration over the phase space defined by the momenta $\mathbf{p}_i$ and the positions $\mathbf{r}_i$ of all the particles ($i = 1, 2, \ldots, N$). The result can be written as

$$Q = Q_{\text{id}} Q_{\text{ex}}\,, \tag{A3.2.1}$$

where

$$Q_{\text{id}} = \frac{1}{N!}\left(\frac{V}{\lambda^3}\right)^N \tag{A3.2.2}$$

is the partition function of the ideal monatomic gas, with $\lambda = (2\pi\hbar^2/mk_BT)^{1/2}$, and

$$Q_{\text{ex}} = V^{-N}\int d\mathbf{r}_1 \cdots \int d\mathbf{r}_N \exp\left[-\frac{\Phi}{k_BT}\right] \tag{A3.2.3}$$

arises from the potential energy $\Phi(\mathbf{r}_1, \ldots, \mathbf{r}_N)$ as a function of the positions of all the particles.

We also report at this point the structure of the grand-canonical partition function Q_G, that will be needed in Appendix 4.1. The number of particles N is allowed to fluctuate in the grand ensemble, and Q_G is obtained from the trace of $\exp[(\mu N - H)/k_B T]$ where μ is the chemical potential. Integration over the momenta for a classical fluid yields an expansion in powers of the fugacity $z = (mk_BT/2\pi\hbar^2)^{3/2}\exp(\mu/k_BT)$, that is $Q_G(z,T,V) = \sum_N z^N Z_N(T,V)$, where the coefficients are determined by $Z_N = (N!)^{-1}\int d\mathbf{r}_1 \cdots \int d\mathbf{r}_N \exp[-\Phi_N/k_BT]$. We also recall that the connection with statistical mechanics is effected through the grand potential $\Omega = -k_BT \ln Q_G$.

Returning to Eq. (A3.2.3), for pairwise additive interactions described by the soft-sphere repulsive model in Eq. (3.55) it takes the form

$$Q_{ex} = V^{-N}\int d\mathbf{r}_1 \cdots \int d\mathbf{r}_N \exp\left[-\frac{\varepsilon\sigma^n}{k_BT}\sum_{i<j} r_{ij}^{-n}\right] . \tag{A3.2.4}$$

The sum in Eq. (A3.2.4) runs over all pairs of particles. With the definitions

$$\rho \equiv \frac{N\sigma^3}{V} \tag{A3.2.5}$$

and

$$\mathbf{s}_i = \left(\frac{N}{V}\right)^{1/3} \mathbf{r}_i\,, \tag{A3.2.6}$$

we then find

$$Q_{ex} = \int d\mathbf{s}_1 \cdots \int d\mathbf{s}_N \exp\left[-\frac{\varepsilon\rho^{n/3}}{k_BT}\sum_{i<j} s_{ij}^{-n}\right] . \tag{A3.2.7}$$

The essential feature to note in Eq. (A3.2.7), as stressed by Hoover and Ross,[62] is that the Helmholtz free energy (in units of k_BT) and other thermodynamic properties depend on the single variable

$$x = \rho\left(\frac{\varepsilon}{k_BT}\right)^{3/n}, \tag{A3.2.8}$$

some consequences of this being discussed in the main text.

Chapter 4

Structure, Forces and Thermodynamics

4.1 Pair Distribution Function $g(r)$

We have emphasised in Chap. 1 that the most basic characteristic of a liquid is that it possesses short-range order, as opposed to the long-range periodicity of a crystalline solid. We now introduce the appropriate tools to describe the short-range order in a monatomic liquid such as argon, or liquid metal sodium.

The idea is simple enough. One selects out an atom and chooses to "sit on" the position of its nucleus as an origin of coordinates while the atom moves through the liquid. Then if the number density of the bulk homogeneous liquid is $\rho = N/V$, where as usual N is the number of atoms in the liquid volume V, we define the density that we see as a function of the distance r from the atom chosen as origin as $\rho g(r)$. A statistical average is, of course, implied. So, more precisely, $g(r)$ is the probability of finding two atoms in the liquid at a distance r from each other and the quantity $4\pi\rho g(r)r^2dr$ is the mean number of atoms inside a spherical shell of radius r and thickness dr, centred on an "average" atom. We are evidently taking the liquid as isotropic at this level of description. Information on preferred bond angles should be sought from the distribution functions of triplets and higher clusters of atoms.

Since atoms cannot come closer together than an atomic diameter σ, then $g(r)$ must be essentially zero in the range $0 \leq r < \sigma$. But then, because of short-range order, there will be a near-neighbour "shell" of atoms around a distance which, for a liquid near the triple point, is close to the corresponding near-neighbour spacing in the crystalline solid prior to melting. In contrast though to the long-range order in a crystal, the next near-neighbour shell in the liquid is much less prominent, and the next outer shell may hardly be visible in $g(r)$. Correlations in atomic positions rapidly die out in a liquid and $g(r)$ tends to unity (corresponding to complete disorder) over a distance of a few atomic diameters.

We are thus led to a form of $g(r)$ as was already shown in Figs. 1.8 and 1.9. This function can actually be obtained by Fourier transform from data coming from a diffraction experiment. Just as the structure of a crystal is determined experimentally by observing the Bragg reflections of X-rays or neutrons, liquid structure is measured *via* diffraction. To describe the experimental data, be it neutron or X-ray, we next introduce the liquid structure factor $S(k)$.

4.2 Definition of Liquid Structure Factor $S(k)$

One direct way to introduce the liquid structure factor $S(k)$ is *via* the pair distribution function $g(r)$ presented above. Essentially, $S(k) - 1$ is the Fourier transform of $g(r)-1$, with a number density factor ρ introduced for dimensional reasons. The precise relation is

$$S(k) - 1 = \rho \int d\mathbf{r}[g(r) - 1]e^{i\mathbf{k}\cdot\mathbf{r}} . \tag{4.1}$$

Notice that with this definition we would have $S(k) = 1$ in a completely disordered system, corresponding to $g(r) = 1$ as already remarked. Actually, as already indicated in Sec. 4.1, it is $S(k)$ itself which is rather directly accessible to neutron or X-ray scattering experiments: from these data $g(r)$ is obtained by Fourier inversion of Eq. (4.1).

Since $g(r)$ is a spherically symmetric function, the volume integral in Eq. (4.1) can be reduced to a single radial integration, the angular integration being completed to yield

$$S(k) - 1 = 4\pi\rho \int_0^\infty r^2 dr[g(r) - 1]\frac{\sin(kr)}{kr} . \tag{4.2}$$

We recognise inside the integrand in Eq. (4.2) the interference function $\sin(kr)/(kr)$ associated with a spherical diaphragm.

The main feature of the structure factor of a simple monatomic liquid, as is for instance reproduced in Fig. 4.1 for liquid sodium[63] at 100°C, is a prominent main peak reflecting a preferred range of first-neighbour distances.

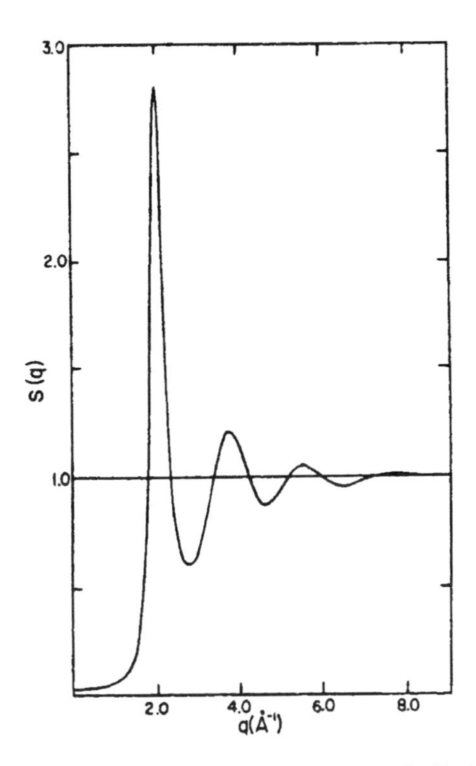

Fig. 4.1. Measured structure factor of liquid sodium at 100° C. (Redrawn from Greenfield *et al.*, Ref. 63.)

This shows up in the corresponding $g(r)$ through the presence of successive shells of neighbours, as we have already noticed in the discussion given in Sec. 4.1. The main origin for this type of short-range order lies in the excluded volume which is associated with the repulsive core of each atom. It is often referred to as topological short-range order, to distinguish it from additional types of short-range order such as arise in molten salts and chemically ordered alloys (see Chap. 8) and in glasses (see Chap. 10).

The short-range order in a liquid is enhanced as its temperature is decreased towards freezing, as is revealed by the increasing height and the narrowing width of the main peak in $S(k)$. We shall discuss the temperature dependence of the main peak in $S(k)$ and its connection with the freezing transition in Sec. 4.4 below. Let us first briefly record how it is measured in neutron or X-ray scattering experiments.

4.3 Diffractive Scattering from a Liquid

To be specific for the latter case, the intensity $I(k)$ of X-rays scattered through an angle 2θ from a liquid sample (say argon) containing N atoms is given by

$$I(k) = N[f(k)]^2 S(k)\,. \tag{4.3}$$

Here, $k = 4\pi \sin\theta/\lambda$, with λ the X-ray wavelength, is the scattering wave number (see Fig. 4.2), while $f(k)$ is the scattering amplitude of a single atom. In a completely disordered system (i.e. for $S(k) = 1$) the intensity results from independent scattering processes on the N atoms: the deviations of $S(k)$ from unity therefore describe the effects of interference between waves scattered by pairs of atoms. Coherence in the scattering arises from short-range order in the liquid.

In Eq. (4.3), since X-rays are scattered dominantly by electrons, the atomic scattering factor $f(k)$ results from the ground-state electron density $\rho(r)$ of a single atom: given again by a Fourier transform relation,

$$f(k) = \int d\mathbf{r}\rho(r)e^{i\mathbf{k}\cdot\mathbf{r}}\,. \tag{4.4}$$

Since for an atom of atomic number Z we have $\int d\mathbf{r}\rho(r) = Z$, it is clear from the definition (4.4) that $f(k = 0) = Z$. The atomic scattering factor $f(k)$ decreases from this value to zero with increasing k, the asymptotic behaviour being proportional to k^{-4}.

An equation of similar form to that in (4.3) holds also for neutron diffractive scattering. However the k-dependent factor $f(k)$ for X-ray scattering is replaced by a k-independent neutron scattering amplitude, since the range of

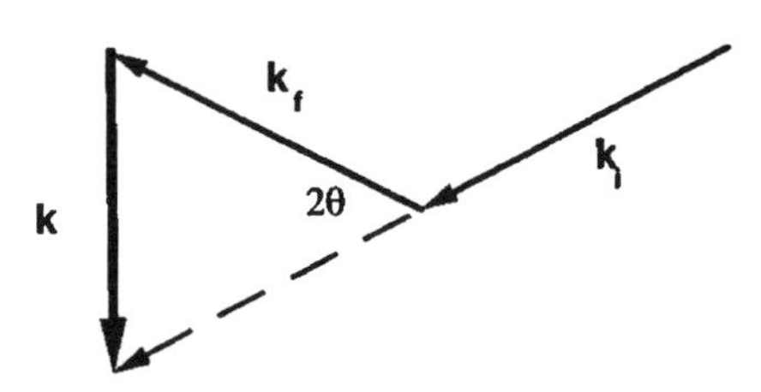

Fig. 4.2. Illustrating the definition of the scattering angle 2θ and of the scattering wave vector $\mathbf{k}$ as the difference between initial wave vector $\mathbf{k}_i$ and final wave vector $\mathbf{k}_f$ of a diffraction probe.

the nuclear force is much smaller than the neutron wavelengths $2\pi/k$ used in these experiments (see, for example the book by Bacon[64]). On the other hand, the neutron scattering cross-section against an atomic nucleus depends on the isotopic state of the nucleus. This fact has been very effectively exploited as a means to enhance the scattering contrast in multi-component liquids such as liquid alloys,[65] by suitable changes in the natural isotopic composition through isotopic enrichment for one of the atomic components. Examples of structural results obtained by the technique of isotopic substitution in neutron diffraction will be shown in Chap. 8, in regard to structure determination for molten salts.

Returning for the moment to monatomic liquids, we proceed below to present the salient features of their structure factor $S(k)$. From time to time throughout this volume, we shall have occasion to refer to specific forms of $S(k)$ extracted from scattering experiments, or sometimes from computer simulation studies.

4.4 Salient Features of Liquid Structure Factor

4.4.1 *Long wavelength limit and connection with thermodynamic fluctuations*

First of all, by arguments from fluctuation theory concerning the long-wavelength density fluctuations in a classical atomic fluid (see the brief account already given in Sec. 3.3), we can obtain the long wavelength limit (wave number $k \to 0$) of $S(k)$ as

$$S(k=0) \equiv S(0) = \rho k_{\mathrm{B}} T K_T \,. \tag{4.5}$$

In Eq. (4.5) ρ is the number density N/V and K_T is the isothermal compressibility. The main point is that $S(0)$ is proportional to the mean square fluctuation in the number of atoms contained in a given volume, and this is in turn measured by K_T.

The latter thermodynamic quantity is usually available from laboratory experiments using thermodynamic or ultracoustic techniques and it proves to be very important in dense liquids, such as argon near its triple point or liquid sodium just above the melting point, to know $S(0)$, for the diffraction measurements of $S(k)$ must eventually join continuously with the long-wavelength limit $S(0)$.

Small-angle (i.e. small k) scattering experiments are very important for a full characterisation of the liquid structure. We shall discuss the small-angle scattering behaviour for liquid argon near the triple point in Sec. 4.6.3 below.

4.4.2 *The Hansen–Verlet freezing criterion*

Less fundamentally based than Eq. (4.6), but nevertheless of importance is the so-called Hansen–Verlet criterion.[66] As a liquid like argon is cooled down, the height of the principal peak in $S(k)$, at $k = k_{\mathrm{m}}$ say, increases and the Hansen–Verlet criterion asserts that the liquid will freeze when this peak reaches a height given by

$$S(k_{\mathrm{m}}) \cong 2.8\,. \qquad (4.6)$$

This criterion for freezing, which was originally proposed for Lennard–Jones liquids, turns out to be of wider applicability. Thus, Ferraz and March[67] noted that a classical plasma would freeze under the same condition and compared this with experiment for the alkali metals Na and K.

The so-called "density wave" theory of freezing (see Kirkwood and Monroe[68] for early work; then see especially Ramakrishnan and Yussouff[69]) leads rather naturally to such a freezing criterion. The essential point is that the fluctuation-theory result given in Eq. (4.5) can be generalised to arbitrary wave number k.

Consider subjecting a classical liquid to an external potential which is periodic in space with a periodicity corresponding to a given wave vector $\mathbf{k}$. If the strength of the external potential is sufficiently weak that the response of theliquid lies in a linear regime, this response will merely be a modulation of the liquid density at the same wave vector $\mathbf{k}$. We may ask about the work done in creating such a density modulation, as a function of the modulus k of the wave vector $\mathbf{k}$. Just as the value of $S(k = 0)$, being proportional in Eq. (4.6) to the isothermal compressibility measures the softness of the liquid against squeezing under uniform pressure, so the value of $S(k)$ at any k measures the softness of the liquid against a density modulation induced by a periodic external potential. By "softness" we mean that the work needed to create the density modulation at given $\mathbf{k}$ is inversely proportional to the height of $S(k)$

at that value of k. This is an exact result in statistical mechanics, following from the general theory of linear response for a fluid subject to static external forces[70] (see Appendix 4.1).

Thus, as the value of $S(k_{\rm m})$ is observed to increase with decreasing temperature towards the freezing point, it signals an increasing softness of the liquid against modulation by density waves having wave vectors of magnitude near $k_{\rm m}$ — until, when the approximate relation (4.7) is satisfied, such density waves are spontaneously locked into the liquid as its transition to the crystalline phase takes place.

The density-wave theory of freezing predicts that the phase transition from liquid to solid in a monatomic system is governed by a balance between a gain of free energy from volume contraction and a loss of free energy incurred in the spontaneous density modulation which changes the uniform density profile of the liquid into the periodic profile of the crystal. A first-order phase transition is predicted in this way, in accord with the behaviour of simple real systems such as argon or sodium: Namely, the bulk liquid phase becomes thermodynamically unstable against crystallisation, but remains mechanically stable and could thus be undercooled by appropriate means. A mechanical instability of the liquid phase would instead correspond to a divergence in the value of $S(k_{\rm m})$ and would prevent supercooling of the liquid state.

According to this interpretation of the liquid–solid phase transition, the position of the main peak in the liquid structure factor should correspond to the main crystalline periodicity (the first set of reciprocal lattice vectors, according to crystallographic terminology). A loose correspondence between liquid-state order in wave number space and the locations of the first one or two sets of Bragg diffraction spots from the crystal is indeed observed in simple systems.

4.4.3 *Relation between the main features of the peak in the structure factor*

We shall elaborate below on the relation between the Lindemann criterion for melting and the Hansen–Verlet criterion for freezing. The work of Bhatia and March,[71] though proposed to relate more generally the height, position and width of the principal peak of the structure factor $S(k)$ of dense monatomic liquids, supplies such a relationship. The aim will simply be to exhibit a correlation between the two different approaches to the phase transition.

Let us first examine, therefore, how the above three main features of the peak in $S(k)$ are related to each other. We start from the fact that in a dense classical liquid, because of excluded volume requirements (see Chap. 2), the pair distribution function $g(r)$ must vanish for values of the interatomic separation r shorter than the atomic diameter. In particular, $g(r)$ must satisfy $g(r=0)=0$. Using the Fourier transform relation (4.1), this condition can be written as

$$2\pi^2\rho = \int_0^\infty k^2[1-S(k)]dk\,. \tag{4.7}$$

An approximate evaluation of the integral in Eq. (4.7) can be achieved as follows. With k_{m} denoting as above the position of the main peak of $S(k)$, let $2\Delta k$ be the peak width as measured by the separation between the two adjacent nodes of $[S(k)-1]$ embracing k_{m}. Furthermore, make the (in general reasonable) assumption that any asymmetry of the peak about k_{m} is sufficiently weak to be neglected. If Eq. (4.7) is now expressed as

$$2\pi^2\rho = \int_0^{k_{\mathrm{m}}-\Delta k} k^2[1-S(k)]dk + \int_{k_{\mathrm{m}}-\Delta k}^{k_{\mathrm{m}}+\Delta k} k^2[1-S(k)]dk + \int_{k_{\mathrm{m}}+\Delta k}^{\infty} k^2[1-S(k)]dk\,, \tag{4.8}$$

then the following approximations prove useful: (i) to take $S(k)=0$ over the range of the first integral in (4.8); (ii) to neglect the third integral in (4.8), due to the oscillations of $S(k)-1$ around zero; and (iii) to estimate the second integral by the triangular area $[S(k_{\mathrm{m}})-1]k_{\mathrm{m}}^2\Delta k$.

Using the above simplifications and writing $\rho = 3/(4\pi R_{\mathrm{A}}^3)$, it is readily verified that $S(k_{\mathrm{m}})k_{\mathrm{m}}^2\Delta k \approx \frac{1}{3}k_{\mathrm{m}}^3[1-\frac{9\pi}{2}(R_{\mathrm{A}}k_{\mathrm{m}})^{-3}]$. Empirically it is found for dense liquids that $R_{\mathrm{A}}k_{\mathrm{m}} \approx 4.4$ and the second term in the brackets contributes 0.15 compared to unity. Thus one has the result

$$S(k_{\mathrm{m}}) \approx \frac{0.3k_{\mathrm{m}}}{\Delta k}\,. \tag{4.9}$$

It is now instructive to compare the approximate prediction (4.9) with the accurate neutron diffraction data of Yarnell *et al.*[72] on liquid argon at 85 K. One finds from their data that $S(k_{\mathrm{m}}) = 2.70$, $k_{\mathrm{m}} = 2.00$ Å^{-1} and $\Delta k = 0.275$ Å^{-1}, yielding the value $S(k_{\mathrm{m}})\Delta k/k_{\mathrm{m}} = 0.37$, which is nearer to 3/8 than the predicted 0.3 in Eq. (4.9). It is also satisfactory that the X-ray data of Greenfield *et al.*[63] on liquid potassium at 65°C yield $S(k_{\mathrm{m}}) = 2.73$, $k_{\mathrm{m}} = 1.62$ Å^{-1} and $k = 0.225$ Å^{-1}, and hence a constant of 0.38, whereas from their experiment on potassium at 135°C the constant is 0.37. Similarly from their experiment on sodium at 100°C and 200°C the measured data yield 0.37 and 0.36. Thus, for the above five examples, it emerges that Eq. (4.9) is quantitative provided 0.3 is replaced by 3/8. The fact that this is greater than 0.3 seems to point to the third integral in Eq. (4.8) having a non-zero negative value.

Turning to the features of the main peak in the pair distribution function $g(r)$, an estimate of $S(0)$ from Eq. (4.1), carried out with similar assumptions to those made above, leads to

$$g(r_{\mathrm{m}})r_{\mathrm{m}}^2 \Delta r = \frac{1}{3} r_{\mathrm{m}}^3 - \frac{1}{3} R_{\mathrm{A}}^3 [1 - S(0)] , \tag{4.10}$$

with definitions paralleling those for $S(k)$. Since $g(r)$ is less readily accessible than $S(k)$, no comparison of Eq. (4.10) with experiment will be attempted. However, again using the data of Yarnell *et al.*[72] on liquid argon at 85 K, one has $g(r_{\mathrm{m}}) = 3.05$, $\Delta r = 0.545$ Å and $r_{\mathrm{m}} = 3.68$ Å. The value of $r_{\mathrm{m}}/\Delta r$ is 6.7, to be compared with $k_{\mathrm{m}}/\Delta k = 7.2$. Thus, quite approximately,

$$\frac{k_{\mathrm{m}}}{\Delta k} \approx \frac{r_{\mathrm{m}}}{\Delta r} . \tag{4.11}$$

4.4.4 *Verlet's rule related to Lindemann's melting criterion*

To return now to the melting and freezing criteria of Lindemann and of Hansen and Verlet respectively, let us use the above estimates. For Ar, Na and K, where freezing at standard pressure involves only minor changes in local coordination, the use of $S(k_{\mathrm{m}})|_{T_{\mathrm{m}}} = 2.8$ at the melting temperature T_{m} yields the estimate $(\Delta r/r_{\mathrm{m}})|_{T_{\mathrm{m}}} \approx 0.11$ from Eqs. (4.9) and (4.10). Lindemann's law of melting gives $(\Delta r/R_{\mathrm{A}})|_{T_{\mathrm{m}}} \approx 0.2$ if one is allowed to identify Δr as the root-mean-square displacement of the atoms in the solid near melting.[73] Since $r_{\mathrm{m}} \cong 1.8 R_{\mathrm{A}}$, the two results are in approximate agreement.

Thus, the main conclusion of the above argument is that there is no difficulty in reconciling the Hansen–Verlet freezing criterion with Lindemann's melting rule.

4.5 Internal Energy and Virial Equation of State with Pair Forces

Whereas the relationship (4.5) for a classical monatomic liquid is independent of the specific model adopted for the interatomic forces, we shall here display two further relations between liquid structure and thermodynamics which are valid only within pair-potential models. They both depend on using the quantity $4\pi\rho g(r)r^2dr$ to count the average number of pairs of atoms at separation r (see Sec. 4.1).

In a classical monatomic liquid at temperature T, the kinetic energy per degree of freedom is $k_BT/2$ (see Sec. 3.6) and the internal energy can be written as

$$U = \frac{3}{2}Nk_BT + \langle\Phi\rangle\,, \tag{4.12}$$

$\langle\Phi\rangle$ being the mean potential energy. If Φ is written as the sum of pairwise interactions, then

$$U = \frac{3}{2}Nk_BT + \left\langle \frac{1}{2}\sum_{i\neq j}\phi(R_{ij})\right\rangle\,, \tag{4.13}$$

where R_{ij} is the separation between atoms i and j and the factor $1/2$ corrects for the fact that each atom pair enters the sum twice. From the definition of $g(r)$ Eq. (4.13) can be rewritten as

$$U = \frac{3}{2}Nk_BT + \frac{1}{2}N4\pi\rho\int_0^\infty r^2g(r)\phi(r)dr\,. \tag{4.14}$$

We thus have a way to evaluate, within a pair-potentials model, the thermodynamic internal energy of a fluid from its pair distribution function.

Similarly, the equation of state can be obtained through the classical virial theorem, going back to Clausius, which relates the average kinetic energy $\langle K\rangle$ to the virial of the forces. The virial of the pressure is $3pV$, yielding for a perfect gas $2\langle K\rangle = 3pV$. More generally,

$$pV = Nk_BT + \frac{1}{3}\left\langle\sum_i \mathbf{R}_i\cdot\mathbf{F}_i\right\rangle\,. \tag{4.15}$$

Here $\mathbf{R}_i$ and $\mathbf{F}_i$ denote respectively the vector position of the ith particle and the total force on this particle due to the remaining particles. Using again the definition of $g(r)$, for a fluid of identical spherical particles interacting *via* a central pair potential Eq. (4.15) takes the form

$$p = \rho k_B T - \frac{2\pi}{3}\rho^2 \int_0^\infty r^3 g(r) \frac{\partial \phi(r)}{\partial r} dr \,. \tag{4.16}$$

This Eq. (4.16) was anticipated in Chap. 3 and will be used in Sec. 4.8 to discuss properties of a fluid such as argon at its critical point.

For the hard sphere fluid, the interatomic force $-(\partial\phi(r)/\partial r)$ is different from zero only at contact, i.e. for r equal to the hard sphere diameter σ. As anticipated in Sec. 3.7, Eq. (4.16) then yields $p = \rho k_B T[1 + 2\pi\rho\sigma^3 g(\sigma^+)/3]$, with $g(\sigma^+)$ the value of the pair distribution at contact.

A form of the equation of state, which is recorded in standard physical chemistry texts,[74] comes into its own in the context of critical behaviour. This is Dieterici equation of state, and we outline in Appendix 4.2 a derivation, following the arguments of Blinder.[1]

4.6 Ornstein–Zernike Direct Correlation Function $c(r)$

A further valuable quantity, going back to the pioneering studies of Ornstein and Zernike[75] on critical phenomena, is the direct correlation function $c(r)$. Though these workers defined $c(r)$ from $h(r) \equiv g(r) - 1$ ($h(r)$ is often called the total correlation function), it is most simply defined in k space where the Fourier transform of $c(r)$, denoted by $\tilde{c}(k)$ below, is related to $S(k)$ by

$$1 - \tilde{c}(k) = \frac{1}{S(k)} \,. \tag{4.17}$$

Why should $\tilde{c}(k)$ be interesting when it is so directly related to $S(k)$? One answer is that it changes the emphasis put on the k-space experimental data for $S(k)$. Thus, $S(0)$ near freezing is often $\cong 0.01$ to 0.06, which means from Eq. (4.17) that $\tilde{c}(0) \cong -100$ to -20. But at k_m, using Eq. (4.6), $\tilde{c}(k_m) \cong 0.7$ and suddenly all emphasis has been moved from an $S(k)$ picture "dominated" by $S(k_m)$ to a $\tilde{c}(k)$ description dominated by $k \ll k_m$ (see Fig. 4.3 from Ashcroft and March[76]). Details on the relevance of $\tilde{c}(k)$ to critical point behaviour will be given in Sec. 4.8.

In fact, according to Eq. (4.5) we have for a classical atomic fluid the result

$$\rho k_B T[1 - \tilde{c}(k \to 0)] = \frac{1}{K_T} \,. \tag{4.18}$$

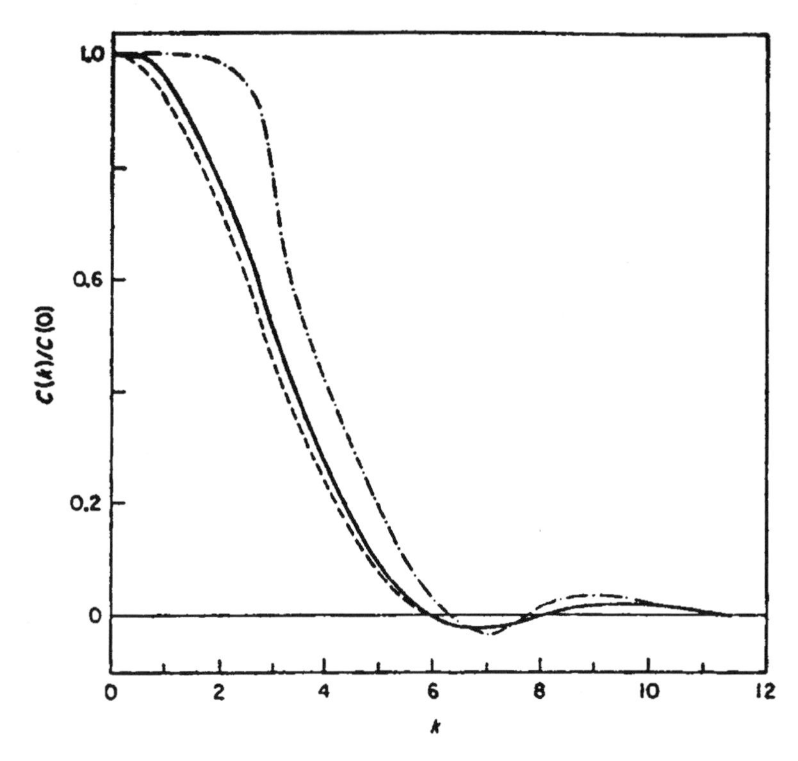

Fig. 4.3. Direct correlation function in **k** space for fluid argon at 84 K. Experimental data (dash-dotted line) are compared with results calculated in slightly different ways for the hard sphere fluid. (Redrawn from Ashcroft and March, Ref. 76.)

Since $\rho k_B T$ is the value of the isothermal bulk modulus for an ideal classical gas, Eq. (4.18) shows that the quantity $-\rho k_B T\tilde{c}(k \to 0)$ measures the contribution of the interatomic forces to the isothermal bulk modulus of the fluid. More generally and recalling the arguments given in Sec. 4.4.2, the function $-\rho k_B T\tilde{c}(k)$ gives the non-ideal term in the mechanical stiffness of the fluid against a modulation of its density by a weak periodic potential having wave vector **k**. It is this meaning as a mechanical-stiffness function, and its connection to the interatomic forces which is implied by the definition (4.17), which makes the function $-\rho k_B T\tilde{c}(k)$ so important in the context of liquid-structure theories.

Ornstein and Zernike, while focusing on critical point behaviour, recognised that in dense monatomic liquids usefully described by a pair potential $c(r)$ is rather directly connected with $\phi(r)$. Indeed, well away from the critical point, it is widely accepted that

$$c(r) \to -\frac{\phi(r)}{k_{\mathrm{B}}T} \tag{4.19}$$

at sufficiently large r. Generally, it can be said that there is rich content on detailed interatomic forces in diffraction measurements of liquid structures. Methods are now emerging by which, in appropriate cases (e.g. argon and sodium) $\phi(r)$ can be extracted by "inverting" the measured liquid structure factor.[77,78] Nevertheless, it is also true that the gross features of the short-range order in a simple monatomic liquid can be mimicked by a hard sphere model: such a structural description based on one particular liquid structural theory, which yields analytic results for the direct correlation function $c(r)$, is set out immediately below.

4.6.1 *Direct correlation function from Percus–Yevick theory for hard spheres*

In the hard sphere model the fluid particles are perfectly impenetrable and therefore the pair distribution function $g(r)$ is exactly zero over the whole range of interatomic distance lying inside the hard sphere diameter σ:

$$g(r) = 0 \quad (\text{for } r < \sigma)\,. \tag{4.20}$$

In the liquid structure theory due to Percus and Yevick,[79] this exact statement for the hard sphere model is combined with an approximate statement on the relationship between the direct correlation function $c(r)$ and the pair potential $\phi(r)$, extending Eq. (4.19) away from the asymptotic ($r \to \infty$) regime. In the hard sphere model, with $\phi(r) = 0$ down to the hard sphere diameter, this approximate relationship is

$$c(r) = 0 \quad (\text{for} \quad r > \sigma)\,. \tag{4.21}$$

It was independently noted by Thiele[80] and by Wertheim[81] that the conditions (4.20) and (4.21) allow a full analytic determination of $c(r)$ when one uses them in the Ornstein–Zernike relation (4.17), written after Fourier inversion as a general relationship[a] between $g(r)$ and $c(r)$:

$$g(r) - 1 = c(r) + \rho \int d\mathbf{r}' c(|\mathbf{r} - \mathbf{r}'|)[g(r') - 1]\,. \tag{4.22}$$

Making use of Eq. (4.20), the result is to supplement the expression (4.21) for $c(r)$ outside the hard sphere diameter σ with a cubic-polynomial expression for $c(r)$ inside σ:

$$c(r) = a_0 + a_1 \left(\frac{r}{\sigma}\right) + a_3 \left(\frac{r}{\sigma}\right)^3 \quad \text{for } r < \sigma\,. \tag{4.23}$$

Here, the coefficients a_i are expressed in terms of the packing fraction $\eta = \pi\rho\sigma^3/6$ as follows:

$$\begin{aligned} a_0 &= -\frac{(1+2\eta)^2}{(1-\eta)^4}\,, \\ a_1 &= \frac{6\eta\left(1+\dfrac{1}{2}\right)^{\eta/2}}{(1-\eta)^4}\,, \\ a_3 &= -\frac{1}{2}\frac{\eta(1+2\eta)^2}{(1-\eta)^4}\,. \end{aligned} \tag{4.24}$$

The Fourier transform $\tilde{c}(k)$ of $c(r)$ in Eqs. (4.21) and (4.23) can also be evaluated analytically, and hence an analytic expression for $S(k)$ follows with the help of Eq. (4.17).

Let us consider in particular the long-wavelength relationship (4.18). After some calculation one finds $1 - \tilde{c}(0) = -a_0$ and hence

$$\left(\frac{\partial p}{\partial \eta}\right)_T = \left(\frac{6k_BT}{\pi\sigma^3}\right)(1+2\eta)^2(1-\eta)^{-4}\,. \tag{4.25}$$

Integration of Eq. (4.25) yields the equation of state of the hard sphere fluid in the form

$$p = \rho k_BT(1+\eta+\eta^2)(1-\eta)^{-4}\,, \tag{4.26}$$

[a]The Fourier transform of the convolution integral in Eq. (4.22) is the simple product of the Fourier transforms: thus Eq. (4.22) yields $S(k) - 1 = \tilde{c}(k) + \tilde{c}(k)[S(k) - 1]$, which is Eq. (4.17).

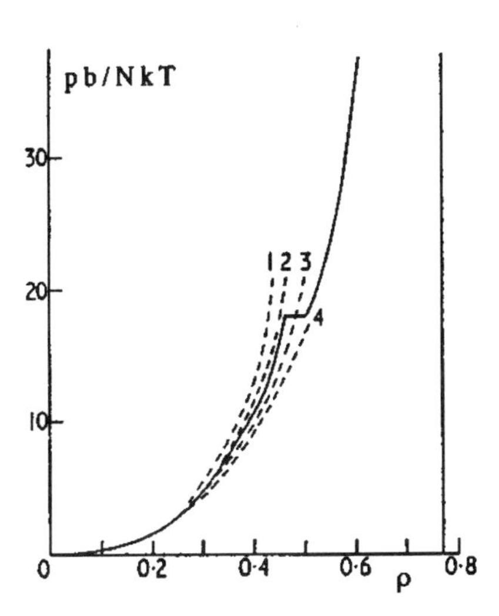

Fig. 4.4. Models for the fluid branch of the isotherm of the hard sphere system. The solid line is from simulation data and is well reproduced by the Carnahan–Starling equation of state. The dashed line marked 2 represents the Percus–Yevick compressibility equation of state, Eq. (4.26). The other dashed lines are alternative evaluations of the fluid isotherm. (Redrawn from Rowlinson, Ref. 83.)

Table 4.1. Values of Ornstein–Zernike function for some liquid metals near freezing.

Metal	$-c(r=0)$	$-\tilde{c}(k=0)$	Ratio $c(r=0)/\tilde{c}(k=0)$
Na	43	41	1.0
K	42	40	1.0
Rb	45	42	1.1
Cs	50	38	1.3
Cu	60	47	1.3
Ag	51	53	1.0
Au	35	38	0.9
Mg	31	39	0.8
Al	45	54	0.8
Ca	34	200	0.2
Pb	44	110	0.4
Sn	40	140	0.3
Fe	46	48	1.0
Ni	41	50	0.8
Co	35	50	0.7

as first obtained by Reiss, Frisch and Lebowitz.[82] This result should be compared with the Carnahan–Starling equation of state in Eq. (2.6), which was obtained from a fit of the simulated isotherms of the hard sphere fluid. There is good numerical agreement between these two analytic forms of the equation of state on the fluid branch,[83] as can be seen from Fig. 4.4.

The relation $1 - \tilde{c}(0) = -a_0$ coming from the Percus–Yevick theory of the hard sphere fluid implies that the ratio $c(0)/\tilde{c}(0)$ is near to unity. A direct test of this approximate theoretical result against experiment for a number of liquid metals is reported in Table 4.1, from Bernasconi and March.[84] More generally, the essentially geometric packing problem for a dense monatomic liquid is well described by the hard sphere model, as is shown by numerous comparisons with experimental data on $S(k)$ e.g. on liquid metals near freezing.

4.6.2 *Softness corrections to the hard sphere potential*

Considerable detailed refinements have been developed in liquid structure theory, taking full advantage of the computer simulation techniques. In particular, much effort has been devoted to the question of how best to correct the thermodynamic properties and the pair distribution function of the hard-core fluid for the finite steepness of the repulsive potential in real fluids. A very successful scheme has been proposed by Andersen *et al.*[85] Defining the function

$$y(r) \equiv g(r)e^{\phi(r)/k_B T}, \tag{4.27}$$

where $\phi(r)$ is the pair potential taken, however, to be purely repulsive, and similarly introducing the function $y_{hs}(r)$ for the hard sphere fluid, these authors show that the relation

$$y(r) = y_{hs}(r) \tag{4.28}$$

holds to first order in the range of the deviation between the two potentials provided that the hard sphere diameter σ is suitably chosen. This simple approximation yields excellent agreement with Monte Carlo data on the equation of state, at the expense of having an effective hard core diameter which is dependent on temperature and density.

The results for $S(k)$ obtained by this approximation are compared with the molecular dynamics results of Verlet[86] for a Lennard–Jones fluid in Fig. 4.5, from work by Chandler and Weeks.[87] The agreement is especially remarkable

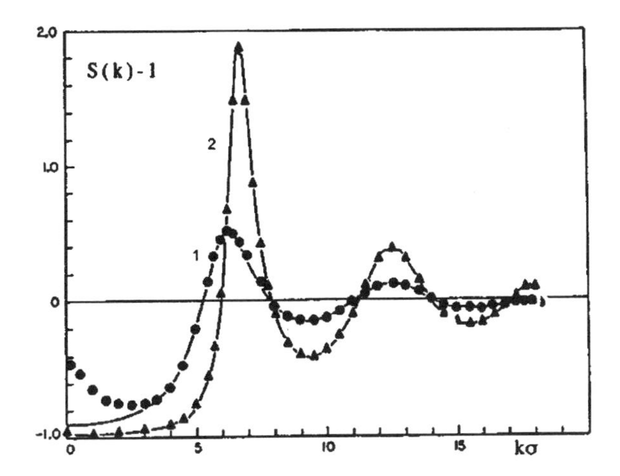

Fig. 4.5. Structure factor of a fluid having the same repulsive interactions as in the Lennard–Jones potential, in an expanded hot-fluid state (state 1) and in a dense cold-fluid state (state 2), compared with simulation data for the *full* Lennard–Jones potential (• and ▲). (Redrawn from Chandler and Weeks, Ref. 87.)

because the theory has omitted the attractive part of the potential, but evidently the latter becomes noticeable at relatively small wave number and that only in low-density states of the fluid.

4.6.3 *Small angle scattering from liquid argon near triple point*

We return to the asymptotic relationship (4.19) between the direct correlation function $c(r)$ and the assumed pair potential $\phi(r)$, to record how it was used by Enderby *et al.*[88] to evaluate the small-angle (neutron or X-ray) scattering from a classical Lennard–Jones liquid such as argon near its triple point. This theory was subsequently brought into contact with the neutron scattering data of Yarnell *et al.*[72] by Matthai and March.[89]

To summarise all this, $\phi(r)$ tends to the van der Waals attraction $-C_6/r^6$ at large r in argon (if retardation effects are neglected: a good approximation here). Use of Eq. (4.19) and simple Fourier transform theory then has the consequence that $\tilde{c}(k)$ has a k^3 term at small k:

$$\tilde{c}(k) = \tilde{c}(0) + c_2 k^2 + c_3 k^3 + \cdots , \tag{4.29}$$

where $c_3 = \pi^2 \rho C_6 / 12 k_B T$. It follows from Eq. (4.17) that $S(k)$ has a similar small-k expansion.

4.7 Thermodynamic Consistency and Structural Theories

As we have seen, three routes to the isothermal compressibility are available for the theory. One is *via* the fluctuation formula in Eq. (4.5), the second is through the first density derivative of the virial pressure in Eq. (4.16), and the third is *via* the second density derivative of the Helmholtz free energy in Sec. 3.2. These three routes to the compressibility, which naturally in an exact theory will yield identical values for K_T, in approximate liquid structural theories allow one to explore "thermodynamic inconsistencies". The following discussion is limited to a density-independent pair potential description, which is a useful starting point for liquid argon (not for a liquid metal such as sodium, where the valence electrons belong to the liquid as a whole and lead to electron-mediated interactions which have an important density dependence).

4.7.1 *Consistency of virial and fluctuation compressibility: Consequences for $c(r)$*

The fluctuation-theory route to the compressibility in Eq. (4.5) is readily re-expressed as

$$(k_B T)^{-1} \left(\frac{\partial p}{\partial \rho} \right)_T = 1 - \rho \int d\mathbf{r} c(r) \,. \tag{4.30}$$

It is now illuminating to compare this equation with the density derivative of the virial equation of state. After some manipulation, following Kumar *et al.*,[90] the equivalence of these two equations is ensured provided the following condition is obeyed:

$$-\rho r^2 c(r) = \frac{\phi(r)}{6 k_B T} \frac{\partial^2}{\partial \rho \partial r} [\rho^2 r^3 g(r)] + F(r; \rho, T) \,, \tag{4.31}$$

where the function $F(r)$ must integrate to zero.

Equation (4.31) suggests[91] that $c(r)$ may be written as the sum of two parts,

$$c(r) = c_p(r) + c_c(r) \,, \tag{4.32}$$

where the "potential" part $c_{\rm p}(r)$ is defined by the first term on the RHS of Eq. (4.31) and the so-called "cooperative" part $c_{\rm c}(r)$ can be anticipated to be short-range compared with $c_{\rm p}(r)$, except when cooperative effects are known from physical arguments to dominate, such as near the critical point. Unfortunately to date $c_{\rm c}(r)$ is only known precisely for a few simple models.

4.7.2 *A route to thermodynamic consistency in liquid-structure theory*

The primary aim of structural theories for classical monatomic liquids is to evaluate the pair distribution function $g(r)$ from a given pair-potential model for the interatomic forces. The availability of computer simulation data obtained with the same model allows useful tests of the statistical mechanical approximations which underlie the theory. In this section we shall briefly present two such approaches: the Percus–Yevick (PY) approximation, that we have met in Sec. 4.6.1 in connection with the structure of the hard sphere fluid, and the hypernetted-chain (HNC) approximation. We shall then return on thermodynamic consistency *via* an interpolation procedure between these approximate theories.

In both the PY and the HNC approximation, the (exact) Ornstein–Zernike relation between $c(r)$ and $g(r)$ in Eq. (4.22) is combined with an approximate "closure" relation bringing the pair potential into the problem. The best choice of closure depends on the character of the pair potential, as we discuss immediately below, and can be justified from the study of diagrammatic expansions for the correlations between atomic positions.[92]

The PY closure for the hard sphere potential is given in Eq. (4.21). More generally, the PY closure relation for a pair potential $\phi(r)$ is

$$c(r) = h(r) + \left[1 - g(r)\exp\left(\frac{\phi(r)}{k_{\rm B}T}\right)\right] \tag{4.33}$$

with $h(r) \equiv g(r) - 1$, yielding back Eq. (4.21) in the case of hard spheres. Experience has shown that Eq. (4.33) is quite a useful approximation when $\phi(r)$ is very short-ranged and leads to a good representation of the main peak in $g(r)$ for a liquid such as argon near its triple point.

The HNC closure relation reads instead as follows,

$$c(r) = h(r) - \ln\left[g(r)\exp\left(\frac{\phi(r)}{k_{\rm B}T}\right)\right] . \tag{4.34}$$

Again, it is a matter of practical experience that Eq. (4.34) works best for systems where the interactions are long-ranged, but underestimates the role of short-range repulsions in determining the shape of the main peak in $g(r)$. A merit of the HNC is that it embodies self-consistency between virial and free energy pressure,[93] although it contains an important internal inconsistency with the fluctuation-theory formula (4.5).

In fact, it is an exact result of statistical mechanics that the pair distribution function $g(r)$ is related to the pair potential $\phi(r)$ by

$$g(r) = \exp\left[\left(-\frac{\phi(r)}{k_B T}\right) + h(r) - c(r) + b(r)\right], \tag{4.35}$$

where the so-called bridge function $b(r)$ is defined through an infinite series of correlation diagrams. The HNC closure in Eq. (4.34) follows from Eq. (4.35) by setting $b(r)$ to zero.

A rather successful proposal in liquid structure theory has been made by Rosenfeld and Ashcroft,[94] who drew attention to the behaviour of the bridge

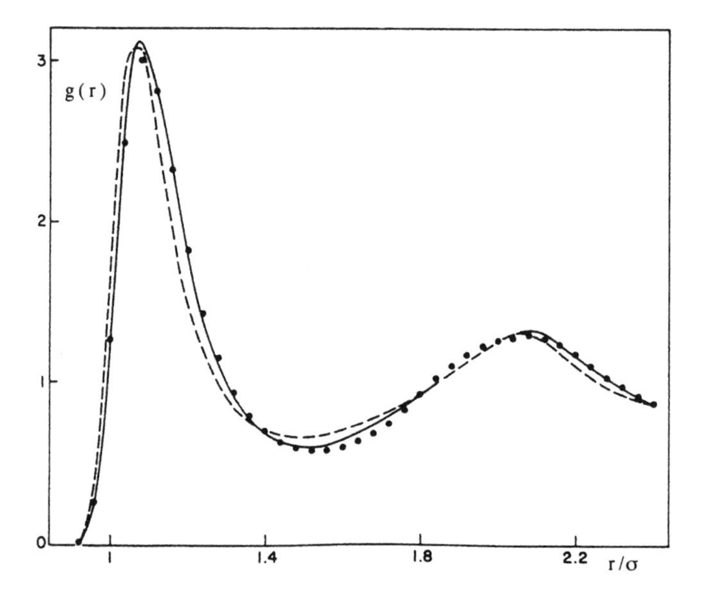

Fig. 4.6. Radial distribution function of the Lennard–Jones fluid near the triple point. The hypernetted-chain result (– – –) and its modification through inclusion of thermodynamic consistency (—) are compared with simulation data. (Redrawn from Rosenfeld and Ashcroft, Ref. 94.)

function in the excluded-volume region at short interatomic distances. They accordingly evaluated $b(r)$ from the PY theory of the hard sphere model. The packing fraction of this model enters as a parameter, and its value may be chosen so as to achieve consistency between virial and fluctuation compressibility. Figure 4.6 gives an illustration of the improvements in the predictions on liquid structure from given pair potentials, which accompany the improved thermodynamic consistency of the theory.

In conclusion, we remark that in a classical monatomic liquid the pair distribution function is exactly related to the distribution function of triplets of particles through an assumed interatomic pair potential. The derivation of this so-called force equation in the grand-canonical ensemble is reported in Appendix 4.3. This approach leads into the Born–Green–Yvon hierarchy of integral equations relating the many-particle distribution functions of successively higher order. Truncations of the hierarchy yield alternative approximate approaches to the theory of liquid structure.

4.8 Liquid–Vapour Critical Point

So far we have been mainly dealing with liquids in the region of the triple point, where liquid and solid densities are comparable. In this section we discuss the region near the critical point, where liquid and gas densities become comparable. Indeed, for the liquid–vapour transition an important thermodynamic variable is the difference $\rho_{\mathrm{l}} - \rho_{\mathrm{v}}$ between the liquid and gas densities. This is zero in the "disordered" single-fluid phase above the critical temperature T_{c} and becomes non-zero below T_{c}. This is the behaviour typical of an order parameter.

4.8.1 *Critical constants for insulating fluids and expanded alkali metals*

The most common definition of the liquid–gas critical point is that state at which the isotherm has a point of inflection satisfying

$$\left(\frac{\partial p}{\partial \rho}\right)_T = 0\,, \quad \left(\frac{\partial^2 p}{\partial \rho^2}\right)_T = 0\,. \tag{4.36}$$

Critical constants for a variety of (mostly insulating) substances[1] are recorded in Table 4.2.

Table 4.2. Measured critical constants for liquid–vapour coexistence.

Substance	$T_c(K)$	p_c (atm)	V_c (cm^3/mol)
Ne	44.7	26.9	44.3
Ar	150.9	48.3	74.6
Kr	209	45.2	92.1
Cl_2	417.1	76.1	123.4
CO	134	34.6	90.0
CO_2	304	72.8	94.2
CS_2	546	72.9	172.7
H_2O	647	218	55.4
D_2O	645	219	55.2
NH_3	406	112	72.0
N_2	126.0	33.5	90.0
O_2	154	50	74
CH_4	190	46	99
CH_3COOH	595	57	171
C_6H_6	562	48	256
n-C_5H_{12}	470	33	310
n-C_6H_{14}	508	30	367
Hg	1735	1042	40.1

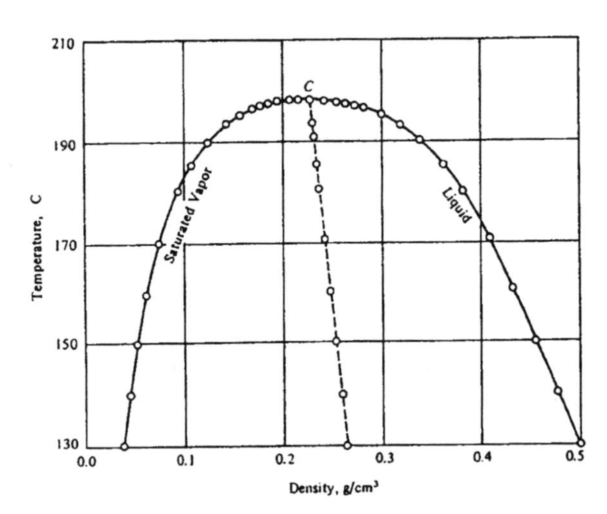

Fig. 4.7. Density of liquid and saturated vapour as function of temperature for n-pentane. Their average (– – –) follows a straight line on the scale of the figure.

Table 4.3. Compressibility ratio at critical point for insulating fluids and alkali metals.

Substance:	H_2	He	N_2	O_2	Ar	Kr	Xe	CO
Z_c	0.33	0.32	0.29	0.29	0.29	0.29	0.29	0.28
Substance:	Li	Na	K	Rb	Cs			
Z_c	0.064	0.132	0.175	0.217	0.203			

For a particular insulating substance, n-pentane, the coexistence curve in the density-temperature plane is plotted in Fig. 4.7. This illustrates the so-called "law of rectilinear diameters", which appears to be traceable back to Cailletet and Mathias in 1886, but which is now known to be restricted in its applicability. The assertion here is that the mean density of the liquid and its saturated vapour is a linear function of the temperature. Major deviations are known, however, for fluid metals near the critical point (see below).

Table 4.3 reports values of the compressibility ratio $Z_c \equiv p_c/\rho_c k_B T_c$ at the critical point for both insulating fluids and alkali metals.[95] As discussed in Appendix 4.2, the Dieterici equation of state gives a good account of the measured value $Z_c = 0.29$ for the heavier rare-gas systems Ar, Kr and Xe, yielding $Z_c = 0.27$.[96] From the virial equation of state (4.16), this fixes for these systems the value of a structural integral at the critical point,[97]

$$\left(\frac{\rho_c}{k_B T_c}\right) \int d\mathbf{r} g_c(r) r \frac{\partial \phi(r)}{\partial r} = 1 - Z_c \,. \tag{4.37}$$

This result is independent of the excluded-volume parameter in the equation of state — a satisfactory result since, as was already known to Ornstein and Zernike (see below), the pair distribution function $g(r)$ becomes very long-ranged as the critical point is approached.

Rather different and system-dependent values are taken by the critical compressibility ratio in the alkali metals (see Table 4.3). The pioneering experiments of Hensel and coworkers[98] on liquid Cs and Rb taken along the liquid–vapour coexistence curve towards the critical point have shown that the main lowering of the density that occurred was reflected in their structure factors measured by neutron scattering through a lowering of the coordination number z, with only a modest increase in the near-neighbour distance. These experiments testify to the value of a "chemical" picture in expanded alkali metals, in which the bond is the basic building block. A coordination-dependent equation of state,[99] which can be justified by glue models of cohesion in metals, is consistent with the variation of Z_c through the alkali series and leads to the relation

$$T_c V_c^{1/3} = \text{constant}\,. \tag{4.38}$$

This result is required from the empirical analysis of Chapman and March[95] who gave the exponent in Eq. (4.38) as 0.3.

4.8.2 *Ornstein–Zernike theory and critical exponents*

From Eq. (4.36) it follows that the isothermal compressibility K_T and hence $S(0)$, the long wavelength limit of the structure factor, diverge at the critical point. The original argument of Ornstein and Zernike[75] for the form of $g(r) - 1 \equiv h(r)$ near the critical point started out from Eq. (4.22) for the direct correlation function $c(r)$. They made in essence two assumptions: (i) that $c(r)$ is short-ranged compared with $h(r)$, which is only true near the critical point; and (ii) that $h(r')$ in the integral in Eq. (4.22) can be Taylor expanded around the point r up to second-order terms. A straightforward calculation leads to the asymptotic solution

$$h(r) \propto r^{-1} \mathrm{e}^{-r/\xi}\,, \tag{4.39}$$

where $\xi \propto [1 - \tilde{c}(0)]^{-1/2} = [S(0)]^{1/2}$ has the meaning of a correlation length between density fluctuations. Evidently, this length diverges as the critical point is approached.

While these results are useful first approximations, the modern theory of criticality[100] requires two independent exponents to describe the long-range correlations between density fluctuations. That is, near the critical point the total correlation function has the asymptotic form

$$h(r) \propto r^{2-d-\eta} e^{-r/\xi}\,, \tag{4.40}$$

where d is the dimensionality of the system and the correlation length has the form $\xi \propto t^{-\nu}$, where $t \equiv |T - T_c|/T_c$ is the reduced temperature. At the same time, the form of K_T at the critical density ρ_c is $K_T \propto t^{-\gamma}$ and the difference between liquid and gas densities is $\rho_l - \rho_\nu \propto t^{\beta}$.

The experimental evidence on the liquid–vapour transition in insulating fluids indicates that the exponent η is small (≈ 0.1) and yields $\nu \approx 0.6$, $\gamma \approx 1.1$ and $\beta \approx 0.35$ (see Sec. 4.8.4).

4.8.3 *Scaling relations*

The definitions of the exponents which characterise the singular behaviours of physical properties near the critical point are usually introduced using a language specific to spin magnetism. In addition to the four exponents introduced above (with K_T becoming the susceptibility χ and $\rho_l - \rho_\nu$ the magnetisation M), two further exponents are introduced. These give the behaviour of the heat capacity in zero field, $C \propto t^{-\alpha}$, and of the order parameter as a function of the field H at $t = 0$, $M \propto H^{1/\delta}$.

Experiments on critical behaviour show that very different systems can have closely similar values of the critical exponents. The exponents have *universal* character in the sense that, while they depend on the dimensionality d of the system and on the number n of components of the order parameter, they do not depend on the details of the interactions. Furthermore, some simple relations exist between the values of the various critical exponents. These facts have led to the concept of *scaling*, which underlies the renormalisation-group theory of critical phenomena.

The crucial aspect of criticality is the divergence of the correlation length ξ on the approach to the critical point. The scale of order in the system, as measured by ξ, increases indefinitely and hence the configuration of the system becomes invariant under a change in scale. In the so-called homogeneity hypothesis[101] one assumes that the correlation function changes like a homogeneous function of the variables r, H and t under a change in scale:

$$h(r; H, t) = b^{2y} h(r/b;\, b^{y_1} H,\, b^{y_2} t)\,, \tag{4.41}$$

for any b, when r is large and H and t small. This hypothesis leads to the following relations:

$$\left.\begin{aligned} \alpha + 2\beta + \gamma &= 2 \\ \gamma &= \beta(\delta - 1) \\ \gamma &= (2 - \eta)\nu \\ \alpha &= 2 - d\nu \end{aligned}\right\} . \tag{4.42}$$

These relations are satisfied by the critical exponents in the exact solution of the Ising model in $d = 2$ given by Onsager[102] ($\alpha = 0$, $\beta = 1/8$, $\gamma = 7/4$, $\nu = 1$, $\delta = 15$, $\eta = 1/4$).

Let us briefly see how the relations (4.42) between the critical exponents can be proved within the homogeneity hypothesis. Taking $H = 0$ and $b = r$ we have $h(r; 0, t) = r^{2y} h(1; 0, r^{y_2} t)$, yielding by comparison with Eq. (4.41) $2y = 2 - \eta - d$ and $y_2 = 1/\nu$. For the scaling of the susceptibility we now have from the fluctuation-theory formula

$$k_B T \chi(H, t) = \int d^d \mathbf{r} h(r; H, t) = \int d^d \mathbf{r} h(r/b; b^{y_1} H, b^{y_2} t)$$
$$= k_B T b^{2y+d} \chi(b^{y_1} H, b^{y_2} t) \,. \quad (4.43)$$

Setting $H = 0$ and $b^{y_2} = 1/t$ we find $\chi(0, t) = t^{-(2y+d)/y_2} \chi(0, 1)$ or $\gamma = (2y + d)/y_2 = (2 - \eta)\nu$. The relations $y_1 = 2 - \eta + \beta/\nu$, $\gamma = \beta(\delta - 1)$ and $\alpha + 2\beta + \gamma = 2$ are obtained by treating in a similar manner the order parameter $M = \int \chi dH$ and the free energy $F = \int M dH$. Finally, by assuming that the singular part F_s of the free energy be extensive and dimensionless, i.e. $F_s/(k_B T) \approx V/\xi^d$ with $\xi \approx t^{-\nu}$, we get the heat capacity $C \approx t^{d\nu - 2}$ namely $\alpha = 2 - d\nu$.

Scaling is justified by the theory of critical phenomena based on the renormalisation-group technique in field theory. The theory allows an approximate evaluation of the critical exponents: for instance, for the Ising model in $d = 3$ it yields $\alpha \approx 0.1$, $\beta \approx 0.33$, $\gamma \approx 1.3$, $\nu \approx 0.6$, $\delta \approx 4.2$ and $\eta \approx 0.07$. The measured values of the critical exponents for the liquid–vapour transition in argon, that we report in Sec. 4.8.4 below, are in fact close to these values. A detailed study of similarities and differences between the critical behaviour of real fluids and critical phenomena of the "broken symmetry" type can be found in the work of Orkoulas *et al.*[103]

4.8.4 *X-ray critical scattering from fluids*

Lin and Schmidt[104] have used X-ray absorption and small-angle scattering to study the equilibrium equation of state of argon near the liquid–vapour critical point. Measurements of the angular dependence of the X-ray scattering at angles less than a few degrees give information about the correlation of density fluctuations in the critical region. Also, the scattering intensity in the zero-angle limit is proportional to $S(0)$ and hence to the isothermal compressibility, both in the one-phase region above the critical temperature and below it in the region where the fluid separates into the liquid and vapour phases. In addition, X-ray absorption measurements can be employed to determine the fluid density in both regions.

The main results of the experiments of Lin and Schmidt on argon are as follows. The exponent of the order parameter is obtained from the temperature variation of the densities of the coexisting phases below T_c and is $\beta = 0.349 \pm 0.005$. The values of the exponents for the correlation length and for the compressibility are $\nu = 0.52 \pm 0.02$ and $\gamma = 1.13 \pm 0.02$ on the coexistence curve in the two-phase region. These exponents instead take the values $\nu = 0.63 \pm 0.02$ and $\gamma = 1.20 \pm 0.03$ at the critical density in the one-phase region above T_c.

4.9 Fluids at Equilibrium in a Porous Medium

The methods of equilibrium liquid-state theory have been extended to investigate the properties of fluids permeating a disordered porous material. Such a material can be viewed as made of many atomic aggregates, which are distributed in space in an essentially random manner to yield a structure of solid regions and voids. This structure can be taken to be homogeneous and isotropic over length scales larger than the characteristic size of the aggregates.

The problem that one faces in the statistical mechanics of a fluid inside a quenched matrix is that the free energy and the correlation and linear response functions are determined by double averages. One must first average over the annealed degrees of freedom keeping the quenched ones fixed, and then average over the quenched degrees of freedom. This problem has been tackled by the replica method, first developed to deal with spin glasses.[105] An outline of the general procedure for the construction of the free energy F, starting from a Hamiltonian function $H(x; y)$ where x stands for the degrees of freedom of the annealed component and y for those of the quenched component, is as follows. The free energy $F(y)$ in a given realisation of the pore network is given by $F(y) = -k_B T \ln Q(y)$ where $Q(y) = \int \exp[-H(x; y)/k_B T] dx$ is the partition function. The free energy F is then obtained by averaging over all realisations of the porous medium through a probability function $P(y)$, i.e. $F = -k_B T \int P(y) \ln Z(y) dy \equiv -k_B T \langle \ln Z(y) \rangle$. The replica method provides a way to carry out this latter average by expressing the logarithm of the partition function in terms of other functions which are more suitable for calculation. This is physically equivalent to creating replicas for the annealed fluid components.

The evaluation of the equilibrium structure taken by a dense fluid inside a quenched disordered matrix has been tackled by a variety of techniques based on the cluster expansion, the replica method, and linear response theory.[106] These results show that the Ornstein–Zernike relations in a classical mixture of annealed and quenched particles can be mapped onto those of a limiting case of a fully multi-component fluid, made from the original fluid and its replicas.

The issue of fluid phase equilibria in such confined geometries has been addressed in a short review by Rosinberg,[107] with regard both to liquid–gas coexistence and to demixing in liquid mixtures. Experimental studies refer to fluids in aerogels, which are very dilute disordered networks of silica strands with porosity well in excess of 90%, and in porous glasses such as Vycor, with porosity of roughly 30%. In high-porosity media the role of confinement is marginal compared to those of disorder and wetting, so that reference can usefully be made to an Ising model of spins subject to a random field as introduced by Brochard and de Gennes.[108] Vycor glass shows instead formation of many microscopic domains rather than macroscopic phase separation. An alternative explanation has been advocated in terms of wetting phenomena occurring inside a single pore.[109] In this view the confinement and the competition between adsorption and interfacial tension are at the origin of a very slow transformation kinetics.

Appendix 4.1 Inhomogeneous Monatomic Fluids

We present in this Appendix some general properties of a monatomic fluid at given temperature T and chemical potential μ, subject to an external potential $V(\mathbf{r})$. The position-dependent potential breaks invariance under translation and makes the average particle density a function of position ($\rho(\mathbf{r})$, say). We shall be concerned with the grand potential $\Omega = -pV$ and the Helmholtz free energy F of such an inhomogeneous fluid.

The discussion requires the mathematical notion of a *functional.* Very briefly, if F is some physical property of a macroscopic body (such as its free energy) and $f(\mathbf{r})$ is a function of position inside the body (such as a density profile), then we say that F is a functional $F[f(\mathbf{r})]$ if its value (a number, at given T and μ) is known once $f(\mathbf{r})$ is known at all points in the body.

In the problem of present interest we first define $u(\mathbf{r}) \equiv \mu - V(\mathbf{r})$. Assuming that we know how the atoms interact with each other, $V(\mathbf{r})$

completely determines the Hamiltonian of the fluid and hence $u(\mathbf{r})$ determines (in principle!) the equilibrium grand-canonical ensemble and the corresponding grand-potential. We express this fact by saying that the grand-potential is a functional $\Omega[u(\mathbf{r})]$ of $u(\mathbf{r})$.

Since the equilibrium ensemble determines the equilibrium density profile $\rho(\mathbf{r})$, we may say that $u(\mathbf{r})$ determines $\rho(\mathbf{r})$. A theorem due to Hohenberg, Kohn and Mermin[70] ensures that also the reverse is true: in principle we can go back from a given density profile $\rho(\mathbf{r})$ to the potential $u(\mathbf{r})$ which determines it.

There is, therefore, a biunivocal relationship between $u(\mathbf{r})$ and $\rho(\mathbf{r})$. This allows us to treat these two microscopic functions in essentially the same way as we treat conjugate thermodynamic quantities, when we pass from one to the other *via* a Legendre transformation. In this case the transformation allows us to introduce a new thermodynamic quantity $\bar{F}$, which is a functional of $\rho(\mathbf{r})$:

$$\bar{F}[\rho(\mathbf{r})] = \Omega[u(\mathbf{r})] + \int d\mathbf{r}\rho(\mathbf{r})u(\mathbf{r})\,. \tag{A4.1.1}$$

By comparing this equation with the relation $F = N\mu - pV$ between the Helmholtz and Gibbs free energies (see Sec. 3.1.3), it is evident that

$$\bar{F} \equiv F - \int d\mathbf{r}\rho(\mathbf{r})V(\mathbf{r})\,. \tag{A4.1.2}$$

Namely, $\bar{F}$ is the so-called "intrinsic" Helmholtz free energy, given by the Helmholtz free energy after subtraction of the mean energy of interaction with the external potential $V(\mathbf{r})$.

A4.1.1 *Equilibrium conditions*

The first-order functional derivative $\delta F[f(\mathbf{r})]/\delta f(\mathbf{r})$ is defined as the change in the value of F due to an infinitesimal change $df(\mathbf{r})$ at any specified position $\mathbf{r}$ in the body. From Eq. (A4.1.1) we find the equilibrium condition which determines $\rho(\mathbf{r})$ at given $u(\mathbf{r})$,

$$\frac{\delta \bar{F}}{\delta \rho(\mathbf{r})} = u(\mathbf{r}) \tag{A4.1.3}$$

as well as the equilibrium condition which determines $u(\mathbf{r})$ at given $\rho(\mathbf{r})$,

$$\frac{\delta \Omega}{\delta u(\mathbf{r})} = -\rho(\mathbf{r})\,. \tag{A4.1.4}$$

Let us at this point consider for an illustration a classical ideal gas of atoms which do not interact with each other. Its equilibrium density profile is given by the Boltzmann distribution,

$$\rho(\mathbf{r}) = \lambda^{-3} \exp\left[\frac{u(\mathbf{r})}{k_B T}\right], \tag{A4.1.5}$$

where λ is the thermal de Broglie wavelength. This result can be obtained from the appropriate equilibrium condition, which is $\delta F_{\text{id}}/\delta\rho(\mathbf{r}) = u(\mathbf{r})$, if the ideal free energy functional is

$$F_{\text{id}}[\rho(\mathbf{r})] = k_B T \int d\mathbf{r}\rho(\mathbf{r})\{\ln[\lambda^3\rho(\mathbf{r})] - 1]\}\,. \tag{A4.1.6}$$

Indeed, by taking $\rho(\mathbf{r}) =$ constant in this expression we recover the free energy of the homogeneous ideal gas (see Eq. (3.50)).

More generally, for a real fluid of interacting particles it is convenient to break $\bar{F}$ into the sum of its ideal part and of the "excess" part due to the interactions between the particles:

$$\bar{F}[\rho(\mathbf{r})] = F_{\text{id}}[\rho(\mathbf{r})] + F_{\text{ex}}[\rho(\mathbf{r})]\,. \tag{A4.1.7}$$

In a classical fluid F_{id} is still given by the expression (A4.1.6). Therefore, the equilibrium density profile takes the form

$$\rho(\mathbf{r}) = \lambda^{-3} \exp\left[\frac{u_{\text{KS}}(\mathbf{r})}{k_B T}\right], \tag{A4.1.8}$$

where $u_{\text{KS}}(\mathbf{r})$ is the potential introduced by Kohn and Sham,[70]

$$u_{\text{KS}}(\mathbf{r}) \equiv u(\mathbf{r}) - \frac{\delta F_{\text{ex}}}{\delta\rho(\mathbf{r})}\,. \tag{A4.1.9}$$

On comparing Eq. (A4.1.8) with Eq. (A4.1.6), we see that Eqs. (A4.1.8) and (A4.1.9) establish a mapping between the fluid of interacting particles and the ideal classical gas. Of course, $u_{\text{KS}}(\mathbf{r})$ is a functional of $\rho(\mathbf{r})$ and a self-consistent evaluation of the equilibrium density is required, after invoking a suitable approximate expression for the functional F_{ex}.

A4.1.2 *Direct correlation function*

The higher derivatives of the functionals Ω and $F_{\rm ex}$ define two hierarchies of correlation functions. In particular, for their second derivatives we have

$$c(\mathbf{r}_1, \mathbf{r}_2) \equiv -(k_{\rm B}T)^{-1} \frac{\delta^2 F_{\rm ex}}{\delta\rho(\mathbf{r}_1)\delta\rho(\mathbf{r}_2)}$$
$$= \frac{\delta(\mathbf{r}_1 - \mathbf{r}_2)}{\rho(\mathbf{r}_1)} - (k_{\rm B}T)^{-1} \frac{\delta u(\mathbf{r}_1)}{\delta\rho(\mathbf{r}_2)} \tag{A4.1.10}$$

and

$$H(\mathbf{r}_1, \mathbf{r}_2) \equiv -k_{\rm B}T \frac{\delta^2 \Omega}{\delta u(\mathbf{r}_1)\delta u(\mathbf{r}_2)} = k_{\rm B}T \frac{\delta\rho(\mathbf{r}_1)}{\delta u(\mathbf{r}_2)}\,. \tag{A4.1.11}$$

In taking the last step in these equations we have used the equilibrium conditions (A4.1.3) and (A4.1.4) as well as the definition (A4.1.7) of the excess free energy. The function $c(\mathbf{r}_1, \mathbf{r}_2)$ defined in Eq. (A4.1.10) is the Ornstein–Zernike direct correlation function for the inhomogeneous fluid, while $H(\mathbf{r}_1, \mathbf{r}_2)$ can be shown to be related to the pair correlation.[b]

The first-order functional derivatives in the RHS of Eqs. (A4.1.10) and (A4.1.11) have a precise physical meaning: $\delta\rho(\mathbf{r}_1)/\delta u(\mathbf{r}_2)$ describes the change in density of the inhomogeneous fluid in response to a change in the external potential, and $\delta u(\mathbf{r}_1)/\delta\rho(\mathbf{r}_2)$ is its inverse. Treating them as matrices with indices $\mathbf{r}_1$ and $\mathbf{r}_2$, from their matrix product we obtain the Ornstein–Zernike relation for the inhomogeneous fluid,

$$H(\mathbf{r}_1, \mathbf{r}_2) = \rho(\mathbf{r}_1)\delta(\mathbf{r}_1 - \mathbf{r}_2) + \rho(\mathbf{r}_1) \int d\mathbf{r}_3 c(\mathbf{r}_1, \mathbf{r}_3) H(\mathbf{r}_3, \mathbf{r}_2)\,. \tag{A4.1.12}$$

Finally we take the limit $V(\mathbf{r}) \to 0$: we recover the homogeneous fluid, in which $c(\mathbf{r}_1, \mathbf{r}_2) = c(r_{12})$ and $H(\mathbf{r}_1, \mathbf{r}_2) = \rho\delta(\mathbf{r}_1 - \mathbf{r}_2) + \rho^2[g(r_{12}) - 1]$. The Ornstein–Zernike relation becomes

$$h(r_{12}) = c(r_{12}) + \rho \int d\mathbf{r}_3 c(r_{13}) h(r_{32})\,, \tag{A4.1.13}$$

[b]The derivation requires starting from the expression of Ω in the equilibrium grand ensemble (see Appendix 3.2), but adding the interaction of the fluid with the external potential: see also Appendix 4.3.

with $h(r) = g(r) - 1$; namely, in Fourier transform

$$S(k) = [1 - \tilde{c}(k)]^{-1}. \tag{A4.1.14}$$

This result is Eq. (4.17) in the main text. The interpretation given above for Eqs. (A4.1.10) and (A4.1.11) shows that a diffraction experiment on a classical liquid gives us directly its static density response susceptibility to a weak external potential varying periodically in space with a wavelength $2\pi/k$.

A4.1.3 *Hypernetted-chain approximation in liquid-structure theory*

A plausible way to obtain an approximate expression for the functional F_{ex} uses an expansion of the inhomogeneous fluid around the homogeneous one at density ρ. With the notation $\Delta\rho(\mathbf{r}) = \rho(\mathbf{r}) - \rho$, the expansion yields for the difference $\Delta\bar{F}$ of the intrinsic free energies of the two fluids, taken at the same temperature and chemical potential, the expression

$$\Delta\bar{F} = (\mu - k_{\mathrm{B}}T)\int d\mathbf{r}\Delta\rho(\mathbf{r}) + k_{\mathrm{B}}T\int d\mathbf{r}\rho(\mathbf{r})\ln\left[\frac{\rho(\mathbf{r})}{\rho}\right]$$
$$- k_{\mathrm{B}}T\iint d\mathbf{r}d\mathbf{r}'c(|\mathbf{r}-\mathbf{r}'|)\Delta\rho(\mathbf{r})\Delta\rho(\mathbf{r}') + \cdots.$$

The higher terms in the expansion involve higher-order direct correlation functions of the homogeneous fluid.

We use the above expansion, truncated at second order terms, to relate the present discussion to the hypernetted-chain (HNC) approximation in liquid-structure theory. To this end we consider the special case in which the "external" potential is the interatomic potential $\phi(r)$ generated by an atom taken at the origin in a homogeneous classical fluid. In this case we have $\rho(\mathbf{r}) = \rho g(r)$, where ρ is the average density of the homogeneous fluid and $g(r)$ is the pair distribution function.

The equilibrium condition in Eq. (A4.1.3) then yields

$$g(r) = \exp\{-\beta\phi(r) + \rho\int d\mathbf{r}'c(|\mathbf{r}-\mathbf{r}'|)h(r') + \cdots\}$$
$$= \exp\{-\beta\phi(r) + h(r) - c(r) + \cdots\}, \tag{A4.1.15}$$

with $\beta = 1/k_BT$. In the last step we have used the Ornstein–Zernike relation (A4.1.13). Equation (A4.1.15) is the HNC closure and corresponds to setting to zero the "bridge function" $b(r)$ in Eq. (4.35) in the main text.

Appendix 4.2 The Dieterici Equation of State

Consider rare gas atoms inside a container. Atoms near one of the container walls will experience an asymmetric distribution of attractive forces, resulting in a reduction of the boundary-layer density ρ_b relative to the bulk density ρ. To estimate the influence of such interatomic forces on the equation of state, let us use the Boltzmann law

$$\rho(\mathbf{r}) \propto \exp\left[-\frac{V(\mathbf{r})}{k_BT}\right], \qquad \text{(A4.2.1)}$$

where $V(\mathbf{r})$ denotes the potential energy of an atom at position $\mathbf{r}$. Let ε represent the average potential energy of an atom in the interior of the fluid, while ε_b is the corresponding quantity near the walls of the container. Neglecting other possible effects, the ratio ρ_b to ρ is

$$\frac{\rho_b}{\rho} = \exp\left[-\frac{\Delta\varepsilon}{k_BT}\right], \qquad \text{(A4.2.2)}$$

where $\Delta\varepsilon = \varepsilon_b - \varepsilon > 0$, the inequality following from the fact that the attractive forces tend to pull the atoms away from the walls.

Introducing excluded volume effects as in the van der Waals equation of state, one is led rather directly from Eq. (A4.2.2) to the Dieterici equation of state:

$$p(v-b) = k_BT \exp\left(-\frac{a}{vk_BT}\right), \qquad \text{(A4.2.3)}$$

where $v = 1/\rho$ is the volume per atom and $a \propto \Delta\varepsilon$. If it is valid to expand the exponential to first-order only, then the van der Waals equation can be recovered by only inessential approximations. However, we stress below that it is very important to retain the exponential form in Eq. (A4.2.3) in modelling the heavier rare gases at the critical point.

Indeed, by standard calculations[96] the critical constants can be derived from Eq. (A4.2.3) as $v_c = 2b$, $k_BT_c = a/4b$ and $p_c = (a/4b^2)\exp(-2)$, yielding for the compressibility ratio

$$Z_c \equiv \frac{p_c v_c}{k_B T_c} = 2\exp(-2) \cong 0.27\,. \tag{A4.2.4}$$

This is in good agreement with the measured value $Z_c \cong 0.29$ for the heavy rare gases Ar, Kr and Xe. In contrast, the van der Waals equation of state leads to $Z_c = 3/8 = 0.375$.

Appendix 4.3 Force Equation and Born–Green Theory of Liquid Structure

We derive in this Appendix a further exact relationship between the pair distribution function $g(r)$ in a classical monatomic liquid and an assumed pair potential $\phi(r)$. This relationship involves the triplet distribution function in the liquid and leads into a hierarchical series of integral equations for many-particle distribution functions that needs truncating by some approximate decoupling procedure in order to yield useful results.

The starting point is the definition $g(r_{12}) \equiv \langle \rho^{(2)}(\mathbf{r}_1, \mathbf{r}_2)\rangle/\rho^2$ for $g(r)$ in terms of the mean two-body density (for $\mathbf{r}_1 \neq \mathbf{r}_2$). Calculating the average over the grand-canonical ensemble and carrying out the classical integration over momenta leads to

$$\rho^2 g(r_{12}) = \sum_{N=2}^{\infty} \frac{\exp[(\Omega + N\mu)/k_B T]}{N!} \left(\frac{mk_B T}{2\pi\hbar^2}\right)^{3N/2} \int d\mathbf{R}_1 \cdots \int d\mathbf{R}_N$$

$$\times \exp\left(-\frac{\Phi_N}{k_B T}\right) \sum_{i\neq j=1}^{N} \delta(\mathbf{R}_i - \mathbf{r}_1)\delta(\mathbf{R}_j - \mathbf{r}_2)\,. \tag{A4.3.1}$$

Here, Φ_N is the potential energy function (depending on all the coordinates $\mathbf{R}_1, \mathbf{R}_2, \ldots, \mathbf{R}_N$ of the atoms), μ is the chemical potential and Ω the grand-potential, related to the grand-canonical partition function Ξ by $\Xi \equiv \exp(-\Omega/k_B T)$. We introduce the fugacity z as

$$z = \left(\frac{mk_B T}{2\pi\hbar^2}\right)^{3/2} \exp\left(\frac{\mu}{k_B T}\right) \tag{A4.3.2}$$

and find

$$\rho^2 g(r_{12}) = \frac{1}{\Xi}\sum_{N=2}^{\infty}\frac{z^N}{(N-2)!}\int d\mathbf{R}_3\cdots\int d\mathbf{R}_N \exp\left[-\frac{\Phi_N(\mathbf{r}_1,\mathbf{r}_2,\mathbf{R}_3,\ldots,\mathbf{R}_N)}{k_\mathrm{B}T}\right], \tag{A4.3.3}$$

where

$$\Xi = \sum_{N=0}^{\infty}\frac{z^N}{N!}\int d\mathbf{R}_1\cdots\int d\mathbf{R}_N \exp\left[-\frac{\Phi_N(\mathbf{R}_1,\ldots,\mathbf{R}_N)}{k_\mathrm{B}T}\right]. \tag{A4.3.4}$$

The force equation follows from Eq. (4.3.3) by taking its gradient with respect to $\mathbf{r}_1$ and setting Φ_N equal to the sum of pair interactions. We get

$$-k_\mathrm{B}T\rho^2\nabla_{\mathbf{r}_1}g(r_{12}) = \frac{1}{\Xi}\sum_{N=2}^{\infty}\frac{z^N}{(N-2)!}\int d\mathbf{R}_3\cdots\int d\mathbf{R}_N \exp\left[-\frac{\Phi_N}{k_\mathrm{B}T}\right] \times\left[\nabla_{\mathbf{r}_1}\phi(r_{12}) + \sum_{i=3}^{\infty}\nabla_{\mathbf{r}_1}\phi(|\mathbf{r}_1-\mathbf{R}_i|)\right]. \tag{A4.3.5}$$

The sum in the square bracket contributes $(N-2)$ equal terms, whose value can be determined by taking $i=3$, say. The result is

$$-k_\mathrm{B}T\nabla_{\mathbf{r}_1}g(r_{12}) = g(r_{12})\nabla_{\mathbf{r}_1}\phi(r_{12}) + \rho\int d\mathbf{r}_3 g^{(3)}(\mathbf{r}_1,\mathbf{r}_2,\mathbf{r}_3)\nabla_{\mathbf{r}_1}\phi(r_{13}), \tag{A4.3.6}$$

where $g^{(3)}(\mathbf{r}_1,\mathbf{r}_2,\mathbf{r}_3)$ the three-body distribution function. This is the force equation.

From Eq. (A4.3.6) it is possible to derive the exact relation (see e.g. Ref. 30)

$$k_\mathrm{B}T\left[\frac{\partial g(r)}{\partial p}\right] = \int d\mathbf{r}_3[g^{(3)}(\mathbf{r}_1,\mathbf{r}_2,\mathbf{r}_3) - g(r)g(r_{23}) - g(r)g(r_{31}) + g(r)]. \tag{A4.3.7}$$

Therefore, the experimental study of the pressure dependence of $g(r)$ gives information on the three-body correlations, integrated over all values of the coordinates of the third particle. Tests of approximate expressions for the triplet function have been made from measurements of the isothermal pressure derivative of the liquid structure factor in rubidium near the triple point and in argon near the critical point.[525]

One such approximate expression is the so-called superposition approximation

$$g^{(3)}(\mathbf{r}_1, \mathbf{r}_2, \mathbf{r}_3) = g(r_{12})g(r_{23})g(r_{31}), \qquad \text{(A4.3.8)}$$

which was proposed in early work by Kirkwood.[526] This is the basis of the Born–Green theory of liquid structure,[527] which is known from experience to work reasonably well for liquids with short-range interactions.

For an application of this general approach in relation to liquid metal Na near freezing, as described by a Fourier-transformable pair potential from the electron theory of metals, the reader is referred to the work of Golden *et al.*[528]

Chapter 5

Diffusion

This and the next two chapters will be concerned with transport processes in dense liquids. The present chapter deals with mass transport (diffusion), the following chapter with momentum transport (viscosity) and the third with transport of energy (thermal conduction).

It has been known for a long time, from kinetic theory arguments, that diffusion in a dilute gas is rapid and depends in a well-defined way on the thermodynamic state. But when we turn to dense liquids, the problem is immediately more difficult because of the complex nature of collisions between particles. A good starting point is Brownian motion, that we have already presented in Sec. 1.3 as part of a largely qualitative introduction to molecular thermal motions. This can be applied to treat a colloidal particle in a liquid, and is also already useful in connection with the irregular zig-zag motions of a molecule in a fluid under the impact of the other molecules.

5.1 Background: Magnitude of Diffusion Coefficients in Gases Contrasted with Liquids

Diffusion, or mass transport, has very distinctive features in the dense liquid state. It is concerned, from a microscopic point of view, with following the motion of a chosen atom, say in liquid argon, which is taken to be at the origin of spatial coordinates ($\mathbf{r} = 0$) at time $t = 0$.

The idea is then to watch the motion of the chosen atom away from the origin as time elapses. More precisely, we shall want to follow how the velocity of the atom at time t correlates with its initial velocity and then take an ensemble average over the initial velocity. In his early work, Einstein recognised that the diffusive motion of a "tagged" atom is such that, at sufficiently long time t, the mean square displacement $\langle r^2 \rangle$ becomes proportional to t, in complete contrast to free particle motion at constant velocity where the distance moved in time t is proportional to t.

In terms of the self-diffusion coefficient D, the diffusive behaviour can be written

$$\langle r^2(t) \rangle = 6Dt \tag{5.1}$$

in the long time limit. As to magnitudes of diffusion coefficients of dense liquids, many of these values fall close to 10^{-5} $cm^2 \cdot s^{-1}$ (see Table 5.1 below for dilute solutes diffusing in water).

Such diffusion coefficients, for orientation, are some ten thousand times slower than the corresponding values in dilute gases. In this limiting case, one may treat the diffusion process by considering a gas of rigid spheres of small molecular dimensions.[110] The result of this model for the diffusion coefficient

Table 5.1. Solute diffusion (at infinite dilution) in water at 25°C.

Solute	$D \times 10^{-5}$ $cm^2 \cdot s^{-1}$
Argon	2.5
Chlorine	1.89
Nitrogen	2.0
Oxygen	2.42
Carbon dioxide	1.91
Ammonia	1.64
Methane	1.84
Benzene	1.02
Methanol	1.28
Ethanol	1.24
Acetic acid	1.29
Acetone	1.28
Glycine	1.05
Haemoglobin	0.07

D is

$$D = \frac{1}{3}\bar{v}\ell\,, \tag{5.2}$$

where $\bar{v}$ represents the average particle velocity while ℓ is the mean free path between successive collisions. Both quantities entering Eq. (5.2) can be calculated in a gas of rigid spheres in equilibrium at temperature T. If the molecular mass is m, then the principle of equipartition of energy yields $\bar{v} = (k_\mathrm{B}T/m)^{1/2}$. In order of magnitude the mean free path ℓ can also be calculated, and we merely quote the result (see also the discussion of a hard sphere fluid below):

$$\ell \approx \frac{k_\mathrm{B}T}{p\sigma^2} = \frac{1}{\rho\sigma^2}\,, \tag{5.3}$$

where p is the pressure, ρ is the number density and σ is the hard sphere diameter. Inserting numbers for dilute gases, one find values of D from Eq. (5.2) of the order of 1 $\mathrm{cm}^2 \cdot \mathrm{s}^{-1}$.

5.1.1 *Practical consequences of "slow" diffusion in dense liquids*

Though, in this basic Introduction to dense liquids most of the quantitative work will take the simplest possible examples (e.g. self-diffusion in liquid argon near its triple point or in liquid metal sodium near freezing), it is important to recognize practical consequences in everyday life of the "slow" rates of diffusion in liquids. These are discussed, for example, in the book by Cussler.[111] As he notes, diffusion frequently limits the overall rate of processes occurring in liquids. He cites some examples, among which are:

(i) in chemistry, diffusion limits the rate of acid-base reactions;
(ii) in chemical industry, diffusion is responsible for the rates of liquid–liquid extraction;
(iii) in metallurgy, diffusion can control the rate of surface corrosion;
(iv) in physiology, the rate of digestion is diffusion limited.

The concept of diffusion-controlled reactions in liquids can be traced as far back as Smoluchowski[112] (see also Chandrasekhar[113]). This concept is applicable whenever the rate of reaction is dominated by the slow process of mutual diffusion of the reaction partners. For example, in some limiting cases one of the reaction partners may be large and can be viewed as immobile. Then it is fruitful to consider an array of static sinks which can absorb particles diffusing independently in the ambient medium. In this type of model, Smoluchowski could already derive an expression for the chemical rate coefficient in the dilute limit by studying the absorption by a single sink. In dense systems, more generally, there will be a competition between such sinks and the rate coefficient will be dependent on the concentration of the sinks. For some account of the theory of this dependence, the interested reader may consult, for instance, the work of Calef and Deutsch.[114]

It is also relevant to note that one method that has been proposed for experimentally testing the concentration dependence of the diffusion-controlled rate coefficient is fluorescence quenching (see again Calef and Deutsch[114]; see also Baird, McCaskill and March[115]).

5.2 Fick's Law and Diffusion Equation

In the above context, pioneering work of Fick (1855) is to be cited. He recognized that diffusion is a dynamical molecular process and developed its law using analogies with the earlier studies of Fourier (1822) on thermal conduction. This led Fick to write, in a one-dimensional case, a flux per unit transverse area (j, say) in terms of a concentration gradient $\partial c/\partial z$ as

$$j = -D\frac{\partial c}{\partial z}\,. \tag{5.4}$$

In fact, the driving force for diffusion of matter is the gradient of chemical potential $\mu(c)$ associated with the concentration gradient, and Eq. (5.4) implies writing $\partial\mu/\partial z = (\partial\mu/\partial c)(\partial c/\partial z)$ and incorporating the factor $\partial\mu/\partial c$ into the definition of D.

Fick also paralleled Fourier's work on heat conduction in using the continuity equation

$$\frac{\partial c}{\partial t} = -\frac{\partial j}{\partial z} \tag{5.5}$$

to obtain from Eq. (5.4) the transport equation

$$\frac{\partial c}{\partial t} = D\frac{\partial^2 c}{\partial z^2}\,. \tag{5.6}$$

Equation (5.6) generalizes in three dimensions to the diffusion equation involving the Laplacian operator $\nabla^2 (\nabla^2 \equiv \partial^2/\partial x^2 + \partial^2/\partial y^2 + \partial^2/\partial z^2$ in Cartesian coordinates):

$$\frac{\partial c}{\partial t} = D\nabla^2 c\,. \tag{5.7}$$

Use will be made of this three-dimensional equation based on Fick's work, later in this chapter.

5.2.1 *Examples of diffusion across a thin film*

As a simple example of Fick's law, let us consider diffusion across a thin film. On either side of the film is a well-mixed solution of one solute. The solute diffuses from the fixed higher concentration, present for $z \leq 0$ into the fixed lower concentration solution, located at $z \geq L$.

The aim now is to employ Fick's law to find the solute concentration profile and also the diffusion flux across the film. To do so, the starting point is to write a mass balance equation on a thin layer with thickness Δz, located in the film at some position z. In order to avoid accumulation of the solute, which is the condition for steady state diffusion, the rate of diffusion into the layer at z must be equated to the same quantity out of the layer at position $z + \Delta z$.

The steady state condition reads $j(z) = j(z + \Delta z)$ and, if we divide by the thickness Δz of the chosen layer, then in the limit as Δz tends to zero we find that the divergence of the current vanishes, $\partial j(z)/\partial z = 0$. The concentration profile is therefore constant in time and, according to Fick's law, is determined by solving the differential equation

$$\frac{d^2 c(z)}{dz^2} = 0\,. \tag{5.8}$$

With the boundary condition $c(0) = c_0$ and $c(L) = c_L$, the solution for the density profile is

$$c(z) = c_0 + \frac{(c_L - c_0)z}{L}\,. \tag{5.9}$$

The steady-state flux is immediately found from Eq. (5.4),

$$j = \frac{D(c_0 - c_L)}{L}\,. \tag{5.10}$$

A richer solution of Eq. (5.7) is given in Sec. 5.4.1.

Other examples related to the above, for instance that of a single solute diffusing through a thin membrane where the membrane is chemically different from the two solutions, are worked out in the book of Cussler[111] and we shall not go into further details here. However, the thrust of the argument is clear enough: in such problems Fick's law (5.4) is to be combined with mass balance considerations to calculate concentration profiles and fluxes.

5.3 Solute Diffusion at High Dilution in Water and in Non-aqueous Solvents

Following the above introduction to Fick's law and the diffusion equation, we want in this section to press further the point as to the "slowness" of diffusion in liquids. We choose to do so by returning to Table 5.1 for solute diffusion coefficients in water at 25°C, in the limit of infinite dilution, from tabulations in the Handbook of Chemistry and Physics.[116] We start off with one of the simplest elements, rare-gas argon. This is followed in the Table by diatomic molecules such as chlorine and carbon dioxide, and by polyatomic hydrocarbons such as methane and benzene. Table 5.1 ends with the example of haemoglobin.

When we change from water to non-aqueous solvents such as chloroform, benzene or ethyl alcohol, solute diffusion coefficients still lie close to 10^{-5} $cm^2 \cdot s^{-1}$. Exceptions arise for solutes having higher molecular weight such as polystyrene or albumin, for which diffusion can be one hundred times slower.

5.3.1 *Stokes–Einstein and semiempirical estimates of solute diffusion*

We turn to enquire as to the way one might begin to make theoretical estimates of such solute diffusion coefficients. A common route for estimating diffusion coefficients in liquids is the so-called Stokes–Einstein relation. As already discussed in Chap. 1, this relation invokes as an essential ingredient another transport property, the shear viscosity of the solvent in which the solute diffusion is occurring. Since the next chapter is devoted to viscosity, the introduction of the Stokes–Einstein relation here is somewhat anticipating the discussion there.

Following the arguments developed in Chap. 1 (see especially Eqs. (1.9) and (1.16)), the Stokes–Einstein relation reads

$$D = \frac{k_B T}{f} = \frac{k_B T}{6\pi\eta a}. \tag{5.11}$$

In the equation, f denotes the "friction coefficient" for the solute, η is the viscosity of the solvent, and a is the "radius" of the solute. The dependence of the diffusion coefficient on the nature of the solute is simply through its "radius". While one has little difficulty in specifying a for the spherical argon atom representing the first entry in Table 5.1, and perhaps a similar situation obtains for the "nearly spherical" molecule of methane, for many of the other entries "molecular shape" is involved, and hence some gross averaging is involved in specifying a in Eq. (5.11).

As Cussler[111] notes in his book, diffusion coefficients do vary inversely with viscosity when the ratio of solute to solvent radius exceeds five. This is encouraging in the sense that the Stokes–Einstein relation (5.11) was (at least initially: but see generalisation eventually effected below) derived by assuming a rigid solute sphere diffusing in a structureless solvent. For a large solute diffusing in a solvent made up of small particles, Eq. (5.11) has a theoretical justification.

As an example in Cussler's book, he uses the Stokes–Einstein relation to estimate the diffusion constant of oxygen in water at 25°C, for which the measured value of D entered in Table 5.1 is $\approx 2 \times 10^{-5}$ cm$^2 \cdot$ s^{-1}. As he notes, the chief difficulty is to estimate a, the "radius" of the oxygen molecule. If one assumes that this is half the collision diameter in the gas, one has $a \approx 1.7$ Å and one is led to $D \approx 1.3 \times 10^{-5}$ cm$^2 \cdot$ s^{-1}: sensible but not quantitative.

It is worthy of note in the above context that a variety of alternatives — usually less theoretically well based than the Stokes–Einstein result — are available for estimating diffusion coefficients in liquids such as recorded in Table 5.1. Thus Scheibel[117] has empirically introduced the ratio of molar volumes of solute and solvent into a modification of the Stokes–Einstein relation: this considerably improves the estimate for oxygen in water, to $D \approx 2.2 \times 10^{-5}$ cm$^2 \cdot$ s^{-1}. Finally, Wilke and Chang,[118] again empirically, have introduced a factor representing solute–solvent interaction into a modification of the Stokes–Einstein relation.

5.4 Summary of Techniques, Including Computer Simulation, for Determining Diffusion Coefficients

In this section, we shall give a quite brief account of some experimental techniques, including computer simulation, which are relevant to the study of diffusion and to the determination of diffusion coefficients. We do not discuss the standard radioactive tracer technique for tracer diffusion measurements. However, we report in Fig. 5.1, from a review by Angell[119] on supercooled

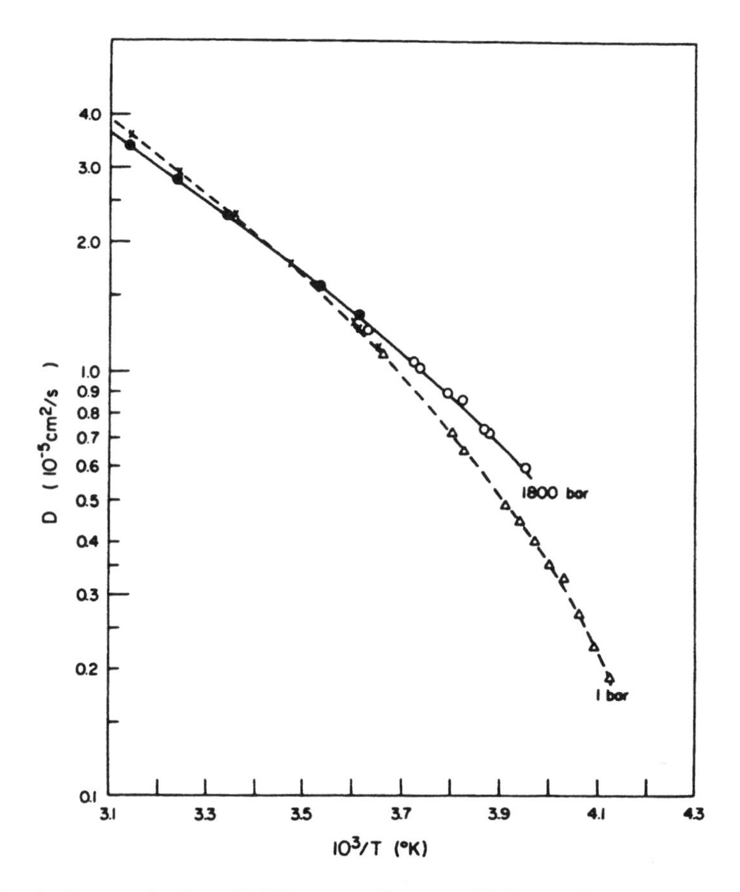

Fig. 5.1. Arrhenius plot for self-diffusion coefficient in H_2O as a function in inverse temperature at two different pressures, showing both the results of tracer diffusion measurements (• and ×) and of NMR measurements (∘ and Δ). (Redrawn from Angell, Ref. 119.)

water, an Arrhenius plot of the self-diffusion constant of H_2O at two different pressures, comparing tracer data with those obtained by the nuclear magnetic resonance technique discussed further below.

5.4.1 *Incoherent neutron scattering*

In Sec. 5.1 we have introduced the atomic diffusive motions in a classical liquid (say in pure liquid argon to be definite) by a vivid picture: we "sit" on a tagged atom as it meanders in time through the liquid, having taken it at time $t = 0$ as being at the origin of coordinates ($\mathbf{r} = 0$). Then it is natural to enquire what is the probability that this tagged atom, at a later time t, is at position $\mathbf{r}$. We denote this probability by $G_s(r,t)$, the subscript s denoting self-motion. The atom during its motion can exchange energy and momentum with its neighbours and we then have only one conservation law to take care of, i.e. that for particle number. This is the reason why the self-diffusion problem is much easier than viscosity and heat conduction.

A natural enough starting point to study some "gross" features of $G_s(r,t)$ is to assume that it obeys the diffusion equation, already introduced in Sec. 5.2 *via* the pioneering work of Fick. Applying this to determine the evolution of $G_s(r,t)$ in time, one can write from Eq. (5.7) in a pure liquid like argon with self-diffusion constant D:

$$\frac{\partial G_s(r,t)}{\partial t} = D\nabla^2 G_s(r,t)\,. \tag{5.12}$$

The physical solution of this equation is readily verified by substitution to be

$$G_s(r,t) = (4\pi Dt)^{-3/2} e^{-r^2/4Dt}\,, \tag{5.13}$$

for $t > 0$. It is easily verified that as $t \to 0$ $G_s(r,t)$ reduces to a representation of the delta function — the specified initial condition. The mean square displacement is

$$\langle r^2 \rangle = \int d\mathbf{r} r^2 G_s(r,t) = 6Dt\,, \tag{5.14}$$

as expected.

Of course, the result (5.13), having been based on Fick's law, is valid only for times long compared with collision times. Without entering microscopic details, however, the important point is that the double Fourier transform of

$G_s(r,t)$ (the spectral function $S_s(k,\omega)$, say) determines the incoherent cross-section in the process of inelastic scattering of a beam of neutrons from the liquid under study. The variables $\hbar k$ and $\hbar\omega$ here have the meaning of the momentum and energy given up to the liquid in the inelastic scattering process. This theorem, which is due to van Hove[120] (also see Marshall and Lovesey[121]) implies that the self-diffusion coefficient can in principle be measured from the half-width of the incoherent scattering spectrum. The spectral function corresponding to Eq. (5.13) is

$$S_s(k,\omega) = \frac{2Dk^2}{\omega^2 + (Dk^2)^2}, \tag{5.15}$$

as expression which is valid at low k and ω. Therefore, the half-width of the spectrum in this limit is Dk^2 (see Fig. 5.2).

More generally, one may express the relationship between the self-diffusion coefficient and the van Hove dynamic structure factor $S_s(k,\omega)$ as

$$D = \frac{1}{2}\lim_{\omega\to 0}\left\{\omega^2 \lim_{k\to 0}\left[\frac{S_s(k,\omega)}{k^2}\right]\right\}. \tag{5.16}$$

The expression, which can be verified from Eq. (5.15), is one of the celebrated Kubo formulae relating transport coefficient to wavenumber and frequency dependent correlation functions.[122]

Complex systems such as polymers exhibit interesting dynamics over a wide temperature range and a broad time scale. By permitting experimental

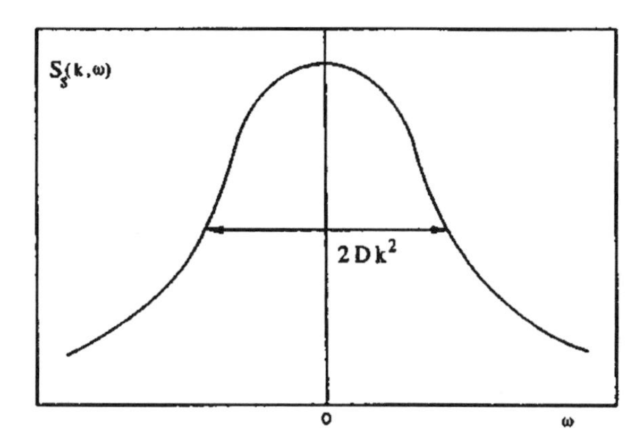

Fig. 5.2. Schematic form of the dynamic structure factor $S_s(k,\omega)$ for self-diffusion in a monatomic liquid.

separation of the incoherent and coherent scattering parts, the use of polarised neutrons with spin polarization analysis enables confident model fitting to scattering data in such systems.[123]

5.4.2 *Dynamic light scattering*

Together with incoherent neutron scattering, dynamic light scattering and dielectric relaxation are widely used experimental tools for the study of molecular motions in polymers, even down into the temperature range approaching the glass transition.[124] To visualise diffusive motions in such macromolecular systems it is useful to focus on the case of a single polymer molecule floating in a neutral solvent of low molecular weight. Then, as proposed by Vrentas and Duda,[125] the molecule can usefully be pictured as a necklace of spherical beads connected by a cord that does not exhibit any resistance to flow. If the solution is very dilute, then the only interaction is with the solvent. In certain cases the solvent will considerably expand the necklace in solution, while in other cases the polymer necklace can shrink into a blob (see Chap. 9).

Between the above extremes, polymer and solvent can interact sufficiently for the segments of the necklace to be essentially randomly distributed. The limit of a "random coil" of polymer is, by convention, chosen as the "ideal" polymer solution. Under such circumstances the polymer diffusion can be estimated from a modified Stokes–Einstein relation:

$$D = \frac{k_B T}{6\pi\eta a_e}, \tag{5.17}$$

where a_e is the effective radius of the polymer. This radius is taken from calculations as

$$a_e = 0.676\langle R^2\rangle^{1/2}, \tag{5.18}$$

where $\langle R^2\rangle^{1/2}$ is the root-mean-square radius of gyration, which is the customary measure of the size of polymer molecules in solution. One method of determining this radius is by light scattering. Equations (5.17) and (5.18) are found to be in agreement with such experiments.

Away from "ideality", the diffusion coefficient is still estimated from the Stokes–Einstein relation, but of course the relation between a_e in Eq. (5.17) and the RMS radius presents more of a problem. Moreover, in certain cases the diffusion coefficient can increase markedly with polymer concentration. Such increase occurs in the face of rapidly increasing viscosity.

Solutes of low molecular weight in a polymer solvent are also of interest as a second limiting case of diffusion in polymer systems, while a third case is that where both solute and solvent are polymers (see, for example Tirrell[126]). This third case has practical relevance in adhesion, in material failure and in polymer fabrication.

5.4.3 *Nuclear magnetic resonance*

Measurements of diffusion coefficients by nuclear magnetic resonance[127] (NMR) allow the achievement of an accuracy of $\approx 5\%$. In the experimental set-up a homogeneous sample is placed in a substantial magnetic field, which aligns the magnetic moments of the nuclei under study. For example, the moments associated with the protons in water molecules may provide the spin magnetisation which is used as the experimental probe. When the magnetic field is then slightly altered, the moments precess and this can induce in an adjacent coil a small time-dependent voltage $V(t) = V_0 \sin(t/\tau)$. The period τ is often the focus of interest in NMR, as it contains information on the immediate chemical environment of the nuclear moment.

For diffusion studies, however, one wants to observe directly the dephasing of the transverse component of the spins, due to the random local magnetic fields arising from molecular motions. This characteristic dephasing time is called the spin–spin relaxation time T_2. When a static magnetic field gradient is imposed, the nuclei during their motions experience different fields and diffusion through the liquid can be followed provided that T_2 is not too short. Typically T_2 in a liquid is of order 1–10 s.

Another NMR method uses instead a pulsed field gradient, which is applied first in one direction and then, after a short interval τ', in the opposite direction. If the solute molecules were fixed, the two perturbations would not change the amplitude V_0. However, these molecules are performing Brownian-line motions, causing a reduction in amplitude which can be measured as a function of the interval τ' between the gradient pulses. The slope of this variation constitutes a direct measure of the Brownian-like motion and hence of the diffusion coefficient.

Figure 5.3 reports pulsed-NMR data[128,129] on diffusion of water molecules as a function of inverse temperature in an Arrhenius plot, both in pure water and in aqueous solutions of $ZnCl_2$ at various concentrations. The activation

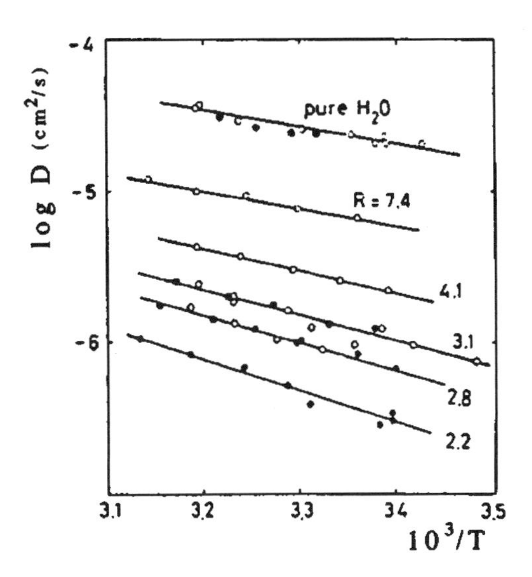

Fig. 5.3. Temperature dependence of the self-diffusion coefficient of water in H_2O–$ZnCl_2$ solutions at various concentrations. (Redrawn from Nakamura *et al.*, Ref. 128.)

energy that can be extracted from this plot essentially doubles in going from pure water to the saturated solution.

5.4.4 *Computer simulation of mean square displacement*

The study of diffusion by the molecular dynamics (MD) method has the advantage that, given an interatomic force law as input, one obtains for a classical liquid reliable results corresponding to a known (though often not fully realistic) force field. The diffusion coefficient D is extracted from the long-time limiting behaviour of the mean square displacement of a particle in the model liquid as a function of the elapsed time, which according to Eq. (5.1) is a linear increase with time: $\langle r^2 \rangle = 6Dt$. Only a few examples will be mentioned here.

An early example of such work, referring to two different pair-potential models for liquid sodium near freezing, is reported in Fig. 5.4.[130] As is seen from the figure, the "long-time" linear behaviour is being established in this model liquid at times ≈ 0.5 ps and the marked difference between the two curves in this regime shows that the diffusion constant is quite sensitive to the interactions. At times below 0.1 ps, on the other hand, an approximately

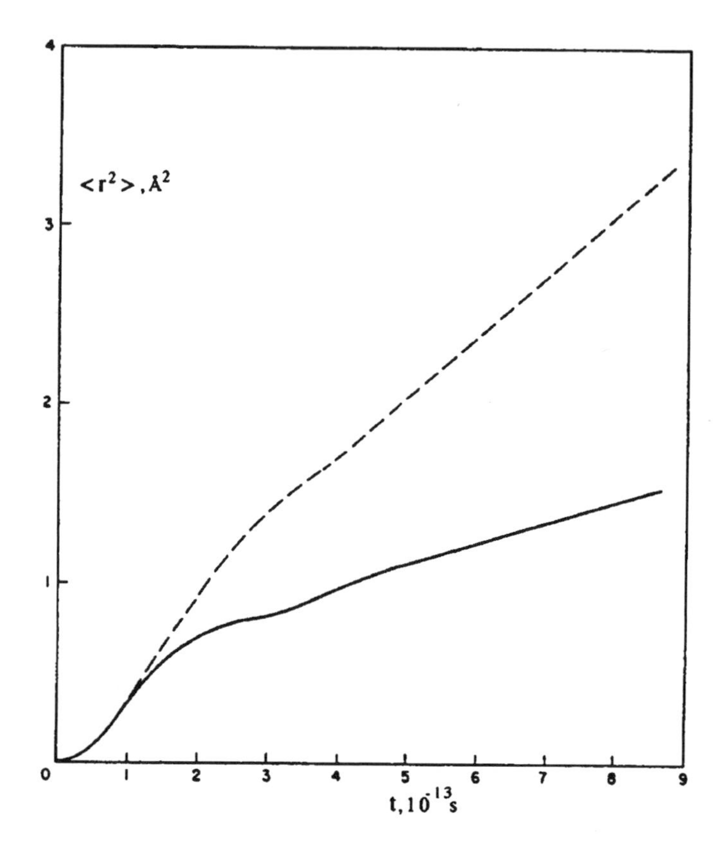

Fig. 5.4. Mean square displacement $\langle r^2 \rangle$ against time t from a molecular dynamics study of two alternative pair-potential models of liquid sodium. (Redrawn from Rahman and Paskin, Ref. 130.)

quadratic increase of $\langle r^2 \rangle$ with time is observed, as for free particle motion. The detailed dynamics of an average particle at low and intermediate times can be better appreciated from the velocity autocorrelation function, which is presented in Sec. 5.5 below.

As a further example of MD studies of diffusion in liquids we refer to the determination of the transport coefficients of the hard-sphere model. These are by now well known over the whole fluid range. A convenient fit of the data is that given by Speedy,[131]

$$D = D_0 \left(1 - \frac{\rho}{1.09}\right) [1 + \rho^4 (0.4 - 0.83\rho^4)] \tag{5.19}$$

where $\rho = N\sigma^3/V$. In this expression D_0 is the infinite dilution value of D, which is given exactly by kinetic theory[132] as

$$D_0 = \frac{3\sigma}{8\rho}\left(\frac{k_B T}{\pi m}\right)^{1/2}, \tag{5.20}$$

where again m is the particle mass and $\rho = N\sigma^3/V$.

An alternative expression for the density dependence of the self-diffusion coefficients for hard spheres has been given by Erpenkeck and Wood,[133] who fitted their simulation data to

$$D = D_E(1 + 0.0382\rho + 3.18\rho^2 - 3.869\rho^3)\,. \tag{5.21}$$

Here, the Enskog value of the self-diffusion constant D_E is given as

$$D_E = \frac{1.01896 D_0}{g(\sigma^+)}, \tag{5.22}$$

where $g(\sigma^+)$ is the value of the pair distribution function at constant (see Sec. 3.7).

Finally, Heyes and Powles[134] have made simulations of fluids with inverse power-law repulsive potentials, $\phi(r) = \varepsilon(\sigma/r)^n$. The scaling of thermodynamics for fluids described by such potentials was set out in Sec. 3.8. The aim of their work was to study how the self-diffusion coefficient $D(n)$ for exponent n converges to the hard sphere limit $D(n = \infty)$, at a packing fraction equal to 0.044. Their results are fitted with remarkable accuracy by the expression

$$D(n) - D(n = \infty) = 7.5n^{-1.71}, \tag{5.23}$$

showing rather slow convergence to the hard-sphere limit. Here D is in units of $\sigma(\varepsilon/m)^{1/2}$.

5.5 Velocity Autocorrelation Function in Pure Dense Liquids

As anticipated in Sec. 5.4.4 above, the technique of MD computer simulation affords a microscopic view of atomic motions in a liquid through the study of the velocity autocorrelation function $Z(t)$. This is defined for a classical fluid by

$$Z(t) = \langle \mathbf{v}(0) \cdot \mathbf{v}(t) \rangle\,, \tag{5.24}$$

$\mathbf{v}(t)$ being the velocity of a particle at time t and the brackets denoting the ensemble average.

The mean square distance $\langle r^2(t)\rangle$ travelled in the time interval t by a particle starting off at the origin $\mathbf{r} = 0$ at time $t = 0$ is related to $Z(t)$ by

$$\langle r^2(t)\rangle = 2\int_0^t dt'(t-t')Z(t')\,. \tag{5.25}$$

This relation is most easily proved by showing first[a] that $d\langle r^2(t)\rangle/dt = 2\int_0^t dt' Z(t')$. From this one recovers at long times the diffusive behaviour given by Eq. (5.1),

$$\lim_{t\to\infty}\langle r^2(t)\rangle = 6Dt\,, \tag{5.26}$$

the diffusion coefficient D being given by

$$D = \frac{1}{3}\int_0^\infty dt' Z(t')\,. \tag{5.27}$$

Such an expression for a transport coefficient as the integral of a time correlation function is known as a Green–Kubo formula.

We have already seen in Fig. 5.4 the behaviour of the function $\langle r^2(t)\rangle$ for two models of liquid sodium near freezing. Figure 5.5 shows the behaviour of $Z(t)$ from MD results of Nijboer and Rahman[135] on a model of liquid argon. The fact that $Z(t)$ becomes negative indicates that each particle on average recoils at short times after hitting its neighbours and before escaping from its coordination cage.

5.5.1 *Frequency spectrum and long-time tails*

The frequency spectrum ($f(\omega)$, say) of the velocity autocorrelation function is introduced from its Fourier representation,

$$Z(t) = \int_0^\infty \frac{d\omega}{2\pi} f(\omega)\cos(\omega t)\,. \tag{5.28}$$

[a]We note that $d\langle r^2(t)\rangle/dt = 2\langle \mathbf{r}(t)\cdot\mathbf{v}(t)\rangle = 2\int_0^t dt'\langle \mathbf{v}(t')\cdot\mathbf{v}(t)\rangle$. The function in brackets is $Z(t-t')$, from the invariance of the equilibrium ensemble under translation in the time: setting $t - t' = t''$, we thus find $d\langle r^2(t)\rangle/dt = 2\int_0^t dt'' Z(t'')$. This agrees with Eq. (5.25), as is easily checked by differentiating the latter with respect to t.

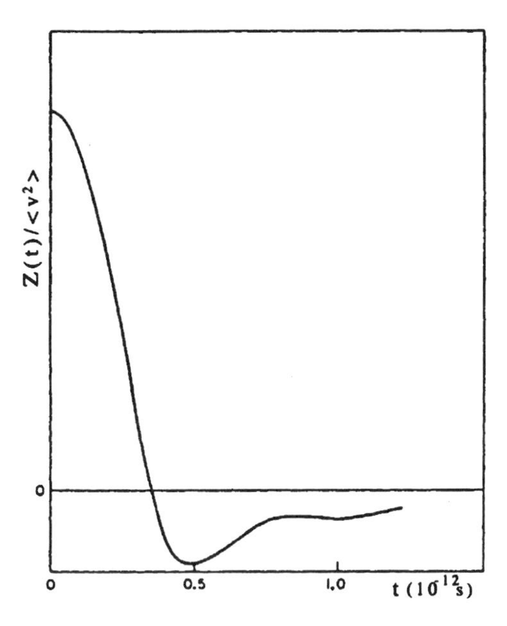

Fig. 5.5. Velocity autocorrelation function $Z(t)/\langle v^2 \rangle$ against time t from a molecular dynamics study of a model for liquid argon. (Redrawn from Nijboer and Rahman, Ref. 135.)

From the definition (5.24) we find $2\pi f(\omega) = \langle \mathbf{v}(\omega)\cdot\mathbf{v}(-\omega)\rangle$, the autocorrelation function of the Fourier transform of the particle velocity. It is evident from this expression that $f(\omega)$ in Eq. (5.28) is an even function of frequency ω.

Figure 5.6 shows the behaviour of $f(\omega)$ for the Nijboer–Rahman model of liquid argon, in comparison with MD data on a model of liquid rubidium near freezing.[136] The frequency range covered by this spectral function is similar to that of the spectrum of harmonic lattice vibrations in the crystal, extending up to $\omega \approx 10^{13}$ s^{-1}. The main new features of $f(\omega)$ in a liquid state are (i) its finite value at zero frequency and (ii) the absence of a high-frequency cut-off. While the latter feature is due to the absence of lattice periodicity and to "strong anharmonicity", the former is related to diffusion, since by inverting Eq. (5.28) and using Eq. (5.27) we find

$$f(0) = 2\int_0^\infty dt Z(t) = 6D\,. \tag{5.29}$$

Thus, the spectrum $f(\omega)$ in a liquid describes not only vibrational motions, but also diffusive motions in the low frequency region.

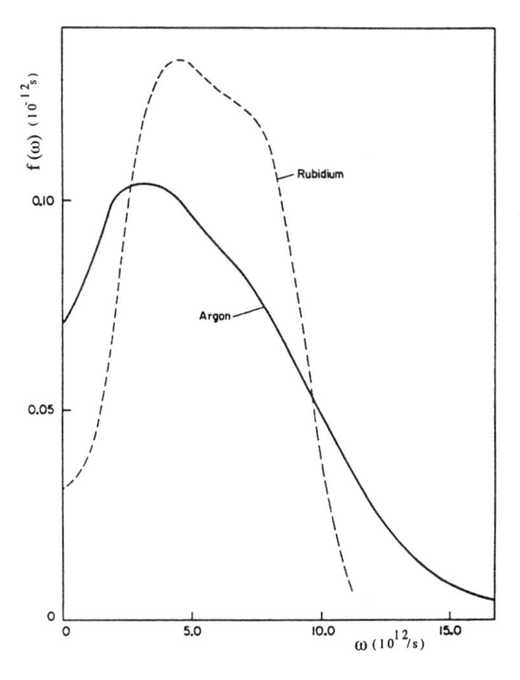

Fig. 5.6. Frequency spectrum $f(\omega)$ against pulsation frequency ω from molecular dynamics studies of models for liquid argon by Nijboer and Rahman and of liquid rubidium by Schommers. (Redrawn from Copley and Lovesey, Ref. 136.)

From Fig. 5.6 we should also notice the different shapes of the frequency spectrum for diffusive motions in argon and in rubidium. We have already remarked in Sec. 3.2, on the basis of specific heat data, that liquid metals such as rubidium are considerably more "harmonic-like" than liquid argon. This observation is confirmed in Fig. 5.6 by the spectrum for rubidium, showing lower diffusivity and a spectral peak which is both higher and narrower. This implies a more marked oscillatory behaviour for each atom on average inside its coordination cage.

The behaviour of the velocity autocorrelations over long times contains some further interesting information. From hydrodynamics, Ernst *et al.*[137] have shown that $Z(t)$ has a long tail, decaying at long time as $t^{-3/2}$. This was first noticed by Alder and Wainwright[138] in the MD study of the hard sphere fluid. As a consequence of the tail, the spectrum acquires an $\omega^{1/2}$ singularity at low frequency[139]: its low frequency expansion is

$$f(\omega) = 6D + d_1\omega^{1/2} + O(\omega)\,, \tag{5.30}$$

where the coefficient d_1 has the value

$$d_1 = -\left(\frac{k_B T}{2^{1/2}\pi m\rho}\right)\left(\frac{D+\eta}{m\rho}\right)^{-3/2}\,. \tag{5.31}$$

Physically, the long-time tail comes about because in a time interval t the diffusing atom shares out its initial momentum with other atoms within a radius $r \approx (Dt)^{1/2}$. This leads to a $t^{-3/2}$ decay of the velocity autocorrelations, although it must be noted that the precise magnitude of the tail involves also the shear viscosity η, through coupling of diffusive motions to collective transverse motions in the liquid.[140] This is evident from the value of d_1 in Eq. (5.31).

Finally, it is also of interest to examine the behaviour of the velocity autocorrelations in the opposite limit of short time. By expanding Eq. (5.28) in powers of t we find

$$Z(t) = \sum_{n=0}^{\infty} \frac{a_n}{(2n)!} t^{2n}\,, \tag{5.32}$$

where

$$a_n = (-1)^n \int_0^{\infty} \frac{d\omega}{\pi} \omega^{2n} f(\omega) \tag{5.33}$$

are known as the moments of the frequency spectrum. It is easily proved that $a_0 = \langle v^2 \rangle = 3k_B T/m$ and that $a_1 = \langle \dot{v}^2 \rangle$. Thus, while a_0 reflects only the free-particle behaviour, a_1 is given by the mean square force acting on a particle in the liquid: in a pair-potential model this can be evaluated from the pair distribution function $g(r)$. In the next section we shall see how the first few spectral moments can be useful in models evaluating $Z(t)$ and the diffusion coefficient.

5.5.2 *The Nernst–Einstein relation*

From its definition through the mean square displacement of a particle over a long time interval, the diffusion coefficient D is the result of atomic motions determined by spontaneous thermal fluctuations in a fluid. Only the correlations of these motions in time are relevant to diffusion, as is evident from Eq. (5.27). The Nernst–Einstein relation, referred to in Sec. 1.3, shows how such spontaneous fluctuations can be revealed through a measurement of driven transport.

To derive the Nernst–Einstein relation we consider the particle motions which are induced by an external, time-independent force field. Common examples may be the motions of charged particles in an electric field or those of massive particles in a gravity field, both in the presence of external scatterers determining a diffusive regime. The mobility μ is defined through the relation $\mathrm{v}_z \equiv -\mu(dV(z)/dz)$ between the mean drift velocity v_z acquired by the particles in an applied potential $V(z)$ and the force exerted by $V(z)$. The particle current density in the z direction is $i_z = \rho \mathrm{v}_z = -\rho\mu(dV(z)/dz)$ if the fluid is in a free-flow configuration with a (constant) particle density ρ. In a blocked-flow configuration, the fluid acquires a density profile $\rho(z)$, which drives a diffusion current according to Fick's law. The particle current density becomes

$$i_z = -\rho\mu \frac{dV(z)}{dz} - D\frac{d\rho(z)}{dz} \tag{5.34}$$

and vanishes in the stationary state provided that

$$\rho(z) \propto \exp\left[-\frac{\mu V(z)}{D}\right] . \tag{5.35}$$

Comparison of Eq. (5.35) with the Boltzmann distribution $\rho(z) \propto \exp[-V(z)/k_{\mathrm{B}}T]$ shows that the diffusion coefficient and the particle mobility are simply related to each other:

$$D = k_{\mathrm{B}}T\mu . \tag{5.36}$$

From Eqs. (5.29) and (5.36), the Nernst–Einstein relation (5.36) is seen to relate the mobility in a constant force field to the zero-frequency value of the velocity spectrum:

$$f(0) = 2m\langle v^2\rangle\mu . \tag{5.37}$$

This relation can be generalised to the case of a time-dependent external field in the linear regime (see e.g. Ref. 30), when it takes the form

$$f(\omega) = 2m\langle v^2\rangle \mathrm{Re}\,\mu(\omega) . \tag{5.38}$$

Here, ω is the angular frequency characterising the time dependence of the (monochromatic) external field and $\mathrm{Re}\,\mu(\omega)$, the real part of the frequency-dependent mobility, measures the dissipation of power from the source of the external field into the system. Equation (5.38) provides an example of the so-called fluctuation–dissipation theorem,[141] according to which a spectrum of spontaneous fluctuations such as $f(\omega)$ is accessible to measurement through the dissipation spectrum of a suitable external probe in the linear regime.

Ideally, the natural dissipative probe for measurements of the frequency spectrum $f(\omega)$ should be provided by the incoherent neutron inelastic scattering experiment referred to in Sec. 5.4.1. It can be proved in full generality that in a classical fluid[142]

$$f(\omega) = \omega^2 \lim_{k\to 0} \frac{S_s(k,\omega)}{k^2}\,, \tag{5.39}$$

with $S_s(k,\omega)$ being the van Hove dynamic structure factor for self-motions as introduced in Sec. 5.4.1. In practice, MD simulations still provide the most effective approach to unravel the microscopic diffusive motions of atoms or molecules in fluids.

5.6 Models of Velocity Autocorrelation Function

The simplest model for the velocity autocorrelations is provided by the Langevin approach described in Sec. 1.3 — which, we recall, applies to a mesoscopic particle diffusing in a molecular fluid medium. From Eq. (1.5), setting $\dot{x}(t) = v_x(t)$ we have

$$m\dot{v}_x = -f v_x(t) + F_r(t)\,. \tag{5.40}$$

We multiply both sides of this equation by $v_x(0)$ and average over the equilibrium ensemble, bearing in mind that the random force $F_r(t)$ does not correlate with the initial velocity (i.e. $\langle v_x(0)F_r(t)\rangle = 0$). We immediately find that Eq. (5.40) integrates to give

$$\langle \mathbf{v}(0)\cdot\mathbf{v}(t)\rangle = 3\langle v_x(0)v_x(t)\rangle = 3\langle v_x^2\rangle e^{-t/\tau}\,, \tag{5.41}$$

for $t \geq 0$, with $\tau \equiv m/f$ being the "relaxation time". The spectrum thus has the Lorentzian form

$$f(\omega) = \frac{2\langle v^2\rangle\tau}{1+\omega^2\tau^2}\,, \tag{5.42}$$

where $\langle v^2\rangle = 3k_{\mathrm{B}}T/m$. Equation (5.29) relates the relaxation time to the diffusion coefficient,

$$D = \left(\frac{k_{\mathrm{B}}T}{m}\right)\tau\,. \tag{5.43}$$

The measured transport coefficient carries information on the time scale of the transport process.

We present in the remaining part of this section some more recent models for diffusive particle motions, which are usefully applied to describe the self-diffusion coefficient and the Stokes–Einstein relation in atomic and molecular liquids.

5.6.1 *The Zwanzig model*

Starting from ideas of Stillinger and Weber,[143] Zwanzig[144] invoked a combination of vibrational and jumping motions in order to derive a Stokes–Einstein relation between the atomic diffusion coefficient and the shear viscosity in a cold, dense liquid directly from the Green–Kubo time correlation formulae. In essence, in the model the configuration space is divided into "cells", each cell being associated with a local minimum on the free energy hypersurface. Some of these minima may correspond to almost crystalline local configurations. The configuration of the liquid remains in one of these minima, performing approximately harmonic rattling motions until it finds a saddle point and jumps through it to another cell. The effects of the jump are (i) to rearrange the positions of the particles in some subvolume V^*, and (ii) to interrupt the oscillations within it, so that the motions inside V^* before and after the jump are uncorrelated.

The realisation of the model introduces a spectrum of vibrational frequencies to reduce the average over atoms, which is involved in Eq. (5.24), to a sum over normal modes localised in the various subvolumes V^* and having a time dependence of the form $\cos(\omega t)$. One also introduces a distribution $w(t/\tau)$ of waiting times, with a characteristic time τ for cell jumps destroying coherence in any subvolume V^*. Evidently, this dynamical picture requires that the waiting time be appreciably longer than the vibrational period. The diffusive jump is then assumed to be essentially instantaneous.

Within a single-frequency Einstein model for the rattling motions one finds

$$Z(t) = \left(\frac{3k_{\rm B}T}{m}\right) w\left(\frac{t}{\tau}\right)\cos(\omega_{\rm E}t)\,, \tag{5.44}$$

$\omega_{\rm E}$ being the Einstein frequency. Here from Eq. (5.27), one obtains with the empirical choice $w(t/\tau) = \mathrm{sec\,h}(t/\tau)$, the expression[145]

$$D = \left(\frac{\pi\tau k_{\rm B}T}{2m}\right)\mathrm{sech}\left(\frac{\pi\omega_{\rm E}\tau}{2}\right), \tag{5.45}$$

for the self-diffusion coefficient D. The early result of Brown and March,[146]

$$D = \left(\frac{k_{\rm B}T}{m\omega_{\rm D}}\right), \tag{5.46}$$

where $\omega_{\rm D}$ is a Debye frequency, follows in the limit $\omega\tau \ll 1$.

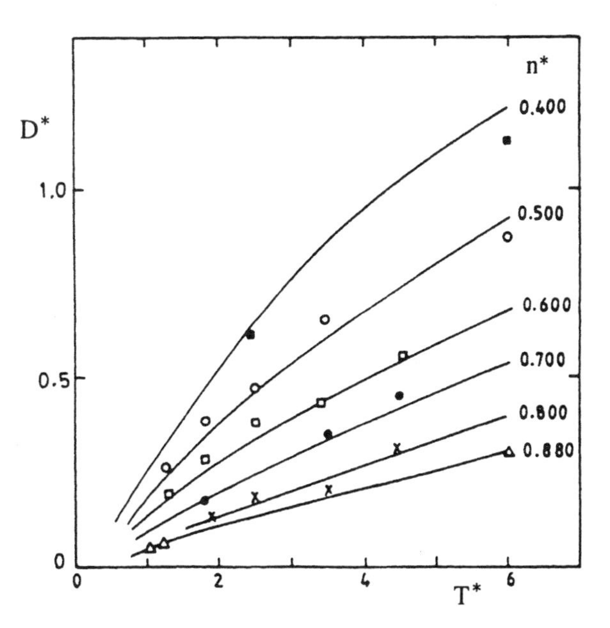

Fig. 5.7. Reduced self-diffusion coefficient $D^* = D(m\varepsilon/\sigma)^{1/2}$ against reduced temperature $T^* = k_{\rm B}T/\varepsilon$ in a Lennard–Jones fluid at various values of the reduced density $n^* = \rho\sigma^3$: theoretical results (solid lines) compared with results from molecular dynamics by Heyes. (Redrawn from Tankeshwar *et al.*, Ref. 145.)

Figure 5.7 reports the work of Tankeshwar *et al.*[145] a comparison of results obtained from Eq. (5.45) with MD data on the temperature and density dependence of the diffusion coefficient of Lennard–Jones fluids. The two disposable parameters ω_E and τ have been adjusted to two low-order moments of $f(\omega)$ (see Sec. 5.5.1).

5.6.2 *Wallace's independent atom model*

In contrast to Zwanzig's model, the model of velocity autocorrelations developed by Wallace[147] focuses on atomic motions within a volume consisting of an atom plus its near neighbours and allows diffusive jumps to occur several times within a vibrational period. The transits that represent the diffusive jumps are then incorporated by Wallace within his description of the atomic motion.

The Wallace model leads to a two-parameter formula for the self-diffusion coefficient

$$D(\omega, \xi) = \left(\frac{4k_B T}{\pi m \omega}\right)\left[\frac{\xi}{(2-\xi)}\right]. \tag{5.47}$$

In Eq. (5.47) ω is linked within the model to a well-defined RMS frequency related to quasi-harmonic phonon frequencies and ξ is a transit parameter, that Wallace extracts from the available data on ten liquid metals (see Table 5.2: with the subscript m denoting the liquid at freezing, the data in Table 5.2 are

Table 5.2. Transport data on liquid metals at the freezing point.

Metal	T_m (K)	ρ_m (g/cm^3)	D_m (10^{-5} cm^2/s)	η_m (cp)	ξ_m	x_m^{-1}
Li	453.7	0.515	5.96	0.60	0.50	53.(0)
Na	371.0	0.925	4.23	0.69	0.55	43.(7)
K	336.4	0.829	3.70	0.54	0.55	43.(5)
Rb	312.6	1.479	2.72	0.67	0.58	44.(0)
Cu	1357.0	8.000	3.98	4.1	0.61	41.(7)
Ag	1234.0	9.346	2.55	3.9	0.50	54.(9)
Pb	600.6	10.68	1.74		0.60	
Zn	692.7	6.58	2.03		0.52	
In	429.8	7.02	1.68	1.9	0.68	52.(6)
Hg	234.3	13.69	0.97		0.92	

the temperature $T_{\rm m}$, the density $\rho_{\rm m}$, the self-diffusion coefficient $D_{\rm m}$ and the shear viscosity $\eta_{\rm m}$, the transit parameter $\xi_{\rm m}$ at $T_{\rm m}$, and the quantity $x_{\rm m}^{-1} = (k_{\rm B}T\rho^{1/3}/D\eta)_{\rm m})$. It will be seen from Table 5.2 that the values of ξ in these liquid metals at the freezing point cluster mostly around $\xi_{\rm m} \approx 0.6$: they fall between about 0.5 and 0.7 for nine metals, with Hg lying outside this range with a value of 0.92.

More generally, Wallace shows that the curve of ξ versus $T/T_{\rm m}$ at atmospheric pressure is approximately universal for nine liquid metals. The simple model of diffusive motions has also proven useful in a description of the glass transition.[148]

5.6.3 *Generalisation of Stokes–Einstein relation*

The model summarised in Sec. 5.6.1 was used by Zwanzig[144] to propose a generalisation of the Stokes–Einstein relation for dense liquids, having the form

$$\frac{D\eta}{k_{\rm B}T\rho^{1/3}} = 0.0658\left(2 + \frac{\eta}{\eta_{\rm l}}\right) . \tag{5.48}$$

In Eq. (5.48), η and $\eta_{\rm l}$ are, respectively, the shear and longitudinal viscosities (see Chap. 6), while ρ is the atomic number density. The RHS of Eq. (5.48) has bounds that can vary between 0.13 and 0.18 and this is in accord with data on a variety of organic liquids. However, these bounds are violated for some liquid metals at freezing.

The transit parameter of the Wallace model has been used[149] to relate shear viscosity and self-diffusion in liquid metals at the freezing temperature. From the result of Brown and March[146] in Eqs. (5.46), Eq. (5.47) can be written in the form

$$D_{\rm m} \propto \left(\frac{T_{\rm m}^{1/2}}{m^{1/2}\rho_{\rm m}^{1/3}}\right)\left[\frac{\xi_{\rm m}}{(2-\xi_{\rm m})}\right] . \tag{5.49}$$

But Brown and March also obtained the shear viscosity from Green–Kubo-type arguments as $\eta_{\rm m} \propto T_{\rm m}^{1/2}m^{1/2}\rho_{\rm m}^{2/3}$, a formula that goes back to Andrade[150] (see Sec. 6.6.3). Multiplying Eqs. (5.49) and (5.50) to get a Stokes–Einstein form immediately yields

$$x_{\rm m} \equiv \left(\frac{D\eta}{k_{\rm B}T\rho^{1/3}}\right)_{\rm m} \propto \left[\frac{\xi_{\rm m}}{(2-\xi_{\rm m})}\right] . \tag{5.50}$$

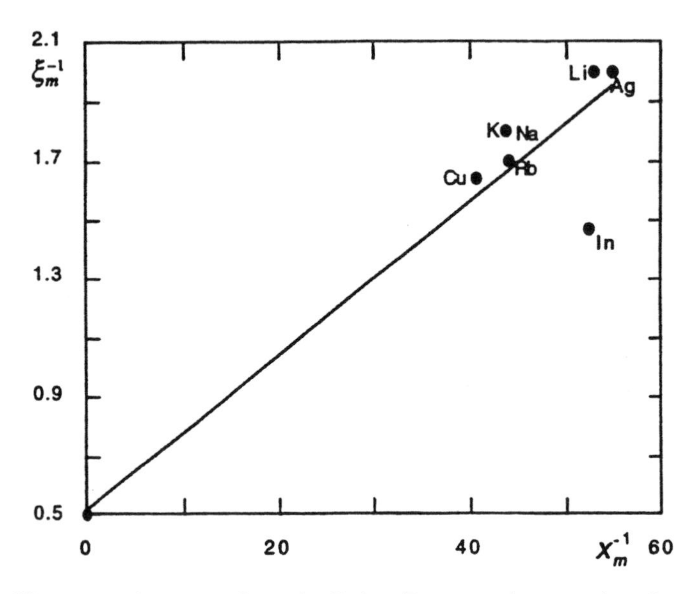

Fig. 5.8. Illustrating deviations from the Stokes–Einstein relation in liquid metals. (Redrawn from March and Tosi, Ref. 149.)

Figure 5.8 shows the extent to which a linear relationship between x_{m}^{-1} and ξ_{m}^{-1}, as predicted by Eq. (5.50), is satisfied from the data reported in Table 5.2. It appears, therefore, that Eq. (5.50) is a step forward in representing deviations from a Stokes–Einstein relation in liquid metals near freezing.

Chapter 6

Viscosity

6.1 Hydrodynamic Variables

Thermodynamic equilibrium in a dense classical fluid is established and maintained by intermolecular collisions occurring with an average time interval τ which in many instances may be of order 10^{-10}–10^{-13} s. The mean free path $\ell = \bar{v}\tau$ is the mean distance travelled by a molecule between successive collisions, with $\bar{v} \approx (k_{\rm B}T/m)^{1/2}$ being the thermal velocity. Consider now a disturbance of the equilibrium which varies periodically in time and space with frequency ω and wave number k. If $\omega\tau \ll 1$ and $k\ell \ll 1$, there are many collisions within each space-time cycle of the perturbation and we may assume that the liquid responds to the perturbation as if at each point in space it were close to equilibrium at each instant in time.

In fact, while almost every degree of freedom in a many-body system will relax to its equilibrium value in a time determined by the system's detailed microscopic interactions, some degrees of freedom are sure to vary slowly in time at long wavelengths. We met an example of such a variable in treating mass diffusion in Chap. 5 (see especially Sec. 5.4.1). Particle conservation implies a continuity equation relating the time derivative of the density of particles to the divergence of the flux (see Eq. (5.5)). As a result, the frequency range relevant to the variation of the particle density in time is Dk^2, vanishing for $k \to 0$ (see Eq. (5.15)).

This argument extends to the densities of all conserved quantities, namely (for a one-component fluid) momentum and energy in addition to the number of particles. Thus, sound waves and thermal conduction in a simple fluid have characteristic inverse times which vanish as the wavelength goes to infinity. In general terms, the time dependence of the density of each conserved quantity is determined by the divergence of its current density and, for slowly varying disturbances, this is a local function of the fields which are thermodynamically conjugate to the conserved quantities. Referring again to Chap. 5, we saw in Fick's law that the particle flux is driven by the gradient in chemical potential (see Eq. (5.4) and the comment following it).

The equations relating currents to fields are known as the constitutive relations and the parities under time reversal of each field and of the currents that it drives are all important. As a general rule, a current has opposite parity to that of the associated density of conserved quantity, since the two are related by a continuity equation. Hence, a coefficient relating a current and a field having the same parity is purely reactive, while the relation between a current and a field having opposite signs under time reversal is necessarily dissipative. Equation (5.4) is an example of such purely irreversible behaviour. Of course, the relations between currents and fields may contain both reactive and dissipative terms — an example being the propagation of sound waves.

The term "hydrodynamics" is commonly used to refer to such dynamics at long wavelength and low frequency.[151,152] As we have already remarked, for a one-component fluid there are five conservation laws and five hydrodynamics modes. This number increases in liquid mixtures because of mass conservation for each component, and also in ordered systems with continuous broken symmetries. As an example of the latter case we may cite a nematic liquid crystal, in which the molecules in the shape of short sticks line up along the so-called director. Since the director can point in any direction (in the absence of anchoring to the container walls), it takes a vanishingly small energy to induce a slow continuous variation in this direction and hence the time rate of change of the variable that describes the "broken symmetry" must be small. For a nematic liquid crystal this argument leads to two additional hydrodynamic modes (see Chap. 11). On the other hand, in a superfluid the additional hydrodynamic variable is the superfluid velocity: the continuous broken symmetry is in this case related to gauge invariance and, since the superfluid velocity can be expressed through the gradient of a phase, there is just one extra hydrodynamic mode (see Sec. 7.7).

In this chapter we shall be concerned with momentum as a conserved quantity in an isotropic one-component fluid and shall defer to Chap. 7 an account of energy conservation. We shall then see that the hydrodynamic theory presented here for sound-wave propagation is correct when energy fluctuations are decoupled from density fluctuations: this occurs in the limit when the specific heat ratio C_p/C_V tends to unity (as for metals near freezing: see Sec. 3.2.2).

6.2 Stresses in a Newtonian Fluid and the Navier–Stokes Equation

The main aims of this section are firstly to introduce the Newtonian law relating the stress in a viscous fluid to the local gradients in fluid velocity, and then to derive the basic equations governing the time variation of the momentum density. The book by Faber[4] gives a good introduction to macroscopic fluid dynamics and will often be cited in the next few sections.

6.2.1 *Viscosity stress tensor*

Consider a fluid flowing along the x_1 direction with velocity $v_1(x_2)$ corresponding to a uniform velocity gradient dv_1/dx_2 (see Fig. 6.1). In a microscopic view the molecules in any given layer are moving more slowly than those in the layer above it. Newton assumed that in such circumstance there is a shearing drag force between adjacent layers, directed along the x_1 axis and acting in the plane orthogonal to the x_2 axis. Denoting the force per unit area (i.e. the stress) as σ_{12} in this case, then Newton's law of viscosity gives its magnitude as

$$\sigma_{12} = \eta \frac{dv_1}{dx_2}, \tag{6.1}$$

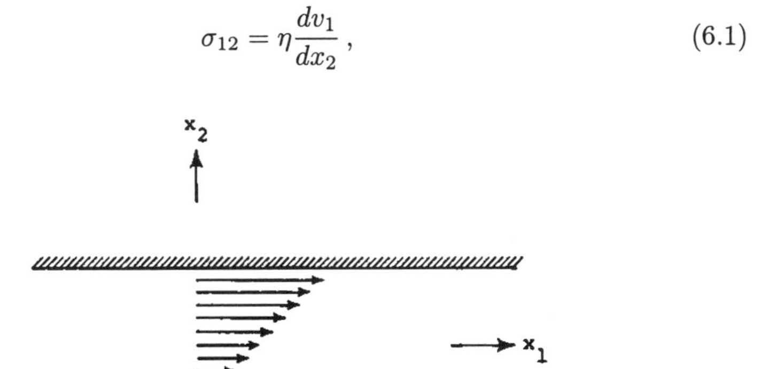

Fig. 6.1. Illustrating the velocity gradient between different layers in a viscous fluid.

at low rates of shear, η being the shear viscosity. This law obeys two basic requirements: (i) the stress changes sign when the flow is reversed; and (ii) if the velocity gradient vanishes everywhere, there is an inertial frame in which the fluid is at rest and the shear stress vanishes. In such a frame only the diagonal elements of the full stress tensor of the fluid are different from zero and are given by the pressure p as $\sigma_{ij} = -p\delta_{ij}$.

As a first step towards the extension of Newton's law to three-dimensional flow, let us consider the case where a velocity gradient dv_2/dx_1 is also present. The quantity

$$\omega_{12} = \frac{1}{2}\left(\frac{dv_1}{dx_2} - \frac{dv_2}{dx_1}\right) \tag{6.2}$$

describes the local rate of rotation of the fluid. No stress can be associated with such vorticity, since local rotation does not change the separation between any two neighbouring points inside the fluid. The stresses σ_{12} and σ_{21} must be equal to each other and proportional to the average of the two velocity gradients — or otherwise each volume element of the fluid would be subject to a couple about the x_3 axis and hence to an angular acceleration.

The conclusion is that stress in a fluid is a symmetric second-rank tensor and may thus be expressed as the sum of an isotropic term and of an anisotropic symmetric term. We may note the analogy with the relation between stress and strain in the theory of elasticity in solids: in the fluid the relative density change (included in the hydrostatic pressure term) and the velocity gradients (i.e. the rates of deformation) take the place of the strain tensor. The most general form of the stress for a Newtonian fluid in the hydrodynamic regime accordingly is

$$\sigma_{ij} = (-p + \eta'\nabla\cdot\mathbf{v})\delta_{ij} + \eta\left(\frac{\partial v_i}{\partial x_j} + \frac{\partial v_j}{\partial x_i}\right), \tag{6.3}$$

where η and η' are known as the first and the second coefficient of viscosity. These are the analogues of the Lamé coefficients $\mu = c_{44}$ and $\lambda = c_{12}$ in the theory of elasticity.

The quantity $\nabla\cdot\mathbf{v}$ in Eq. (6.3) is the divergence of the velocity field and by particle conservation a continuity equation relates it to the total time derivative of the particle density ρ,

$$\nabla\cdot\mathbf{v} = -\frac{1}{\rho}\frac{D\rho}{Dt}. \tag{6.4}$$

By the operator D/Dt it is meant that the rate of change is being taken following the fluid, i.e.

$$\frac{D\rho}{Dt} = \frac{\partial \rho}{\partial t} + \mathbf{v} \cdot \nabla \rho \,. \tag{6.5}$$

Thus, Eq. (6.4) can also be written in the more transparent form

$$\frac{\partial \rho}{\partial t} = -\nabla \cdot (\rho \mathbf{v}) \,, \tag{6.6}$$

$\rho \mathbf{v}$ being the particle current density $\mathbf{j}$. For incompressible flow we set $\nabla \cdot \mathbf{v} = 0$ in Eq. (6.3).

6.2.2 *Bulk and shear viscosity*

Let us introduce, as usual, the mean pressure $\bar{p}$ through the trace of the stress tensor:

$$\bar{p} = -\frac{1}{3}(\sigma_{11} + \sigma_{22} + \sigma_{33}) = p - \eta_{\mathrm{b}} \nabla \cdot \mathbf{v} \,, \tag{6.7}$$

where we have defined $\eta_{\mathrm{b}} = \eta' + 2\eta/3$. The parameter η_{b} is known as the bulk viscosity. The second term on the RHS of Eq. (6.7) may be rewritten in terms of the total time derivative of the particle density through Eq. (6.4). The pressure p in Eq. (6.7) will instead be evaluated by the thermodynamic rules from the instantaneous free energy density of the fluid.

Bulk viscosity is often irrelevant in the dynamics of simple fluids, but enters explicitly in the attenuation of sound waves. Let us consider a planar sound wave of small amplitude propagating along the x_1 direction, with a velocity field v_1 depending sinusoidally on x_1. From Eq. (6.3) the normal component of the stress tensor is $-\sigma_{11} = p - (\frac{4}{3}\eta + \eta_{\mathrm{b}})(\partial v_1/\partial x_1)$. This expression provides the basis for a discussion of sound-wave attenuation (see Sec. 6.7).

6.2.3 *The Navier–Stokes equation*

We now write the equation of motion of the fluid in the form of a continuity equation relating the time derivative of the momentum density $g_i = m\rho v_i$ to the divergence of the momentum current density π_{ij},

$$\frac{\partial g_i}{\partial t} = -\sum_j \frac{\partial \pi_{ij}}{\partial x_j} \,. \tag{6.8}$$

Assuming for a moment that dissipation may be neglected, we set $\pi_{ij} = p\delta_{ij} + g_i v_j$. Equation (6.8) may then be written with the help of Eq. (6.6) in the form

$$\frac{\partial \mathbf{v}}{\partial t} + (\mathbf{v} \cdot \nabla)\mathbf{v} = -\frac{1}{m\rho}\nabla p\,. \tag{6.9}$$

This is the equation first proposed in 1755 by Euler to treat inviscid flow. With the definition given earlier for the total time derivative, it may be written in the form

$$m\rho\frac{D\mathbf{v}}{Dt} = -\nabla p\,, \tag{6.10}$$

showing that Euler's equation is nothing but Newton's dynamical law relating to the pressure force the acceleration of a mass element having instantaneous velocity $\mathbf{v}(\mathbf{r}(t), t)$. A density of external force may be added to the RHS of Eq. (6.10) if needed.

The same considerations hold, of course, for Eq. (6.8) when viscous dissipation is taken into account (as already noted in Sec. 6.1, we neglect thermal dissipation in the present chapter). We only need to include in the momentum current density the full stress tensor from Eq. (6.3):

$$\pi_{ij} = -\sigma_{ij} + g_i v_j\,. \tag{6.11}$$

Using Eqs. (6.11) and (6.3) in Eq. (6.8) then leads us to the Navier–Stokes equation,

$$m\rho\left[\frac{\partial \mathbf{v}}{\partial t} + (\mathbf{v} \cdot \nabla)\mathbf{v}\right] = -\nabla p + \left(\frac{1}{3}\eta + \eta_b\right)\nabla(\nabla \cdot \mathbf{v}) + \eta\nabla^2\mathbf{v}\,. \tag{6.12}$$

This is to be combined with the continuity Eq. (6.6). Because of the non-linear terms on the LHS of Eq. (6.12), the solutions of this equation can be very complex and rich.

6.2.4 *Viscous dissipation*

The viscous forces supplementing the Euler terms in the Navier–Stokes equation are frictional forces causing irreversible loss of energy. Let us consider the amount of work δW done on the fluid per unit volume in giving to the molecules quasi-static displacements $\delta\mathbf{s}(\mathbf{r}) = \mathbf{v}(\mathbf{r})\delta t$ in a time interval δt. From the definitions of stress and strain we have

$$\frac{\delta W}{\delta t} = \frac{1}{2}\sum_{ij}\sigma_{ij}\left(\frac{\partial v_i}{\partial x_j}+\frac{\partial v_j}{\partial x_i}\right)$$

$$= -p\nabla\cdot\mathbf{v}+\eta'(\nabla\cdot\mathbf{v})^2+\frac{1}{2}\eta\sum_{ij}\left(\frac{\partial v_i}{\partial x_j}+\frac{\partial v_j}{\partial x_i}\right)^2 . \qquad (6.13)$$

The first term on the RHS is the pressure work done in changing the density (see Eq. (6.4)) and may have either sign depending on the sign of $\nabla\cdot\mathbf{v}$. The viscosity terms instead cannot be negative and are responsible for energy dissipation.

In particular, if only the component $\partial v_1/\partial x_1$ of the velocity gradient is different from zero we find from Eq. (6.13)

$$\left.\frac{\delta W}{\delta t}\right|_{\mathrm{visc}} = \left(\frac{4}{3}\eta+\eta_{\mathrm{b}}\right)\left(\frac{\partial v_1}{\partial x_1}\right)^2 . \qquad (6.14)$$

Such a dissipative term has already been met in Sec. 6.2.2. On the other hand, if only the component $\partial v_1/\partial x_2$ is non-vanishing we get

$$\left.\frac{\delta W}{\delta t}\right|_{\mathrm{visc}} = \eta\left(\frac{\partial v_1}{\partial x_2}\right)^2 . \qquad (6.15)$$

As we shall see in Sec. 6.7 below, these two expressions are directly relevant to determining the attenuation of longitudinal sound waves and of transverse motions, respectively.

6.3 Laminar Flow and the Measurement of Shear Viscosity

A presentation of the various methods by which the shear viscosity of a fluid may be measured can be found in the book of Kestin and Wakeham.[153] Illustrative values for some materials at atmospheric pressure are shown in Table 6.1.[116] An Arrhenius plot of the shear viscosity of water, extending down into the supercooled region, is reported in Fig. 6.2.[119]

Here we pause on the equations governing two of these methods of measurement (the oscillating-disk viscometer and the Couette viscometer) as examples of laminar flows described by the Navier–Stokes equation (see also Faber[4]). The term "laminar flow" is used in a restricted sense, implying that the fluid

Table 6.1. Shear viscosity of some fluids at atmospheric pressure.

Fluid	T (C)	η (cp)
Air	15	1.789×10^{-2}
Argon	-188	0.28
Bromine	20	1.252
	50	0.746
Water	0	1.793
	25	0.890
	100	0.282
Mercury	25	1.526
	100	1.245
Glycerol	25	934.0
	100	14.8
Glass		$> 10^{15}$

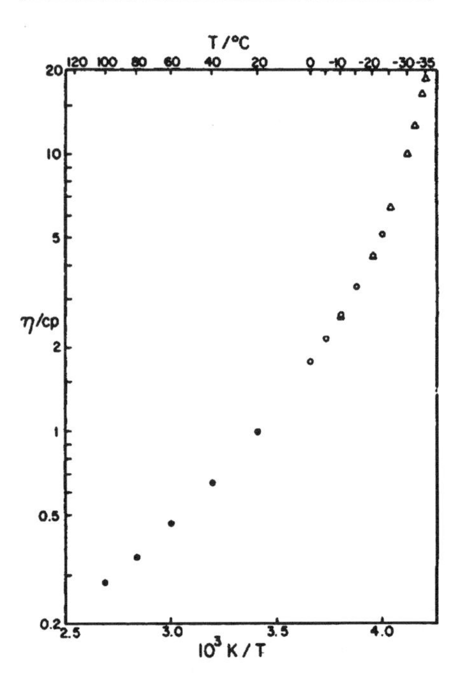

Fig. 6.2. Arrhenius plot of the shear viscosity of water, extending into the supercooled region. (Redrawn from Angell, Ref. 119.)

is being viewed as an assembly of thin laminae of uniform thickness, whose boundaries remain fixed as the fluid moves through them. The two cases of present interest are when the laminae are (i) plane sheets, or (ii) concentric cylindrical shells.

6.3.1 *Oscillating disk viscometer*

For viscous flow in planar laminae we take $v_1(x_2, t)$ as the only non-zero component of the fluid velocity. Equation (6.12) then yields

$$m\rho \frac{\partial v_1}{\partial t} = -\frac{\partial p}{\partial x_1} + \eta \frac{\partial^2 v_1}{\partial x_2^2} \,. \tag{6.16}$$

This equation also governs the case in which the laminar flow is induced in a large body of fluid by moving in it a flat solid plate, in the absence of a pressure gradient. Taking the surface of the plate at $x_2 = 0$ and the fluid in the region $x_2 > 0$, we have in this region

$$m\rho \frac{\partial v_1}{\partial t} = \eta \frac{\partial^2 v_1}{\partial x_2^2} \,, \tag{6.17}$$

with the non-slip boundary condition that $v_1(x_2 = 0, t)$ equals the velocity of the plate $u(t)$.

Let us take $u(t)$ as an oscillating function of time with angular frequency ω and look for travelling-wave solutions of the form $v_1(x_2, t) \propto \exp[i(kx_2 - \omega t)]$, with the understanding that only the real part of this expression is significant. From Eq. (6.17) we find $k = \pm(1 + i)/\delta$ with

$$\delta = \left(\frac{2\eta}{m\rho\omega} \right)^{1/2} \,. \tag{6.18}$$

The length δ may be viewed as the boundary layer thickness in the fluid, within which the waves induced by the oscillating plate are attenuated. In practice, the viscosity is measured from the logarithmic decrement of a system of horizontal disks undergoing torsional oscillations.

6.3.2 *Couette viscometer*

Viscous flow in concentric cylindrical shells is best described by adopting cylindrical coordinates (r, θ, x_3). The fluid may either flow along the x_3 axis or circulate about this axis. The former case leads back to the Poiseuille equation (see Sec. 1.4.3), while the latter is appropriate to the Couette viscometer (for a full discussion see the book of Faber[4]).

The viscometer consists of two coaxial cylinders, with the outer one rotating at constant angular velocity ω_b. In such steady state the fluid rotates with tangential velocity $v(r)$ increasing with the radius r. At a point in the fluid, the rate of rotation of a short line embedded in it is v/r if it lies at right angles to the radius vector but dv/dr if it lies along the radius vector. We set $\partial v_1/\partial x_2 = -dv/dr$ and $\partial v_2/\partial x_1 = -v/r$ to find

$$\sigma_{12} = -\eta r \frac{d(v/r)}{dr} . \tag{6.19}$$

The torque per unit length is $g_3(r) = -r(2\pi r \sigma_{12})$. The viscosity may be measured from the torque transmitted from one cylinder to the other after a steady state has been established.

6.3.3 *Hydrodynamic lubrication*

Following Tabor,[3] let us consider a shaft of radius a rotating with angular velocity ω inside a bearing of radius $(a + c)$. If the shaft carries a light load and the bearing is filled with oil, they will remain concentric. The velocity gradient across the lubricant is $\omega a/c$ and the tangential force f per unit length that must be exerted on the shaft to maintain steady rotation is

$$f = (2\pi a) \frac{\eta \omega a}{c} . \tag{6.20}$$

The equation was derived by Petroff in 1883 and is adequate for a lightly loaded bearing.

However, as was first realised by Reynolds in 1886, a much larger load can be supported if the shaft rotates in an eccentric position, so that the oil is squeezed through the converging gap between the two surfaces. Rotation of the shaft causes its centre to shift, inducing a pressure difference which supports the weight.

6.4 Creeping Flow Past an Obstacle

Following again the book by Faber,[4] we consider in this section cases in which an effectively incompressible fluid is in quasi-steady flow at very slow velocity past an obstacle of size σ. This requires values of the Reynolds number Re $= \rho\sigma v/\eta$ which are much less than unity. In these cases the LHS of the Navier–Stokes equation can be neglected and, setting $\nabla\cdot\mathbf{v} = 0$ in its RHS, the equation reduces to

$$\nabla^2\mathbf{v} = \eta^{-1}\nabla\delta p\,. \tag{6.21}$$

Here, $\delta p(\mathbf{r}) = p(\mathbf{r}) - p$ is the excess local pressure. Equation (6.21) is to be solved under appropriate asymptotic conditions on the flow far away from the obstacle.

6.4.1 *Stokes' law revisited*

The problem that we have posed above is that solved by Stokes in 1851 for a spherical obstacle of diameter σ. We have already indicated its solution in Sec. 1.4.4 (see Eq. (1.16)). Let us pause here to see how Eq. (6.21) is solved.

In general, the solution of such a differential equation is the sum of a particular integral $\mathbf{v}_1(\mathbf{r})$, obeying the equation $\nabla^2\mathbf{v}_1 = \eta^{-1}\nabla p$, and of a complementary function $\mathbf{v}_2(\mathbf{r})$ satisfying the equation $\nabla^2\mathbf{v}_2 = 0$ and enforcing asymptotic conditions at large distance from the sphere. With regard to $\mathbf{v}_2(\mathbf{r})$, in the absence of vorticity we may introduce a single-valued scalar potential function $\phi(\mathbf{r})$ such that $\mathbf{v}_2(\mathbf{r}) = \nabla\phi(\mathbf{r})$. Evidently, the condition $\nabla\cdot\mathbf{v}_2 = 0$ is satisfied if $\phi(\mathbf{r})$ obeys the Laplace equation $\nabla^2\phi(\mathbf{r}) = 0$. By applying the divergence operator $(\nabla\cdot)$ to Eq. (6.21) it is seen that the pressure must also obey the Laplace equation $\nabla^2 p(\mathbf{r}) = 0$.

In the problem of creeping flow past a sphere, the solution of the Laplace equation which satisfies the asymptotic condition of uniform flow at velocity $\mathbf{u}$ along the x_1 direction, say, is

$$\phi(r,\theta) = (ur + Ar^{-2})\cos\theta \tag{6.22}$$

in polar coordinates. Similarly, taking the excess pressure to vanish at infinity we have

$$\delta p(r,\theta) = Br^{-2}\cos\theta\,. \tag{6.23}$$

Expressions for the radial and tangential velocity fields follow at once from these expressions, and the two coefficients A and B are determined by asking that $\mathbf{v}(r,\vartheta = 0)$ should vanish at contact with the sphere (i.e. for $r = \sigma/2$).

The result for the radial and tangential components of the velocity field is

$$\begin{cases} v_r(r,\theta) = u\left(1 - \frac{3}{2}x + \frac{1}{2}x^3\right)\cos\theta \\ v_\theta(r,\theta) = -u\left(1 - \frac{3}{4}x + \frac{1}{4}x^3\right)\sin\theta\,, \end{cases} \tag{6.24}$$

with $x = \sigma/2r$. This solution is easily checked by substitution into Eq. (6.21). The no-slip condition ($v_r = 0$) is satisfied on the surface of the sphere.

The excess stress acting on the surface of the sphere has normal and shear components which can be obtained from Eqs. (6.3) and (6.19). Taken together, these are equivalent to a uniform force per unit area in the direction of $\mathbf{u}$, of magnitude $3\eta u/\sigma$. The total drag force is

$$F = \pi\sigma^2\left(\frac{3\eta u}{\sigma}\right) = 3\pi\eta\sigma u\,. \tag{6.25}$$

This is Stokes' law (see Eq. (1.16)).

If, on the other hand, the sphere is itself a liquid of viscosity η', one has to account for circulating currents arising within it and modifying the external pattern of flow. In this case the surface force per unit area becomes

$$F = 2\pi\eta\sigma u\frac{\eta + 3\eta'/2}{\eta + \eta'}\,, \tag{6.26}$$

giving back Eq. (6.25) for $\eta \ll \eta'$ but yielding $F = 2\pi\eta\sigma u$ when $\eta \gg \eta'$.

As an application let us consider a small solid sphere of radius a and mass density ρ_s, falling under gravity down the axis of a vertical cylinder which is filled with liquid of mass density ρ_ℓ — the essence of the falling-sphere viscometer. In the conditions under which it was derived, Eq. (6.25) can be used in the reference frame in which the liquid is stationary and the sphere is moving. The falling sphere may thus be expected to reach a thermal velocity given by

$$3\pi\eta\sigma u = \frac{1}{6}\pi\sigma^3 g(\rho_s - \rho_\ell)\,. \tag{6.27}$$

In the case of a gas bubble in a liquid, on the other hand, we need to use Eq. (6.26) with $\eta \gg \eta'$ instead of Eq. (6.25) and to replace ρ_s in Eq. (6.27) by $\rho_{\mathbf{gas}} \ll \rho_\ell$. The result is that the bubble *rises* with a terminal velocity given by $u = \rho_\ell\sigma^2 g/12\eta$.

6.4.2 *The viscosity of suspensions*

As a preliminary to evaluate the effective viscosity of a fluid containing a dilute system of solid particles in suspension, Faber[4] considers again the problem of the velocity field around a solid sphere, but now under the assumption that the vorticity-free shear be uniform far away from the sphere. That is, the velocity field $\mathbf{v}_2(\mathbf{r})$ is given asymptotically by

$$\mathbf{v}_2(\mathbf{r}) = \zeta \left(x_1, -\frac{1}{2}x_2, -\frac{1}{2}x_3 \right), \tag{6.28}$$

with ζ being the rate of extension of the fluid.

By comparing Eq. (6.28) with Eq. (6.13) it is easily shown that the rate of energy dissipation approaches asymptotically the amount $3\eta\zeta^2$ per unit volume. In the presence of the foreign sphere, on the other hand, the energy dissipation is calculated to be

$$\delta W = 3\eta\zeta^2 \left[1 + \frac{5}{2} \left(\frac{\sigma}{2R_0} \right)^3 + \cdots \right] V \tag{6.29}$$

over a spherical volume V of radius $R_0 \gg \sigma$. Thus, the presence of the foreign sphere increases the shear viscosity of the incompressible fluid by the factor shown in the brackets in Eq. (6.29).

The above result immediately suggests an effective-medium description of the shear viscosity of a dilute dispersion of solid particles. The effective viscosity of a fluid containing a large number of solid particles in suspension can be estimated as

$$\eta_{\text{eff}}^{(\text{solid})} \approx \eta \left(1 + \frac{5}{2} f \right), \tag{6.30}$$

where f is the fraction of the total volume of the suspension which is occupied by solid matter. A similar argument applied to the suspension of spherical gas bubbles leads to the result

$$\eta_{\text{eff}}^{(\text{gas})} \approx \eta (1 + f), \tag{6.31}$$

for its effective viscosity. To such a suspension one may, in fact, also attribute an effective compressibility $K_{\text{eff}} \approx f K_{\text{gas}}$ and an effective viscosity $\eta_{\text{b,eff}} \approx 4\eta/3f$.

6.4.3 *Percolation*

By invariance under a Galilean transformation the problem of a liquid flowing through a set of stationary solid particles is equivalent to the fall of a set of solid particles through a stationary liquid. Assuming the solid particles to be spheres of diameter σ occupying a fraction f of the total volume and the liquid to have viscosity η, an estimate of the terminal velocity can be obtained from Eq. (6.27) by using the effective viscosity (6.30) of the suspension in place of η.

More generally, the semiempirical expression

$$u = \frac{g\sigma^2}{18\eta}(\rho_s - \rho_\ell)\mathrm{fn}(f) \qquad (6.32)$$

is often used in this connection by engineers, the first factor being taken from Eq. (6.27) and the function $\mathrm{fn}(f)$ being taken in the form $\mathrm{fn}(f) = (1-f)^3/10f$. This expression can be used in situations where the particles are in contact and are falling as an essentially solid permeable block, but disregards effects of the geometry of the packing and the shape of the particles.

The problem is of great practical importance in situations where a liquid is percolating through a porous medium under a pressure gradient, as for water through soil or oil through shale.[4] The velocity of flow is proportional to the pressure gradient and inversely proportional to the viscosity of the liquid, through a permeability coefficient k which depends on the properties of the porous medium but not on the properties of the liquid. This is known as Darcy's law. Equation (6.32) gives an estimate of the coefficient as $k = \sigma^2\mathrm{fn}(f)/18f$.

6.5 Vorticity

In deriving Stokes' law for the drag on a solid sphere, the Reynolds number was assumed to be much less than unity. The effects that arise when this assumption does not apply depend critically on the behaviour of the boundary layers, where the flow is contaminated by vortices.

The velocity $\boldsymbol{\omega} = \boldsymbol{\nabla} \times \mathbf{v}$ described the local rate of rotation of the fluid (see Sec. 6.2). With this definition, the divergence of $\boldsymbol{\omega}$ is necessarily zero everywhere — a property that it shares with the electromagnetic fields $\mathbf{E}$ and $\mathbf{B}$ in the free space and with all other so-called solenoidal vectors which are defined as the curl of another vector. Thus, its variation in space can be described by continuous field lines whose direction coincides with the local direction of $\boldsymbol{\omega}$ and whose density is proportional to the magnitude of $\boldsymbol{\omega}$. Every free vortex line can be described as a bundle of such lines of vorticity.

6.5.1 *Vorticity diffusion*

The analysis of vorticity dynamics is usefully based on the Navier–Stokes Eq. (6.12). Standard vectorial calculus yields the relation $\boldsymbol{\nabla}(\boldsymbol{\nabla}\cdot\mathbf{v}) = \nabla^2\mathbf{v} + \boldsymbol{\nabla}\times\boldsymbol{\omega}$ and hence Eq. (6.12) can be rewritten as

$$m\rho\left[\frac{\partial\mathbf{v}}{\partial t} + (\mathbf{v}\cdot\boldsymbol{\nabla})\mathbf{v}\right] = -\boldsymbol{\nabla}p^* - \eta\nabla\times\boldsymbol{\omega}\,, \tag{6.33}$$

where $p^* = p - (\eta_{\mathrm{b}} + 4\eta/3)\boldsymbol{\nabla}\cdot\mathbf{v}$. We discuss below the case of uniform fluid density and incompressible flow, when by taking the curl of Eq. (6.33) we find the equation of motion for vorticity in the form

$$\frac{\partial\boldsymbol{\omega}}{\partial t} + (\mathbf{v}\cdot\boldsymbol{\nabla})\boldsymbol{\omega} = (\boldsymbol{\omega}\cdot\boldsymbol{\nabla})\mathbf{v} + \frac{\eta}{m\rho}\nabla^2\boldsymbol{\omega}\,. \tag{6.34}$$

The LHS of this equation is $D\boldsymbol{\omega}/Dt$, the rate of change of $\boldsymbol{\omega}$ taken while following the fluid.

The last term on the RHS of Eq. (6.34) vanishes for inviscid flow described by the Euler Eq. (6.9). In this case the lines of vorticity can be regarded as embedded in the fluid and forced to move with it. Furthermore, they are conserved in number so that the magnitude of $\boldsymbol{\omega}$ increases when the fluid is stretched in the direction of $\boldsymbol{\omega}$.

The viscous term on the RHS of Eq. (6.34), on the other hand, implies the presence of three-dimensional diffusion for each of the three components of the vector $\boldsymbol{\omega}$ (see Chap. 5 for the diffusion equation in the context of matter transport). Thus, vorticity is not permanently embedded if the fluid has viscosity: rather, it spreads by diffusion with a diffusivity coefficient given by the kinematic viscosity $\nu = \eta/m\rho$.

Of course, as in any other diffusive process the vorticity is still conserved. However, if it is positive in some regions and negative in others, some cancellation will occur as the lines of vorticity diffuse. In particular, a line of vorticity forming a closed loop can disappear by collapsing into a point. In other situations a line of vorticity may diffuse towards the liquid surface and disappear there. Surfaces may act both as sinks or as sources of vorticity.

The above elementary facts are at the basis of the behaviour of fluids in boundary layers adjoining the surfaces of solid bodies at increasing Reynolds numbers, leading to the formation of eddies and turbulent wakes — a subject of crucial importance in fluid dynamics and its applications. We refer once again the interested reader to the excellent introduction of these topics given in the book of Faber[4] (see also Chap. 12 below concerning turbulence).

6.5.2 *The Magnus force*

Here we briefly refer to the so-called Magnus effect, which concerns a uniformly rotating solid cylinder embedded in a fluid moving relative to it with a transverse velocity $\mathbf{v}(\mathbf{r})$ tending to the value $\mathbf{u}$ asymptotically far away. The electromagnetic analogue is the current-carrying wire inside a uniform magnetic field $\mathbf{B}$, which experiences a force $\mathbf{f} = \mathbf{i} \times \mathbf{B}$ per unit length.

Similarly, the so-called Magnus force tends to uplift the rotating cylinder and is given by $\mathbf{f} = -m\rho\mathbf{K} \times \mathbf{u}$, where $\mathbf{K} = \oint \mathbf{v} \cdot d\mathbf{l}$ is the circulation integral or in essence the strength of the vortex line associated with the rotating cylinder in terms of the local fluid velocity. This lift force which acts on a bound vortex line is independent of the cross-section of the solid body to which the line is attached.

6.6 Models of Viscosity

As noted in the short review by Heyes and March,[154] transport of momentum and energy occurs in a liquid not only by the bodily movement of molecules as in the transport of matter, but also by the action of intermolecular forces at a distance. Of these two independent mechanisms, the former is predominant at low density while the latter is the important one at high densities. Transport in monatomic fluids at low densities is well understood through the work of Chapman and Enskog.[132] The theory of transport in liquids is less well developed: as the inter-molecular potential varies continuously with distance, collisional momentum transport is associated with a distortion of the radial distribution function $g(r)$, as emphasised in the early work of Irving and Kirkwood.[155] Much progress has been made by exploiting the simplicity of the hard sphere model and by computer simulation of transport in Lennard–Jones fluids.

6.6.1 *Shear and bulk viscosity of hard sphere fluid*

Collins and Raffel[156] and Longuet-Higgins and Pople[157] emphasised that in a liquid of rigid molecules the singular nature of the interactions permits a finite flux of momentum and energy even when the radial distribution function is momentarily isotropic, as it is in equilibrium.

The earliest study of transport in a dense hard sphere fluid goes back to Enskog. He used a Boltzmann-like equation to determine the evolution of the single-particle distribution function as for a dilute gas, but modified the collision term to take account of effects which become important as the density increases: the molecules cannot be treated as point-like and the free volume that they can occupy is reduced. The calculation of Longuet-Higgins and Pople tackled directly the collisional contributions to the shear and bulk viscosities on the assumptions that (i) the pair distribution function does not depend on the rate of strain, but only on density and temperature; and (ii) the velocity distribution function has a Maxwellian form peaking at the local hydrodynamic velocity and with a spread determined by the local temperature. Some details of their calculation can be found in Appendix 6.1.

The results of Longuet-Higgins and Pople can be expressed in terms of the deviation of the equation of state from the ideal-gas form, through the quantity $\alpha \equiv (p/\rho k_{\mathrm{B}}T - 1)$ where ρ is the number density, as follows:

$$\eta = \frac{2}{5}\rho\sigma\left(\frac{mk_{\mathrm{B}}T}{\pi}\right)^{1/2}\alpha \tag{6.35}$$

and

$$\eta_{\mathrm{b}} = \frac{5}{3}\eta\,, \tag{6.36}$$

σ being the hard sphere diameter and m the particle mass. The factor α reflects the probability of finding two spheres at contact, which is given by the contact value of the pair distribution function (see Eq. (3.53)). Use of the virial expansion for α, which for the hard sphere fluid is $\alpha = b\rho[1 + 5b\rho/8 + 0.2869(b\rho)^2 + \cdots]$ with $b = 2\pi\sigma^3/3$, emphasises the strong dependence of the viscosities on density. The same model yields the Stokes–Einstein relation in the form

$$D\eta = \frac{1}{10}\sigma^2\rho k_{\mathrm{B}}T \tag{6.37}$$

and relates the thermal conductivity λ to the shear viscosity in the form

$$\lambda = \left(\frac{5k_B}{2m}\right)\eta . \tag{6.38}$$

Correlations between these transport coefficients and the excess entropy from studies of simple fluids by molecular dynamics will be discussed in Sec. 7.2.

Later work on transport in the hard sphere fluid is discussed in the books of Kestin and Wakeham[153] and of Ferziger and Kaper.[158] On the phenomenological side, Fig. 6.3 reports from Dymond and Brawn[159] a plot of the volume dependence of the reduced viscosity η^* for fluid argon and some monatomic

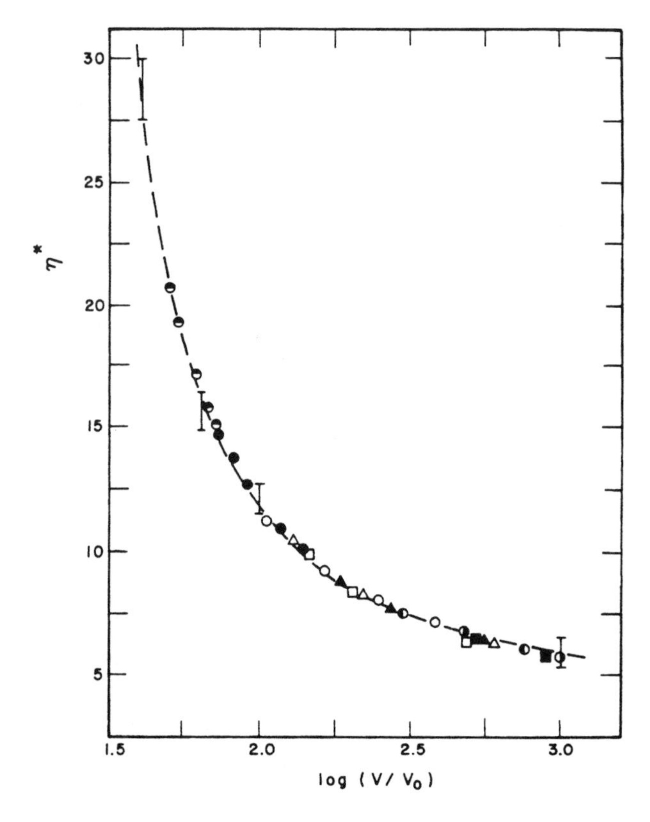

Fig. 6.3. A plot of the reduced viscosity η^* *versus* reduced volume V/V_0 for argon (○), krypton (△) and xenon (□) at various temperatures. The vertical bars show hard sphere results. (Redrawn from Dymond and Brawn, Ref. 159.)

gases at several temperatures. This quantity is defined as

$$\eta^* = \frac{\eta}{\eta_0}\left(\frac{V}{V_0}\right)^{2/3}, \tag{6.39}$$

where η_0 refers to the low density limit and V_0 is the atomic core volume. Figure 6.3 includes as vertical bars results for the hard sphere fluid, showing that this model supplies us with a good description of the data for these simple fluids.

More generally, a rapid increase of shear viscosity with pressure is the rule for liquids. This can be a very significant factor in lubrication: many sliding mechanisms may operate successfully because local high pressures, which would normally be expected to squeeze out the lubricant, so increase its viscosity that it remains trapped between the surfaces. This type of elasto-hydrodynamic lubrication is of substantial importance in lubrication practice.

6.6.2 *Temperature dependence of shear viscosity*

We have introduced in Sec. 1.4.2 the Eyring model for shear viscosity. We present it here in some detail to show how it leads to an Arrhenius behaviour for the temperature dependence of the shear viscosity, as a first approximation for dense liquids (see for example Fig. 6.2).

The model introduces an energy barrier E for the relative sliding of local neighbouring sheets of the liquid. The frequency with which such sliding will occur in the absence of applied stress will be $\nu_0 = (k_B T/h)\exp(-E/k_B T)$. Of course, jumps to the right and to the left are equally probable in this situation. Under an applied force F, however, the work associated with sliding over a distance $\pm d$ is $\pm Fd/2$ and the activation energy is $E \pm Fd/2$. The frequency of jumps in the direction of the force, assuming $Fd \ll k_B T$, becomes

$$\nu = \left(\frac{Fd}{h}\right)\exp\left(-\frac{E}{k_B T}\right) \tag{6.40}$$

and this is soon seen to imply $\eta \propto \exp(E/k_B T)$. This relation is in accord with the observed marked decrease in viscosity with increasing temperature. For a very large range of liquids the activation energy E is of order 0.3–0.4 times the latent heat of vaporisation L.

However, for liquid metals it is found that $E \ll 0.4L$. Brown and March[160] have drawn attention to the role of the temperature dependence of $g(r)$, as first

Table 6.2. Activation energy for shear viscosity and melting temperature of liquid metals.

	Hg	Bi	Pb	Sn	Zn	Cd	Sb	Cu	Ag	Na	K
E (eV)	0.027	0.069	0.086	0.060	0.141	0.089	0.136	0.154	0.211	0.065	0.050
$T_{\rm m}$ (K)	234.3	544	600.6	505	692.7	594	903	1357.0	1234.0	371.0	336.4

noted by Green.[160] On this basis they proposed that the activation energy for shear viscosity is $E \approx |\Phi_{\rm min}|$, $\Phi_{\rm min}$ being the value on interatomic pair potential at its main minimum. In a metal $|\Phi_{\rm min}| \ll L$, because of the large contribution to cohesion from the sea of valence electrons.

Table 6.2 reports the values of the activation energy for shear viscosity in a number of metals[162] and shows that it roughly correlates with the melting temperature $T_{\rm m}$.

6.6.3 *Green–Kubo formulae for viscosity*

As was the case for the diffusion coefficient in Chap. 5 (see Eqs. (5.16) and (5.27)), the shear and bulk viscosities can be expressed through the limits taken by appropriate spectral functions in the hydrodynamic regime (see Sec. 6.7). The corresponding space-time functions express the autocorrelations of transverse currents for shear viscosity and of the dissipative component of the longitudinal current for bulk viscosity. For instance, the shear viscosity can be written as the time integral of a function $\eta(t)$,

$$\eta = \int_0^\infty \eta(t)dt \tag{6.41}$$

and $\eta(t)$ is conveniently calculated in computer simulation runs by molecular dynamics on simple monatomic fluids[163] from the autocorrelations of the off-diagonal components of the microscopic stress tensor (see Sec. 7.2.4).

In this viewpoint Brown and March,[146] by arguments involving the assumption of a Debye frequency, obtained an appropriate formula relating the shear viscosity of a liquid metal at freezing to the melting temperature $T_{\rm m}$ and the atomic number density ρ (see Sec. 5.6.3):

$$\nu_{\rm m} \propto T_{\rm m}^{1/2} m^{1/2} \rho_{\rm m}^{2/3} \,. \tag{6.42}$$

Choosing the proportionally constant in accord with the work of Andrade[150] there is excellent agreement between Eq. (6.42) and experiment, as is shown in Table 6.3.

Table 6.3. Shear viscosity of liquid metals near freezing (in cp).

	Li	Na	K	Rb	Cs	Cu	Ag	Au	In	Sn
Expt.	0.60	0.69	0.54	0.67	0.69	4.1	3.9	5.4	1.9	2.1
Eq. (6.42)	0.56	0.62	0.50	0.62	0.66	4.2	4.1	5.8	2.0	2.1

6.6.4 *Computer simulation of shear viscosity in a Lennard–Jones fluid*

Computer simulation of space-time autocorrelation functions in Lennard–Jones fluids by the methods of molecular dynamics was first carried out by Rahman.[164] Levesque *et al.*[163] used this technique in long-duration runs on large samples of such fluids near the triple point and evaluated the transport coefficients from the Green–Kubo formula. They reported a long-time tail for shear viscosity autocorrelations, but none for the bulk-viscosity function.

Since then the technique has been applied numerous times to cover essentially the whole of the Lennard–Jones fluid phase diagram[165] and to study the dependence of shear viscosity on strain rate.[166] The data for all transport coefficients fit reasonably well to a range of analytical approaches and the success with respect to kinetic theories such as Enskog's is quite acceptable.

6.7 Transverse Currents and Sound Propagation in Isothermal Conditions

We report in this section the solution of the linearised Navier–Stokes equation describing isothermal momentum transport in a one-component fluid in the hydrodynamic regime. The reactive coupling between particle density and momentum density leads to two propagating sound modes, while two purely diffusive modes are due to transverse currents. The effect of coupling with thermal fluctuations, which shifts the velocity of sound propagation from isothermal to adiabatic and contributes to sound wave attenuation, will be discussed in Chap. 7.

6.7.1 *Linearised Navier–Stokes equation*

Linearisation is carried out with respect to the deviations from the equilibrium state, which are $\delta\rho(\mathbf{x},t) = \rho(\mathbf{x},t) - \rho$, $\delta p(\mathbf{x},t) = p(\mathbf{x},t) - p$ and $\mathbf{v}(\mathbf{x},t)$. To first order in these quantities, Eq. (6.13) is expressed in terms of the current density $\mathbf{j} = \rho\mathbf{v}$ as follows:

$$\frac{\partial \mathbf{j}}{\partial t} = -\nabla p + \frac{1}{m\rho}\left(\frac{1}{3}\eta + \eta_b\right)\nabla(\nabla \cdot \mathbf{j}) + \frac{\eta}{m\rho}\nabla^2 \mathbf{j}. \tag{6.43}$$

It is convenient to separate $\mathbf{j}$ into the sum of its longitudinal and transverse parts, $\mathbf{j} = \mathbf{j}_l + \mathbf{j}_t$, defined so that $\nabla \times \mathbf{j}_l = 0$ and $\nabla \cdot \mathbf{j}_t = 0$. The pressure force is purely longitudinal and the density is coupled only to $\mathbf{j}_l$ *via* the mass conservation equation.

Equation (6.43) then yields a diffusive equation for each of the two components of $\mathbf{j}_t$:

$$\frac{\partial \mathbf{j}_t}{\partial t} = \nu \nabla^2 \mathbf{j}_t\,, \tag{6.44}$$

where $\nu = \eta/m\rho$ is the kinematic viscosity. The attenuation length L can be estimated from this equation as $L \approx (\omega/\nu)^{1/2}$, yielding $L \approx 10^{-5}$ cm for water (with $\eta \approx 1$ cp and $m\rho = 1$ g/cm^3) at frequency $\omega = 2\pi \times 10^5$ Hz — a high rate of damping.

By analogy with Eqs. (5.12) and (5.15) we easily get the spectral function for transverse-momentum fluctuations in the linear hydrodynamic regime,

$$S_T(k,\omega) = \frac{2\nu k^2}{\omega^2 + (\nu k^2)^2} \tag{6.45}$$

and the Kubo formula for shear viscosity,

$$\eta = \frac{1}{2} m\rho \lim_{\omega \to 0}\left\{\omega^2 \lim_{k \to 0}\left[\frac{S_T(k,\omega)}{k^2}\right]\right\}. \tag{6.46}$$

Again by analogy with diffusive motions (see Eqs. (5.27)), the shear viscosity can be expressed through the time integral of the transverse-momentum autocorrelation function. As already noted in Sec. 6.6.3, this function reflects the autocorrelation of the off-diagonal microscopic stress tensor.

We turn to the longitudinal part of Eq. (6.43). Taking its divergence and using the mass conservation equation,

$$m\frac{\partial \delta\rho(\mathbf{x},t)}{\partial t} = -\nabla \cdot \mathbf{j}_l(\mathbf{x},t)\,, \tag{6.47}$$

we find a second-order differential equation in time for $\delta\rho(\mathbf{x},t)$,

$$\left[-m\frac{\partial^2}{\partial t^2} + \frac{1}{\rho}\left(\frac{4}{3}\eta + \eta_b\right)\frac{\partial}{\partial t}\nabla^2\right]\delta\rho(\mathbf{x},t) = -\nabla^2 \delta p(\mathbf{x},t)\,. \tag{6.48}$$

However, thermodynamics relates the pressure and density fluctuations in Eq. (6.48) through

$$\delta p(\mathbf{x},t) = \left(\frac{\partial p}{\partial \rho}\right)_T \delta\rho(\mathbf{x},t)\,. \tag{6.49}$$

Thus, after taking the space-time Fourier transform of Eq. (6.48) we get from it the eigenvalue equation $\omega^2 - c^2k^2 + i\omega k^2\Gamma = 0$, where $c^2 \equiv m^{-1}(\partial p/\partial\rho)_T$ and $\Gamma \equiv (\frac{4}{3}\eta + \eta_b)/m\rho$. The complex solutions of the eigenvalue equation are

$$\omega = \pm ck - \frac{1}{2}i\Gamma k^2\,, \tag{6.50}$$

and describe two longitudinal sound waves which (i) propagate with speed c determined by the isothermal compressibility, and (ii) are attenuated by bulk and shear viscosity. The value of the attenuation coefficient Γ was anticipated in Sec. 6.2.2. Notice that the attenuation vanishes with wave number faster than the real part of the sound frequency.

For typical sound-wave intensities in the laboratory, corresponding to a power of ≈ 0.1–0.3 Watt/cm^2 transmitted per unit area normal to the direction of propagation, the mean particle velocity is much smaller than the wave velocity. In these conditions the linearisation of the Navier–Stokes equation is justified. However, in highly viscous liquids the omission of higher-order derivatives of the velocity in the derivation of the stress tensor is no longer permissible.

6.7.2 *Bulk viscosity*

We pause at this point to reflect on the molecular origin of bulk viscosity. From the way in which this friction term enters Eq. (6.43), its presence can be viewed as equivalent to having an enhanced local pressure in a fluid element where $\nabla \cdot \mathbf{j} < 0$ (or a reduced one when $\nabla \cdot \mathbf{j} > 0$).

Such an effect can be quite significant in the dynamics of polyatomic fluids with internal degrees of freedom. The partition of energy between translational and internal degrees of freedom need not be the same as in equilibrium: thus, if an element of the fluid is suddenly compressed the extra energy may initially reside in the translational degrees of freedom and only later the internal ones may receive part of it.

This argument suggests that η_b may be much larger than η in some polyatomic fluids. Available estimates indicate $\eta_b/\eta \approx 2.5$ in water and as large as 10^2 in benzene. Molecular dynamics on the Lennard–Jones fluid near its triple point[163] yields instead $\eta_b/\eta \approx 0.25$.

6.7.3 *Brillouin light scattering*

Light scattering measures fluctuations in the local complex refractive index, which to a good approximation in isotropic fluids is a function of the local particle density. The spectral function for density fluctuations is obtained from the results reported in Sec. 6.7.1 as follows:

$$S(\mathbf{k},\omega) = \frac{2k_{\mathrm{B}}T\Gamma k^4/m}{(\omega^2 - c^2k^2)^2 + (\Gamma\omega k^2)^2}\,. \tag{6.51}$$

This is the value taken in the isothermal hydrodynamic regime by the van Hove dynamic structure factor of a one-component fluid, that we shall present in full generality in Sec. 6.8 below.

It is evident from Eq. (6.51) that this form of the spectrum for density fluctuations, and hence of the light scattering spectrum, presents two peaks at frequencies $\omega = \pm ck$ with widths given by Γk^2. This double-peak structure is known as the Brillouin doublet and its measurement determines the speed of sound in a frequency range definitely higher than that covered by standard ultracoustic techniques. We must, however, defer a full discussion of light scattering from a liquid in the hydrodynamic regime to the next chapter.

6.8 Microscopic Density Fluctuations and Inelastic Scattering

We have seen in Sec. 6.7.1 that propagating sound waves are the manifestation of hydrodynamic density fluctuations in a simple liquid. The relationship between longitudinal particle currents and the driving field — the gradient of the stress tensor — involves both a reactive term and a dissipative term: the former leads to wave propagation at a finite velocity and the latter to damping by viscosity. On the other hand, a liquid offers no resistance to slow shears and there is no reactive term for transverse currents in the hydrodynamic regime.

In the last section of this chapter we aim to extend the above picture outside the hydrodynamic regime, by treating microscopic density fluctuations in isothermal conditions. We shall first give a general introduction to the inelastic scattering of neutrons and X-rays as the basic techniques for the experimental study of the microscopic dynamics of liquids. We shall then emphasise the wave-vector and frequency dependence of both restoring and friction forces in the collective particle dynamics. With regard to transverse currents we shall merely recall at this point that, as already introduced in Sec. 1.5 on the rigidity of liquids, restoring forces may arise in high-frequency transverse motions.

6.8.1 *Inelastic neutron scattering from liquids*

In introducing the microscopic space-time correlations between density fluctuations in a simple liquid we follow the line of argument developed by van Hove[167] in evaluating the differential cross-section for inelastic neutron scattering from condensed matter.

Let $\mathbf{R}_i(t)$ be the position of the ith atom at time t in a classical fluid of mean density ρ containing N particles in a volume $V = N/\rho$. The function $\rho(\mathbf{r}, t) = \sum_i \delta(\mathbf{r} - \mathbf{R}_i(t))$ can be used to describe the actual density of atoms at point $\mathbf{r}$ at time t. We construct the correlation function

$$G(|\mathbf{r} - \mathbf{r}'|, t - t') = \frac{1}{\rho}\langle \rho(\mathbf{r}, t)\rho(\mathbf{r}', t')\rangle \tag{6.52}$$

or

$$G(r, t) = \frac{1}{N}\left\langle \sum_{ij} \delta(\mathbf{r} - \mathbf{R}_i(t) + \mathbf{R}_j(0)) \right\rangle. \tag{6.53}$$

As usual, the brackets $\langle \cdots \rangle$ denote the average over the equilibrium ensemble. In each term of the double sum we are asking what is the probability of finding atom i in $\mathbf{r}$ at time t if atom j (which could be the same atom or another atom) was in the origin at time $t = 0$.

It is in fact useful to consider separately the terms with $i = j$. Van Hove wrote

$$G(r, t) = G_s(r, t) + G_d(r, t), \tag{6.54}$$

with

$$G_s(r, t) = \frac{1}{N}\sum_i \langle \delta(\mathbf{r} - \mathbf{R}_i(t) + \mathbf{R}_i(0))\rangle \tag{6.55}$$

and

$$G_{\mathrm{d}}(r,t) = \frac{1}{N}\sum_{i\neq j}\langle\delta(\mathbf{r} - \mathbf{R}_i(t) + \mathbf{R}_j(0))\rangle\,. \tag{6.56}$$

Of course, $G_{\mathrm{s}}(r,t)$ describes single-particle motions as already introduced in Sec. 5.4.1. Setting $t = 0$ we get $G_{\mathrm{s}}(r,0) = \delta(\mathbf{r})$ and $G_{\mathrm{d}}(r,0) = \rho g(r)$, with $g(r)$ the pair distribution function introduced in Chap. 4. On the other hand, in the limit $t \to \infty$ the correlations should vanish and hence $G_{\mathrm{s}}(r,\infty) = 1/V$ and $G_{\mathrm{d}}(r,\infty) = \rho$. The sketch shown in Fig. 6.4 indicates how $G_{\mathrm{s}}(r,t)$ and $G_{\mathrm{d}}(r,t)$ may be expected to evolve as time passes.

The space Fourier transform of the particle density, defined by

$$\rho_{\mathbf{k}}(t) = \int d\mathbf{r}\rho(\mathbf{r},t)e^{i\mathbf{k}\cdot\mathbf{r}} = \sum_i e^{i\mathbf{k}\cdot\mathbf{R}_i(t)}\,, \tag{6.57}$$

describes for $\mathbf{k} \neq 0$ a wave-like density fluctuation of given wave vector $\mathbf{k}$. Hence, the function

$$F(k,t) = \int d(\mathbf{r}-\mathbf{r}')G(|\mathbf{r}-\mathbf{r}'|, t-t')e^{i\mathbf{k}\cdot(\mathbf{r}-\mathbf{r}')} = N^{-1}\langle\rho_{\mathbf{k}}(t)\rho_{-\mathbf{k}}(0)\rangle \tag{6.58}$$

(which is known as the *intermediate scattering function*) gives the probability amplitude that, starting from the liquid at equilibrium and having created a density fluctuation with wave vector $-\mathbf{k}$ at time $t = 0$ and a density fluctuation with wave vector $\mathbf{k}$ at time t, the liquid is found at the end in the same equilibrium state. If such a density fluctuation corresponds to an excited eigenstate with frequency $\omega_{\mathbf{k}}$, we expect that $F(\mathbf{k},t)$ should depend on time through the

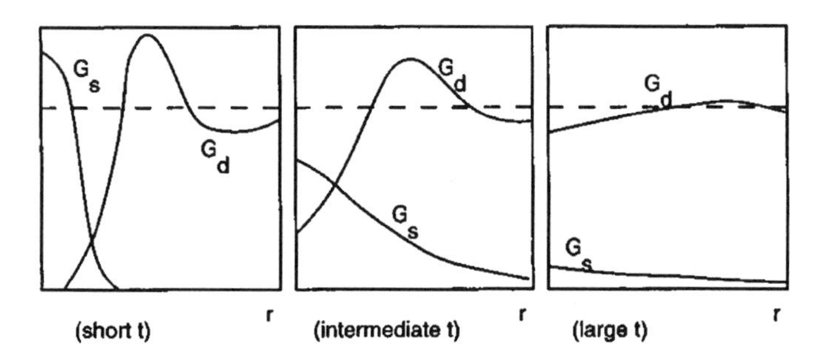

Fig. 6.4. Schematic drawing of the spreading of the van Hove correlation functions with increasing time.

phase factor $\exp(\pm i\omega_{\mathbf{k}} t)$. This is, for instance, the case of harmonic lattice vibrations in crystals. More generally, the decay of $F(\mathbf{k}, t)$ in time reflects the lifetime of a collective excited state, at least in ranges of $\mathbf{k}$ where a wave-like density fluctuation is not simply an overdamped oscillation.

A further Fourier transform leads us to van Hove's dynamic structure factor $S(\mathbf{k}, \omega)$:

$$S(\mathbf{k}, \omega) = \int dt F(k, t) e^{-i\omega t} = \iint d\mathbf{r} dt G(\mathbf{r}, t) e^{i(\mathbf{k}\cdot\mathbf{r} - \omega t)} \tag{6.59}$$

expressing the correlations between the wave-like fluctuations with frequencies ω and $-\omega$. We have

$$\int_{-\infty}^{\infty} \frac{d\omega}{2\pi} S(k, \omega) = F(k, t = 0) = S(k) \ : \tag{6.60}$$

we recover the structure factor $S(k)$ measured in a diffraction experiment (see Chap. 4).

Starting from these definitions and using time-dependent second-order perturbation theory (equivalent to the first Born approximation in scattering theory), van Hove showed that the differential cross-section for inelastic scattering of neutrons from a liquid is determined by $S(k, \omega)$ and by its analogue $S_s(k, \omega)$, obtained by Fourier transform from $G_s(r, t)$. More precisely, the probability $P(\mathbf{k}, \omega)$ of an inelastic scattering event in which momentum $\hbar\mathbf{k}$ and energy $\hbar\omega$ are exchanged between the neutron beam and the liquid is the sum of a coherent contribution and an incoherent one, which are given by

$$P_{\text{coh}}(\mathbf{k}, \omega) = N|\langle f\rangle|^2 S(\mathbf{k}, \omega) \tag{6.61}$$

and

$$P_{\text{inc}}(\mathbf{k}, \omega) = N[\langle |f|^2\rangle - |\langle f\rangle|^2] S_s(\mathbf{k}, \omega) \, . \tag{6.62}$$

Here, $\langle f\rangle = (1/N)\sum_i f_i$ and $\langle |f|^2\rangle = (1/N)\sum_i |f_i|^2$, f_i being the scattering amplitude for a neutron from nucleus i. The scattering amplitude depends on the chemical nature and on the isotopic state of the nucleus, but is independent of $\mathbf{k}$ at the wave vectors of interest in neutron scattering experiments from condensed matter, which correspond to wavelengths of order 0.5–50 Å (i.e. much larger than the radius of the nuclear forces). The corresponding excitation frequencies are of order 10^{13} s^{-1}. These values are in match with the peak momentum and energy parameters of a beam of thermal neutrons from a nuclear reactor, making this probe a particularly useful one for inelastic scattering studies of atomic motions. Exhaustive presentations of this topic can be found in the books of Lovesey[168] and Egelstaff.[169]

Of special interest in liquids and amorphous solids is the region of scattering momentum and energy which is intermediate between the hydrodynamic behaviour, where a density fluctuation can be regarded as a small perturbation in an isotropic continuum in thermal equilibrium, and the regime at very short length and time scales, where the dynamics can be viewed as that of independent particles between successive collisions. The intermediate region reflects the properties of the correlation function at distances of the order of the mean first-neighbour separation and at times characteristic of microscopic collective motions.

In Fig. 6.5 we report, from the experiments of Copley and Rowe[170] on liquid Rb at 370 K, the dispersion relation for collective density fluctuations as measured from the position of an observed side peak in the inelastic scattering spectrum *versus* the scattering wave vector. The side peak is visible up to

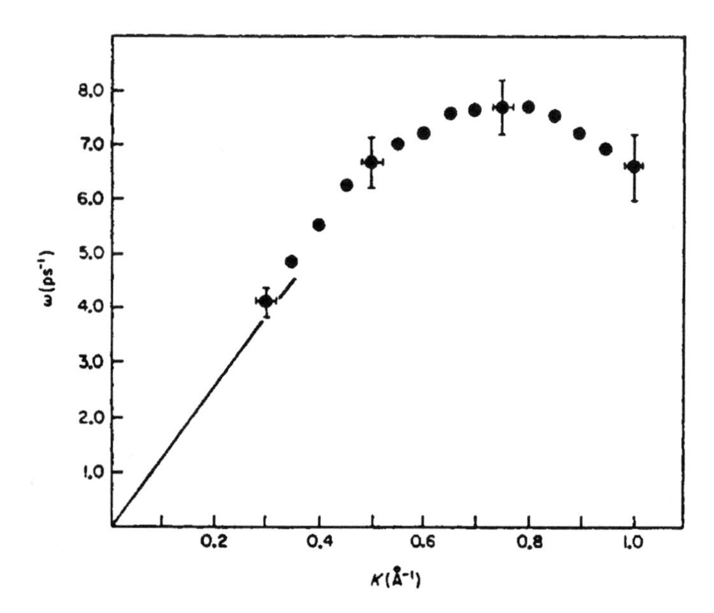

Fig. 6.5. Frequency of side peak in the scattering function of liquid rubidium at 320 K, as a function of wave number (•). The solid straight line is obtained from the measured speed of sound. (Redrawn from Copley and Rowe, Ref. 170.)

$k \approx 1$ Å^{-1} in the spectrum of this liquid metal and is further evidence for the "harmonicity" of liquid alkali metals as compared to liquid argon, say. Figure 6.5 also includes a linear dispersion relation obtained from the velocity of sound as measured in the same liquid by ultrasonic techniques. These results demonstrate how the hydrodynamic and the microscopic regime merge together in this liquid.

For a review of chemical applications of diffractive and inelastic neutron scattering to a variety of systems (molecular solids and liquids, hydrogen-bonded materials, zeolites, polymers, adsorbates) the reader is referred to the article of Trouw and Price.[171] We shall return on neutron inelastic scattering in discussing excitations in superfluid Helium in Chap. 7.

6.8.2 *Inelastic photon scattering from liquids*

Traditionally, the use of photons for the study of the microscopic dynamics of condensed-matter systems has been based on the Raman inelastic scattering technique. The cross-section for these scattering processes[172] is determined by the Fourier transform of the autocorrelation of the polarizability tensor $\alpha(\mathbf{r}, t)$ and is therefore not simply related to the van Hove function $S(k, \omega)$ in the appropriate range of momentum and energy (corresponding to $\omega = ck$ of order 50 to 1000 cm^{-1} say, with c the speed of light). Thus, Raman scattering data on liquid argon have been interpreted in terms of a second-order mechanisms,[173] in which a density fluctuation is polarized by the incident light and interacts *via* a dipolar field with another density fluctuation: the additional time-varying polarization scatters the light, the cross-section being approximated by the convolution integral of the product of two van Hove functions.

In fact, the Raman scattering techniques is most useful for the study of internal vibrations of structural units in complex-forming molten salts (see Chap. 8) and in molecular liquids (see Chap. 9). These excitations show no dispersion and are often insensitive to the surrounding liquid medium. Their study gives mainly structural information on the molecular units that may be present in the liquid and on how these may change with variables such as temperature and chemical composition.

On the other hand, with the development of high-flux synchrotron radiation sources and of high-energy resolution techniques, the inelastic scattering of X-ray photons has become a very powerful tool for the study of microscopic atomic motions in solids and liquids.[174] Collective excitations have been revealed in all investigated liquids and glasses at wavelengths approaching the mean interparticle distance. The wave number dependence of their energy shows that they are the short-wavelength evolution of the hydrodynamic sound mode, while their k-dependent broadening contains information on microscopic relaxation processes.

An example of these spectra and of their analysis is shown for liquid Li at 475 K in Fig. 6.6, from the work of Scopigno *et al.*[175] The two side peaks observed in this spectrum at $k = 0.7$ Å^{-1} provide clear evidence for propagating microscopic excitations. The figure also reports tests of various fits based on extensions of Eq. (6.51) into the microscopic regime. A very good fit of the data, shown by the full line in Fig. 6.6, is obtained by means of a visco-elastic model, as will be introduced immediately below in connection with the findings on molecular dynamics in water. In brief, the best fit in Fig. 6.6 involves the use of two relaxation times, reflecting damping of collective density fluctuations *via* mechanisms associated with interatomic collisions and with couplings to other collective modes of motion.

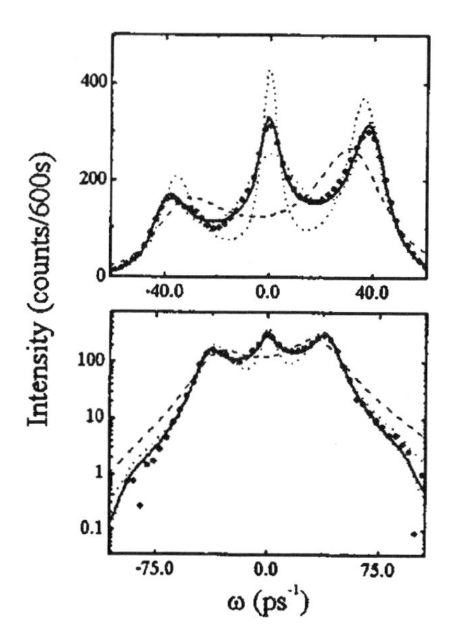

Fig. 6.6. Intensity of inelastic X-ray scattering spectrum *versus* frequency ω in liquid lithium at $T = 475$ K and $k = 0.7$ Å^{-1}. The various lines report fits of the data by alternative microscopic approaches. (Redrawn from Scopigno *et al.*, Ref. 175.)

6.8.3 *Fast sound in water*

The propagation of sound in water has received much attention since a molecular dynamics study[176] showed that the spectrum of density fluctuations in a mesoscopic region of wave number exhibits a secondary maximum at a much higher frequency than that appropriate to hydrodynamic sound. Sette *et al.*[177] have used inelastic X-ray scattering to follow the evolution of acoustic-like longitudinal excitations in water as their speed of propagation goes from 2000 m/s at $k = 0.1$ Å^{-1} to the fast-sound value $v_\infty \approx 3200$ m/s at $k \geq 0.4$ Å^{-1}. Hydrodynamic sound propagates in water at $v_0 \approx 1500$ m/s. The experiment thus provides a mapping of the cross-over from hydrodynamic to high-frequency sound. The data for the dispersive behaviour of the speed of sound in this molecular liquid are shown in the LHS panel in Fig. 6.7.

A simple visco-elastic model[178] yields for the relevant susceptibility the expression

$$\chi(k,\omega) = \frac{nk^2}{m}\left[\omega^2 - \omega_0^2(k) + i\omega\frac{\omega_\infty^2(k) - \omega_0^2(k)}{-i\omega + \tau^{-1}(k)}\right]^{-1}, \tag{6.63}$$

the scattering function $S(k,\omega)$ being simply proportional to the imaginary part of $\chi(k,\omega)$. To describe the data on water we should in Eq. (6.61) set $\omega_0(k) = v_0 k$ and $\omega_\infty(k) = v_\infty k$, and interpret $\tau(k)$ as a phenomenological relaxation time governing the cross-over from a resonance at frequency ω_0 to

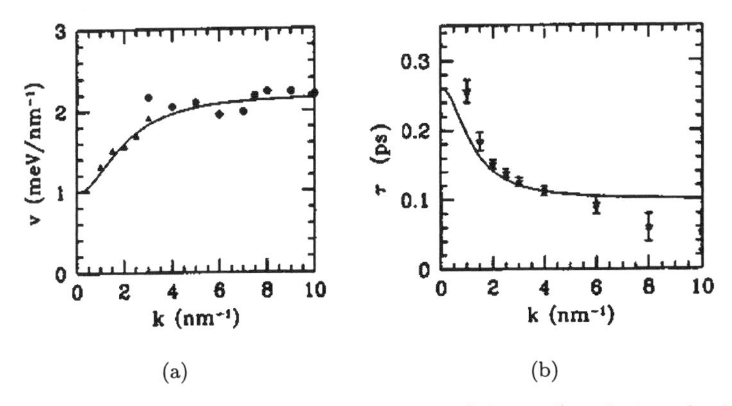

Fig. 6.7. Velocity of acoustic-like excitations in water (left panel) and visco-elastic relaxation time (right panel) *versus* wave number k. (Redrawn from Tozzini and Tosi, Ref. 178.)

one at frequency ω_∞. The results of the analysis of the data on the basis of this simple model are shown in Fig. 6.7.

The dispersive behaviour of sound in water is qualitatively similar to that observed in glass-forming liquids. There the transition between the two dynamical regimes is determined by the coupling of the propagating density waves with the dynamics of structural rearrangements. The transition occurs at $\omega\tau \approx 1$, the system having a viscous behaviour for $\omega\tau \ll 1$ and a solid-like behaviour for $\omega\tau \gg 1$. This transition is accompanied by the appearance of propagating transverse currents at high frequencies.

Appendix 6.1 Kinetic Calculation of Shear Viscosity for Hard Spheres

Longuet-Higgins and Pople[157] consider the transfer of x-momentum inside a hard sphere fluid, occurring in a very short time interval through unit distance in the y direction. The transfer is taken to occur *via* collisions of pairs of spheres which are already close together at the beginning of the time interval. If $\mathbf{p}_1 = p_1\mathbf{l}$ and $\mathbf{p}_2 = p_2\mathbf{l}$ are the momenta of the two particles, with $\mathbf{l}$ the unit vector along the interparticle separation vector $\mathbf{r}_{12}$, then the relative momentum of approach is $p = p_1 - p_2$ and the relative rate of approach is p/m. The spheres will soon collide if $p > 0$ and when this happens a momentum $p\mathbf{l}$ will be transferred from one to the other. The x component of this momentum is pl_x and the y component of $\mathbf{r}_{12}$ is σl_y, where σ is the sphere diameter. Therefore, the total flux of x-momentum in the y direction from two-body collisions is

$$P_{xy}^{(c)} = \oint \int_0^\infty h(p,\omega) \frac{\sigma p^2 l_x l_y}{m} d\omega dp \,. \tag{A6.1.1}$$

Here $h(p,\omega)d\omega$ is the number of pairs per unit volume within distance of contact, having relative momentum p and such that $\mathbf{l}$ lies in the solid angle $d\omega$. The first integration is over all orientations and the second is over the positive range of p.

At this point the two approximations already noted in Sec. 6.6.1 are made to evaluate $h(p,\omega)$: it is written as

$$h(p,\omega) = (2\pi\rho^2\sigma^2)\left(\frac{g_0}{4\pi}\right)\phi(p,\omega) \,, \tag{A6.1.2}$$

where g_0 is the value of the equilibrium pair distribution function at contact and $\phi(p,\omega)$ has a local Maxwellian form. If the fluid velocity is zero at the point of collision of the two spheres, the mean values of p_1 and p_2 are $(\sigma/2)p'l_x l_y$ and $-(\sigma/2)p'l_x l_y$, so that the distribution of p is

$$\phi(p,\omega) = (4\pi m k_B T)^{-1/2} \exp\left[-\frac{(p - \sigma p' l_x l_y)^2}{4\pi m k_B T}\right] . \tag{A6.1.3}$$

In evaluating the integral in Eq. (A6.1.1) it is a valid procedure to expand to the first power of p', since velocity gradients are assumed small. The result for the integral in Eq. (A6.1.1) is

$$P_{xy}^{(c)} = \frac{4\pi}{15}\rho^2\sigma^4 g_0 p' \left(\frac{k_B T}{\pi m}\right)^{1/2} . \tag{A6.1.4}$$

The shear viscosity is given by $P_{xy}^{(c)}/(p'/m)$ and the result in Eq. (6.35) in the main text follows by using Eq. (3.53) for the contact value of the pair distribution function.

Chapter 7

Heat Transport

7.1 Fourier's Law

Of all transport coefficients in condensed matter, thermal conductivity has some notable features that distinguish it from the others, such as self-diffusion and viscosity treated in the previous two chapters. Unlike these other two, the thermal conductivity coefficient does not exhibit any major discontinuity at the liquid–solid phase boundary. Also, there can be a substantial electronic contribution in the case of metallic systems (see Sec. 7.3).

In the linear response regime thermal conduction is described by Fourier's law,

$$\mathbf{j}_Q(\mathbf{r},t) = -\lambda \cdot \nabla T(\mathbf{r},t)\,, \tag{7.1}$$

where $\mathbf{j}_Q(\mathbf{r},t)$ is the local heat flux and $\nabla T(\mathbf{r},t)$ the temperature gradient driving the heat flow. Strictly, λ is a second-rank tensor, which is important when considering anisotropic solids and liquids (e.g. liquid crystals). Here, we will confine our discussion to isotropic liquids, which over time have no preferred direction for the molecules. The thermal conductivity is then a scalar, λ.

The thermal conductivity λ in a liquid varies with density and temperature in much the same way as does the shear viscosity η, except near the critical point where thermal conduction may become really large. In normal liquids η and λ are typically 10 to 100 times larger than in a dilute gas. This enhancement is dominantly due to collisional transfer: this being an efficient way of transporting momentum and energy. This is seen clearly in the hard sphere model (see the kinetic calculation of shear viscosity for hard spheres in Appendix 6.1), where a collision leads to an instantaneous transfer over the distance between the centres of two molecules. For the same essential reason, the speed of sound in a liquid metal near freezing is substantially larger than the mean thermal velocity of the atoms. In contrast, self-diffusion involves transport of the molecules themselves and does not involve collisional transfer of mass.

The methods for the measurement of thermal conductivity are reviewed in the book of Kestin and Wakeham,[153] already referred to in Chap. 6 in connection with the measurements of shear viscosity. A steady-state parallel plate method and a concentric cylinder method parallel the viscometers presented in Sec. 6.3. In Fig. 7.1 we report from the work of Dymond[179] the analogue of Fig. 6.3 for the reduced shear viscosity η^* by showing the reduced thermal

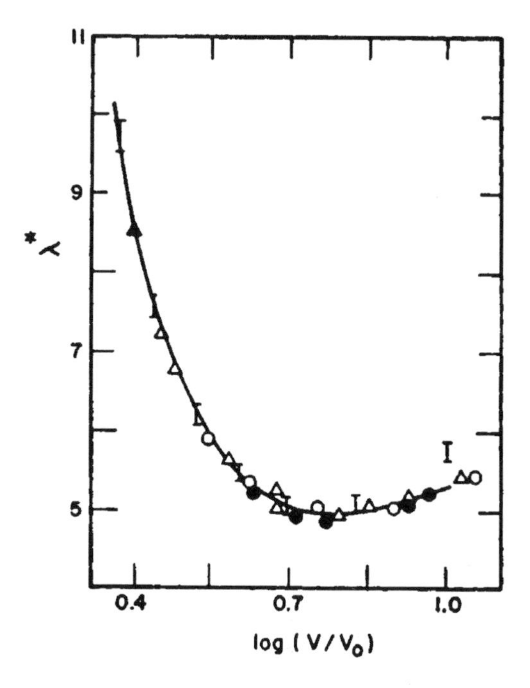

Fig. 7.1. The reduced thermal conductivity λ^* defined in Eq. (7.2) in dense gases *versus* reduced volume V/V_0. The vertical bars show hard sphere results. (Redrawn from Dymond, Ref. 179.)

conductivity λ^* in dense gases of Neon, Argon and Krypton as a function of the reduced volume V/V_0. Here, λ^* is defined as

$$\lambda^* = \frac{\lambda}{\lambda_0}\left(\frac{V}{V_0}\right)^{2/3}, \tag{7.2}$$

as in Eq. (6.39) for η^*. In fact, the core volume V_0 in Fig. 7.1 has been taken from viscosity measurements. Again, hard sphere results (shown in Fig. 7.1 by vertical bars) give good estimates for real monatomic fluids.

In Fig. 7.2 we report from the work of Basu and Levelt Sengers[180] quoted by Kestin and Wakeham[153] the observed behaviour of the thermal conductivity of carbon dioxide as a function of mass density ρ at various temperatures in the neighbourhood of the critical point. As anticipated, thermal conductivity becomes very large in the critical region.

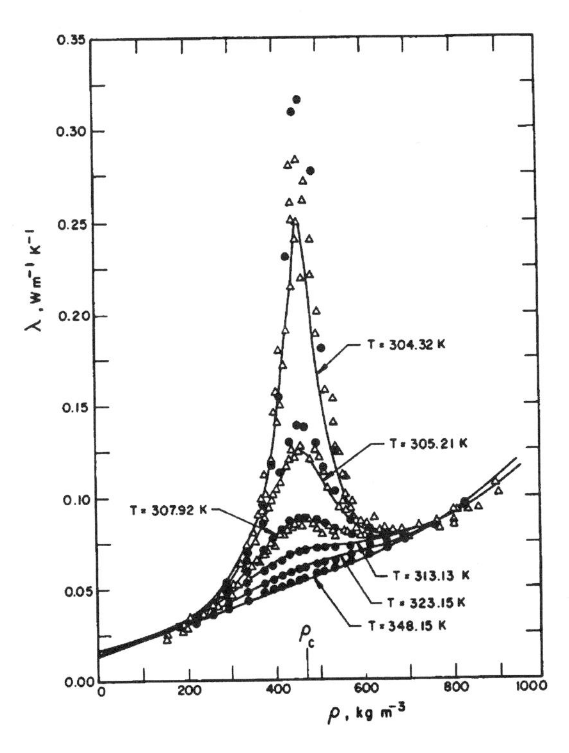

Fig. 7.2. The thermal conductivity of carbon dioxide in the critical regime. (Redrawn from Kestin and Wakeham, Ref. 153.)

According to the general criterion that we have presented in Sec. 6.1, Fourier's law in Eq. (7.1) tells us that the relation between thermal current and driving field is purely dissipative. An important theme of this chapter will therefore be to discuss the thermodynamic laws in the presence of mass flow and to explicitly show how dissipative forms for the constitutive relations are associated with entropy production (see Sec. 7.4). As a consequence, heat flow in normal systems shows diffusive-type behaviour and contributes to sound wave damping *via* coupling to density fluctuations (see Sec. 7.5). On the other hand, heat propagation can occur in a superfluid and gives rise to novel phenomena in this state, as will be discussed in Sec. 7.7. Before addressing these themes, however, we shall dwell on the microscopic background to heat conduction.

7.2 Studies of Heat Conduction by Molecular Dynamics

Accurate theoretical predictions of transport coefficients are still lacking and currently we rely to a large extent on computer simulations. The transport coefficients in simple fluids appear to obey a corresponding-states behaviour which correlates fairly well with the configurational entropy S_{E} in excess of the ideal-gas value.[181] Hundreds of simulation results can be fitted to a form which is suggested by elementary kinetic theory for a dense fluid of particles with thermal velocities and a mean free path between collisions of the order of the mean interparticle distance:

$$\begin{aligned} D &\approx 0.6\rho^{-1/3}\bar{v}e^{-0.8s}\,, \\ \eta &\approx 0.2m\rho^{2/3}\bar{v}e^{0.8s}\,, \\ \lambda &\approx 1.5k_{\mathrm{B}}\rho^{2/3}\bar{v}e^{0.5s}\,. \end{aligned} \tag{7.3}$$

Here, $\bar{v} = (kT/m)^{1/2}$ and s is a positive quantity defined as $s = -S_{\mathrm{E}}/Nk_{\mathrm{B}}$.

The heat flux has components that are purely kinetic and components that represent energy transfer through the interaction forces. Molecular dynamics (MD) simulation solves the many-body problem for a representative region of the liquid by numerical integration of the equations of motion of the interacting molecules, and can be used to calculate the heat flux from appropriate microscopic expressions. A range of model systems have been studied: here we focus on one-component monatomic and molecular liquids.[182]

7.2.1 *Green–Kubo formula*

Two classes of MD methods have been proposed to compute the thermal conductivity for an arbitrary molecular system, specified solely in terms of pair potentials and imposed thermodynamic conditions. In equilibrium MD methods the molecules are allowed to interact in the absence of any perturbing field (e.g. a temperature gradient). The other approach is to employ a non-equilibrium MD method, which is closer in spirit to Fourier's law.

In the limit of a small applied field, the Green–Kubo method relates heat conductivity to the integral of the correlation function of fluctuations in the heat-flux vector in a chosen direction x:

$$\lambda = \frac{V}{k_B T^2} \lim_{t\to\infty} \int_{t_0}^{t_0+t} dt' \langle J_{Qx}(t') J_{Qx}(0)\rangle \, . \tag{7.4}$$

For a fluid consisting of N atoms in volume V, we have

$$J_{Qx} = V^{-1}\left[\sum_{i=1}^{N} e_i \mathrm{v}_{ix} + \sum_{i=1}^{N-1}\sum_{j=i+1}^{N} (\mathbf{r}_{ij}\cdot\mathbf{v}_i)\phi'_{ij} r_{ijx}\right] , \tag{7.5}$$

for the x-component of the heat flux, where $e_i = m_i v_i^2/2 + \sum_j \phi_{ij}$ is the energy of a molecule with index i in the fluid, ϕ_{ij} being the pair interaction potential. The equilibrium Green–Kubo approach has been applied to Lennard–Jones liquids by a number of groups since the pioneering work of Levesque *et al.*[163]: see for example the work of Hoheisel *et al.*[183]

This treatment has been generalized to a single-component *molecular* fluid of N molecules containing a number n of interaction sites (see Heyes and March[182] for details). Simulations of the thermal conductivity of compact near-spherical molecules such as SF_4 and CF_4 have been carried out by this method.[184] The heat flux is given in this case by

$$\mathbf{J}_Q = \frac{1}{V}\left[\sum_{i=1}^{N} e_i^m \mathbf{v}_i + \sum_{i=1}^{N-1}\sum_{j=i+1}^{N} \mathbf{r}_{ij}(\mathbf{v}_i\cdot\mathbf{f}_{ij} + \boldsymbol{\omega}_i\cdot\boldsymbol{\Gamma}_i)\right] , \tag{7.6}$$

where i and j refer to center-of-mass quantities and e_i^m is the total energy of molecule i. The other quantities entering Eq. (7.6) are the force $\mathbf{f}_{ij}$ between the centres of mass of molecules i and j, the principal angular velocity $\boldsymbol{\omega}_i$ of molecule i and the corresponding principal torque $\boldsymbol{\Gamma}_i$.

7.2.2 *Non-equilibrium methods*

It has proved rather difficult to devise accurate non-equilibrium MD approaches to thermal conduction. The transport coefficient is straightforwardly measured in the laboratory by direct application of Fourier's law. This approach can be mimicked in computer simulations by sandwiching the sample between two "thermal" walls at a finite separation in the z direction.[185] The problem with all such "wall-based" methods for transport coefficients is that the liquid becomes spatially inhomogeneous (the molecules form layers against each wall) and very large temperature gradients have to be imposed. These techniques also tend to be statistically quite poor. Nevertheless, thermal-wall MD techniques have proved useful in studying other non-equilibrium phenomena such as Rayleigh–Benard convection rolls.[186]

The application of a constant field across a simulation cell must be compatible with periodic boundary conditions. This is achieved for shear viscosity by the use of the Lees–Edwards boundary conditions.[187] The nearest that can be achieved for heat transport is to impose an oscillatory temperature profile in the cell, with a finite wave vector compatible with periodic boundary conditions. Another procedure that has been applied to Lennard–Jones fluids and ionic systems involves dividing the MD cell into a number of layers parallel to one of its faces. One layer at each end of the cell is heated and two layers in the middle are used at heat sinks. Two heat fluxes in opposite directions are induced towards the centre of the cell and the thermal conductivity can be obtained by direct application of Fourier's law in the two regions.[188]

Evans[189] and Gillan and Dixon[190] have devised ingenious synthetic non-equilibrium equations of motion which are free of gradients. Some of the underlying ideas were contained in an earlier external field perturbation method invented by Jacucci *et al.*[191] The theory for this approach is described below using the formalism of Morriss and Evans,[192] who extended the Green–Kubo formula to apply to non-equilibrium systems.

7.2.3 *Transient time correlation formula*

We consider an external force field $\mathbf{F}$ being switched at time $t = 0$ onto a system at equilibrium. The response of any time-dependent property of the system, $B(t)$ say, is given by

$$\langle B(t)\rangle = \langle B(0)\rangle + \frac{1}{k_{\mathrm{B}}T}\int_0^t dt'\langle B(t')\dot{H}(0)\rangle\,. \tag{7.7}$$

The time derivative of the Hamiltonian is viewed as "dissipative heat" in the phenomenological form of a flux times a perturbing thermodynamic force, i.e. $\dot{H} = -\mathbf{J}_Q \cdot \mathbf{F}$ where $\mathbf{J}_Q$ is the dissipative flux. In the case of thermal conductivity, the fictitious force or "heat field" $\mathbf{F}$ replaces the temperature gradient but still induces an additional heat flux $\Delta\mathbf{J}_Q(t)$ in the system. Setting $B = \mathbf{J}_Q$ and substituting for $\dot{H}$ in Eq. (7.7) gives the so-called transient time correlation function formula for the thermal conductivity at *arbitrary* applied field. We then have

$$\lambda(\mathbf{F}) = \lim_{t\to\infty} \frac{\Delta J_Q(t)}{F} = \frac{V}{3k_\mathrm{B}T^2} \int_0^t dt' \langle \Delta\mathbf{J}_Q(t') \cdot \mathbf{J}_Q(0) \rangle \,. \tag{7.8}$$

In order to satisfy the equation $\dot{H} = -\mathbf{J}_Q \cdot \mathbf{F}$ at the microscopic level, the equations of motion for a monatomic system are

$$\dot{\mathbf{r}}_i = \frac{\mathbf{p}_i}{m} \,, \tag{7.9}$$

$$\dot{\mathbf{p}}_i = \mathbf{F}_i + (e_i - \langle e \rangle)\mathbf{F} - \sum_{j=1}^{N} \mathbf{f}_{ij}\mathbf{r}_{ij} \cdot \mathbf{F} + \frac{1}{2N} \sum_{j=1}^{N-1} \sum_{k=j+1}^{N} \mathbf{f}_{ij}\mathbf{r}_{ij} \cdot \mathbf{F} - \alpha\mathbf{p}_i \,. \tag{7.10}$$

Here, $\mathbf{F}_i$ is the instantaneous force on particle i, $\langle e \rangle$ is the average of the instantaneous energy e_i taken over all atoms in the system, and as before $\mathbf{f}_{ij}$ is the pair force between atoms i and j. The last term on the RHS of Eq. (7.10) serves to prevent a rise in the temperature of the system from the imposed heat force: a constant-kinetic-energy condition is enforced by applying a thermostatic control in the form of a multiplier α acting on the chosen momentum $\mathbf{p}_i$ and being evaluated at each time step according to

$$\alpha = \frac{1}{\sum_{i=1}^{N} p_i^2} \sum_{i=1}^{N} \mathbf{p}_i \cdot \left[\mathbf{F}_i + (e_i - \langle e \rangle)\mathbf{F} - \sum_{j=1}^{N} \mathbf{f}_{ij}\mathbf{r}_{ij} \cdot \mathbf{F} \right.$$
$$\left. + \frac{1}{2N} \sum_{j=1}^{N-1} \sum_{k=j+1}^{N} \mathbf{f}_{ij}\mathbf{r}_{ij} \cdot \mathbf{F} \right] \,. \tag{7.11}$$

Equations (7.9)–(7.11) are homogeneous and conserve the momentum of the MD cell. Through the term $(e_i - \langle e \rangle)\mathbf{F}$ they originate a heat current, since the atoms with energy greater than the average $\langle e \rangle$ will be driven by the external force in the opposite direction to those atoms which instantaneously have an energy below the mean.

The thermal conductivity is obtained from such simulations by extrapolation of the field-dependent conductivity to zero **F**. Even when linear response transport coefficients are of sole interest, several simulations at different field strengths still need to be made in order to carry out the extrapolation to zero field. The simulations at finite heat field have no experimental analogue. In contrast, in the case of shear viscosity the finite-field simulations correspond to finite shear rates, which are experimentally realizable and can lead to important phenomena in non-Newtonian flows (see Chap. 11).

We conclude by remarking that, away from equilibrium and from the linear-response regime, the decoupling of transport coefficients such as shear viscosity and thermal conductivity no longer holds. Simulations of thermal conduction in a strongly sheared Lennard–Jones fluid[193] reveal that the symmetry-breaking due to the shear field causes (i) the diagonal elements of the thermal conductivity tensor to become unequal and (ii) non-diagonal elements to emerge.

7.3 Electronic Contribution to Heat Conduction in Liquid Metals

In earlier chapters we have often referred to liquid metals as examples of *atomic* liquids. This is correct in regard to phenomena in which the valence electrons follow adiabatically the motions of the ions in the metal. In heat conduction the nature of metal as a two-component fluid of ions and valence electrons emerges, since the two components separately contribute to thermal transport and indeed the electronic contribution is usually dominant. A clear indication of this fact comes from the Wiedemann–Franz law relating thermal and electrical conductivity. Electronic conduction in a metal is, of course, due to the motion of the valence electrons against the background of positively charged ionic cores under an applied voltage drop.

The most elementary approach to the Wiedemann–Franz law involves two steps.[194] In the first step, one uses the kinetic theory of gases to express the thermal conductivity λ and the electrical conductivity σ of the valence electron gas in terms of the respective relaxation times (see Appendix 7.1). These expressions are $\lambda = C_e u^2 \tau_\lambda / 3$ and $\sigma = ne^2 \tau_\sigma / m$, where n is the number density of electrons (having charge e and mass m), C_e is the electronic contribution to the heat capacity of the metal per unit volume and u is the velocity of an electron on the Fermi surface. In evaluating these expressions one must take account of the quantum degeneracy of an electron gas in simple liquid metals. The second essential step is to assume that the relaxation times for the two transport processes are equal. The result is

$$\frac{\lambda}{\sigma T} = \frac{\pi^2 k_B^2}{3e^2} \equiv L\,. \tag{7.12}$$

The RHS is the so-called Lorenz number, $L = 2.45 \times 10^{-8}\ \mathrm{W \cdot \Omega \cdot K^{-2}}$. Equation (7.12) does not depend on such details as the shape of the Fermi surface and the density of electron states, but assumes elastic scattering. The validity of this assumption is discussed, for example, by Rice.[195]

Experimentally, the measured value of $\lambda/\sigma T$ for the liquid alkali metals is between 2.1 and 2.6×10^{-8}, but somewhat larger deviations from the prediction (7.12) are met in other liquid metals (Table 7.1). Both inelastic scattering processes and electron–electron interactions may play a role in determining these deviations.[196] We shall return in Chap. 14 on the microscopic background to the properties of conduction electrons in metals.

An evaluation of the viscosity of liquid alkali metals by electron plasma theory[197] indicates that the electronic contribution to this transport coefficient is instead quite small, of order 10% at most. Momentum transport in a liquid metal mostly occurs through ionic motions and ion–ion collisions, with the valence electrons following in an almost adiabatic manner.

March and Tosi[198] have related the thermal conductivity λ_m of the liquid alkali metals at freezing to their electrical resistivity ρ_m and shear viscosity

Table 7.1. Values of $\lambda/\sigma T$ for a number of liquid metals (in units of $10^{-8}\ \mathrm{W \cdot \Omega \cdot K^{-2}}$).

Metal		Metal		Metal	
Li	2.6	Pb	2.4	Ta	2.4
Na	2.2	Sb	2.6	Re	1.75
K	2.1	Bi	2.5	Os	1.75
Cs	2.4	Ti	2.9	Pt	2.3
Cd	2.5	Zr	2.25	In	2.7
Hg	2.75	Hf	2.7	La	2.65
Zn	3.2	Mo	2.6	Ce	2.56
Al	2.4	W	2.5	Pr	2.89
Ga	2.07	Ru	2.45	Nd	2.27
Tl	3.2	Ir	1.95	Gd	1.83
Sn	2.9	Nb	2.6	Dy	2.34

$\eta_{\rm m}$ by

$$\lambda_{\rm m} = \frac{L\eta_{\rm m}^2}{A\rho_{\rm m} M n^{4/3}}, \tag{7.13}$$

with M the ionic mass and $A = 5.1 \times 10^{-4}$. This expression agrees with the data well within 10%.

7.4 Thermodynamics with Mass Motion and Entropy Production

Like viscosity, heat flow is an important source for the attenuation of sound waves in fluids. We have seen in Sec. 6.7 that propagation of sound waves is accompanied by compressions and rarefactions of the medium. The compressed regions experience a rise in temperature and heat flows out of them into the rarefied regions. To treat these phenomena we need to couple the equations of motion for density fluctuations with the time dependence of the entropy.

7.4.1 *Thermodynamic relations*

As a first step we have to extend the thermodynamic relations that we have presented in Sec. 3.1 to the case of a fluid moving with velocity $\mathbf{v}$ (for a more detailed presentation see for example the book of Chaikin and Lubensky[199]). The internal energy U is the sum of the rest-frame energy $U_0(S, N, V)$ and of the kinetic energy $P^2/(2Nm)$ associated with the center-of-mass motion, with $\mathbf{P} = Nm\mathbf{v}$ being the total momentum:

$$U(S, N, V, \mathbf{P}) = U_0(S, N, V) + \frac{P^2}{2Nm}. \tag{7.14}$$

We thus have $\mathbf{v} = (\partial U/\partial \mathbf{P})_{S,N,V}$ and hence Helmholtz free energy is

$$F(T, N, V, \mathbf{v}) = U - TS - \mathbf{P}\cdot\mathbf{v} = F_0(T, N, V) - \frac{Nmv^2}{2}. \tag{7.15}$$

From the identity (3.9) for the differential of the free energy F_0 in the rest frame, we find

$$dF = -SdT - pdV + \mu dN - \mathbf{P}\cdot d\mathbf{v}, \tag{7.16}$$

so that the pressure $p = -(\partial F/\partial V)_{T,N,\mathbf{v}}$ and the entropy $S = -(\partial F/\partial T)_{N,V,\mathbf{v}}$ do not depend on $\mathbf{v}$ when expressed in terms of T, N and V. The chemical potential is instead given by

$$\mu = \left(\frac{\partial F}{\partial N}\right)_{T,V,\mathbf{v}} = \mu_0 - \frac{1}{2}mv^2\,, \tag{7.17}$$

where $\mu_0 = (\partial F_0/\partial N)_{T,V}$ is the chemical potential in the rest frame.

In terms of intensive densities of conserved quantities at fixed volume and of the entropy density $s = S/V$, we can obtain the pressure from the relation $pV = N\mu - F$ as in Sec. 3.1.3:

$$p(\mu, T, \mathbf{v}) = \rho\mu - u + Ts + \mathbf{g}\cdot\mathbf{v}\,, \tag{7.18}$$

where $\rho = N/V$, $u = U/V$ and $\mathbf{g} = \mathbf{P}/V \equiv m\mathbf{j}$. Finally, by differentiating Eq. (7.14) and using Eq. (3.5) for dU_0, we find the relation

$$Tds = du - \mu d\rho - \mathbf{v}\cdot d\mathbf{g}\,, \tag{7.19}$$

for the differential of the entropy density in terms of the differentials of the densities of conserved quantities. These equations are valid for uniform translations of the whole fluid, but will extend to situations in which there are only slow spatial variations.

7.4.2 *Entropy production*

Equation (7.19) is crucial for the derivation of hydrodynamic equations, since we can use it to get the form of the constitutive relations between the currents of conserved quantities and the fields T, μ and $\mathbf{g}$. We only need to combine it with the conservation laws for (i) particle number (Eq. (6.6)), (ii) total momentum (Eq. (6.8)), and (iii) total energy. The latter is

$$\frac{\partial u}{\partial t} = -\nabla\cdot\mathbf{j}_u\,, \tag{7.20}$$

where $\mathbf{j}_u$ is the energy current density. We aim to get an equation for the rate of change of the entropy, bearing in mind that the entropy current contains both a term $s\mathbf{v}$ and a heat-current term.

From Eq. (7.19) we find

$$T\frac{\partial s}{\partial t} = \frac{\partial u}{\partial t} - \mu\frac{\partial \rho}{\partial t} - \mathbf{v}\cdot\frac{\partial \mathbf{g}}{\partial t} = -\nabla\cdot\mathbf{j}_u + \mu\nabla\cdot\mathbf{j} + \sum_{i,j} \mathrm{v}_i\nabla_j\pi_{ij} \tag{7.21}$$

and

$$T\mathbf{v}\cdot\nabla s = \mathbf{v}\cdot\nabla u - \mu\mathbf{v}\cdot\nabla\rho - \sum_{i,j} \mathrm{v}_i\mathrm{v}_j\nabla_i g_j\,. \tag{7.22}$$

By combining these two equations, and using the fact that the vector $\mathbf{g}-m\rho\mathbf{v}$ must vanish since it is by definition the momentum density in the fluid rest frame, we find that the rate of change of the entropy density obeys the equation

$$T\left[\frac{\partial s}{\partial t} + \nabla\cdot\left(s\mathbf{v} + \frac{1}{T}\mathbf{j}_Q\right)\right] = -\frac{1}{T}\mathbf{j}_Q\cdot\nabla T - \sum_{i,j}(\pi_{ij} - p\delta_{ij} - \mathrm{v}_i g_j)\nabla_i\mathrm{v}_j\,. \tag{7.23}$$

Here, the pressure p is given by Eq. (7.18) and the heat current density $\mathbf{j}_Q$ is defined as

$$j_{Qi} = j_{ui} + \sum_j(\mathrm{v}_i\mathrm{v}_j g_j - \mathrm{v}_j\pi_{ji})\,. \tag{7.24}$$

Equation (7.23), combined with the requirement that entropy production must be non-negative, will give the constitutive relations for the currents.

7.4.3 *Constitutive relations*

In the absence of dissipation the entropy must obey a continuity equation $(\partial s/\partial t) = -\nabla\cdot(s\mathbf{v}))$ and Eq. (7.23) can be satisfied by setting $\mathbf{j}_Q = 0$ and $\pi_{ij} = p\delta_{ij} + g_i\mathrm{v}_j$. We recover, of course, the inviscid flow of the Euler fluid presented in Sec. 6.2.3. The corresponding energy current density is $\mathbf{j}_u = (u+p)\mathbf{v}$.

When there is dissipation, the RHS of Eq. (7.23) must be positive and this restricts the form of the dissipative couplings between currents and fields having opposite parity under time reversal (see Sec. 6.1). Considering each term in turn, we can see that (i) the dissipative heat current $\mathbf{j}_Q$ must flow against the temperature gradient, thus giving back Fourier's law; and (ii) there is a dissipative term in the momentum current density (σ'_{ij}, say) which is defined through

$$\pi_{ij} = p\delta_{ij} + m\rho\mathrm{v}_i\mathrm{v}_j - \sigma'_{ij}\,. \tag{7.25}$$

In the Newtonian regime σ'_{ij} is linearly related to the velocity gradients $\nabla_k\mathrm{v}_l$ by a fourth-rank tensor, which must have the same symmetry as the elastic-constants tensor in the theory of linear elasticity. We recover the results already presented in Sec. 6.2.1: namely, the tensor $\sigma_{ij} \equiv -p\delta_{ij} + \sigma'_{ij}$ has the form given in Eq. (6.3).

7.5 The Effect of Heat Flow on Sound Wave Propagation

We can now return to the discussion of sound wave propagation given in Sec. 6.7 and complete it by including heat flow. The Navier–Stokes equation still holds and its linearisation yields Eq. (6.46), that we report here for convenience:

$$\left[-m\frac{\partial^2}{\partial t^2} + \frac{1}{\rho}\left(\frac{4}{3}\eta + \eta_b\right)\frac{\partial}{\partial t}\nabla^2\right]\delta\rho(\mathbf{x},t) = -\nabla^2\delta p(\mathbf{x},t)\,. \tag{7.26}$$

From Eq. (7.24) we get the linearised energy current density as

$$\mathbf{j}_u = (u+p)\mathbf{v} - \lambda\nabla T\,. \tag{7.27}$$

Inserting this expression in Eq. (7.20) and using the linearised continuity equation for the particle density $[\nabla\cdot\mathbf{v} = -\rho^{-1}(\partial\delta\rho/\partial t)]$ we obtain

$$\frac{\partial}{\partial t}\left[\delta u(\mathbf{x},t) - \frac{u+p}{\rho}\delta\rho(\mathbf{x},t)\right] = \lambda\nabla^2\delta T(\mathbf{x},t)\,. \tag{7.28}$$

Here, from the linearised forms of Eqs. (7.18) and (7.19) and with $\tilde{s} \equiv s/\rho$ we have

$$\delta u(\mathbf{x},t) - (u+p)\frac{\delta\rho(\mathbf{x},t)}{\rho} = \rho T\delta\tilde{s}(\mathbf{x},t)\,. \tag{7.29}$$

7.5.1 *Hydrodynamic modes*

In solving the coupled Eqs. (7.26) and (7.28) it is convenient to adopt particle and entropy fluctuations as the independent variables. The relevant transformations are $\delta p = (\partial p/\partial\rho)_S\delta\rho + (\partial p/\partial\tilde{s})_\rho\delta\tilde{s}$ and $\delta T = (\partial T/\partial\rho)_S\delta\rho + (\partial T/\partial\tilde{s})_\rho\delta\tilde{s}$, where $\rho(\partial p/\partial\rho)_S \equiv K_S^{-1}$ is the adiabatic bulk modulus and $\rho T(\partial\tilde{s}/\partial T)_\rho \equiv C_V$ is the isochoric heat capacity per unit volume. We also need the thermodynamic relation for the difference between isobaric and isochoric heat capacities, $C_p - C_V = -T(\partial p/\partial V)_T[(\partial V/\partial T)_p]^2$ (see Eq. (3.19)).

Equations (7.26) and (7.28) may now be written as

$$\left[-\frac{\partial^2}{\partial t^2}+\left(c_S^2+\Gamma_\eta\frac{\partial}{\partial t}\right)\nabla^2\right]\delta\rho(\mathbf{x},t)=-m^{-1}\left(\frac{\partial p}{\partial\tilde{s}}\right)_\rho\nabla^2\delta\tilde{s}(\mathbf{x},t) \tag{7.30}$$

and

$$\left(\frac{\partial}{\partial t}-\Gamma_\lambda\nabla^2\right)\delta\tilde{s}(\mathbf{x},t)=\frac{\lambda}{\rho T}\left(\frac{\partial T}{\partial\rho}\right)_S\nabla^2\delta\rho(\mathbf{x},t)\,. \tag{7.31}$$

We have set $c_S^2=m^{-1}(\partial p/\partial\rho)_S$, $\Gamma_\eta=(\frac{4}{3}\eta+\eta_b)/m\rho$ and $\Gamma_\lambda=\lambda/C_V$.

Equations (7.30) and (7.31) are easily solved by taking Fourier–Laplace transforms. The main results to be noted are as follows. From (7.31) we get a diffusive heat mode at frequency

$$\omega=-iD_T k^2\,, \tag{7.32}$$

where D_T is the thermal diffusion coefficient. In an incompressible fluid the RHS of Eq. (7.31) vanishes and $D_T=\Gamma_\lambda$ in this case. More generally, we can set

$$\delta\rho(\mathbf{x},t)\cong-\left(\frac{\partial\rho}{\partial p}\right)_S\left(\frac{\partial p}{\partial\tilde{s}}\right)_\rho\delta\tilde{s}(\mathbf{x},t) \tag{7.33}$$

in Eq. (7.31), to order k^2, with the result

$$D_T=\Gamma_\lambda\left[1-C_V\left(\frac{\partial\rho}{\partial p}\right)_S\left(\frac{\partial p}{\partial\tilde{s}}\right)_\rho\left(\frac{\partial T}{\partial\rho}\right)_S\right]=\frac{\lambda}{C_p}\,. \tag{7.34}$$

With regard to density fluctuations, we see from Eq. (7.30) that coupling to entropy fluctuations has shifted the speed of sound to the adiabatic value c_S from the isothermal value c_T obtained in Sec. 6.7.1. The complex sound frequencies are given by

$$\omega=\pm c_S-\frac{1}{2}i\Gamma k^2\,, \tag{7.35}$$

where

$$\Gamma=\Gamma_\eta+D_T\left(\frac{C_p}{C_V}-1\right)\,. \tag{7.36}$$

Coupling to heat fluctuations contributes to the damping of sound waves. Recalling that $c_S/c_T = (C_p/C_V)^{1/2}$, we see that all effects of this coupling vanish in the limit $C_p/C_V \to 1$.

According to Eq. (7.36), the contributions to sound wave damping from viscosity and from heat flow are additive. For monatomic fluids the theory is in good agreement with experimental observations, but in polyatomic fluids the observed attenuation is substantially larger than predicted. A sound wave can disturb the distribution of energy between translational and internal degrees of freedom of the molecules as well as their spatial arrangement. These two types of phenomena are called thermal relaxation and structural relaxation, respectively, and contribute to sound wave damping in molecular liquids.

7.5.2 *Light scattering*

We have already noted in Sec. 6.7.3 that density fluctuations can be observed by light scattering through the fluctuations that they induce in the complex refractive index of the medium. Following the pioneering work of Landau and Placzek,[200] the light scattering spectrum is evaluated from the hydrodynamic equations and is proportional to the van Hove function $S(k,\omega)$ in the appropriate region of long wavelength and low frequency (see for example the review by Mountain[201]). For a plane-polarized incident wave with frequency $\omega_i = ck_i$, the intensity of scattered light at position $\mathbf{R} = (R, \phi)$ from the scattering site is[202]

$$\frac{I(\mathbf{R},\omega)}{I_0} = \left(\frac{N\alpha^2\omega_1^4}{2\pi R^2 c^4}\right) \sin^2\phi S(k,\omega)\,, \tag{7.37}$$

where N is the number of (spherically symmetric) molecules of polarizability α in the scattering volume, ω is the shift in angular frequency and $k = 2k_i \sin(\theta/2)$ with θ the scattering angle.

From the calculations that we have reported in Sec. 7.5.1, it is clear that there are three peaks in $S(k,\omega)$ as a function of angular frequency ω (see Fig. 7.3). The central component of the spectrum, peaking at $\omega = 0$, arises from the coupling of density fluctuations to thermal fluctuations (see Eq. (7.33)) and is known as the Rayleigh peak. There also are two "Doppler-shifted" peaks, referred to as the Brillouin components, centred at $\omega = \pm c_S k$ with width Γk^2 in accord with Eq. (7.35). Non-Lorentzian parts of the

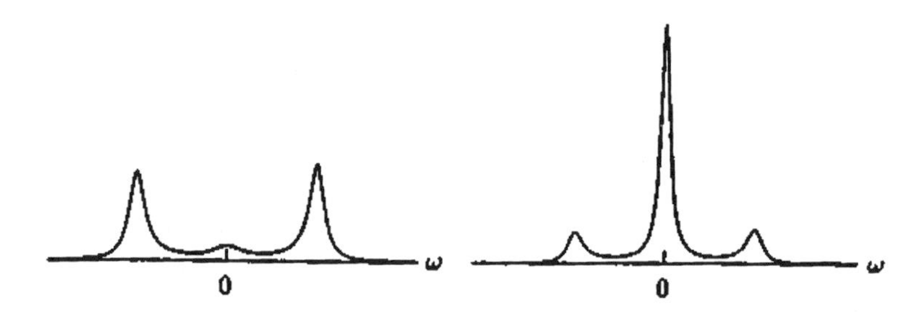

Fig. 7.3. A sketch of the light scattering intensity from density fluctuations in a simple fluid in the hydrodynamic regime. Left, far from the critical point; right, approaching the critical point.

spectrum are also present but are usually neglected, since they do not contribute significantly to the intensity near the three main peaks.

The ratio of integrated intensities for the Rayleigh (I_{R}) and Brillouin peaks ($2I_{\mathrm{B}}$) is

$$\frac{I_{\mathrm{R}}}{2I_{\mathrm{B}}} = \frac{C_p}{C_V} - 1\,. \tag{7.38}$$

The ratio becomes very large on the approach to the critical point, where $(\partial\rho/\partial p)_T$ and hence C_p are diverging (see Chap. 4). This is the phenomenon of critical opalescence, referred to earlier. At the same time the width of the Rayleigh peak narrows, being proportional to $\lambda/C_p \to 0$.

The light scattering technique was used in the work of Zollweg *et al.*[203] to measure the effective kinematic viscosity of Xenon along the coexistence curve up to $T_{\mathrm{c}} - T \approx 0.070$ K. Very precise data on shear viscosity at various frequencies for Xenon approaching its critical point have become available from experiments performed in microgravity condition.[204]

7.5.3 *Sound propagation in the critical region*

As was shown theoretically by Kadanoff and Swift[205] and by Kawasaki,[206] the dynamical behaviour of systems near criticality is intimately related to the anomalous static properties as the critical point is approached. A valuable method for investigating this behaviour is to use an acoustic-beam probing technique to measure the speed and attenuation of sound.[207]

In a pure fluid, sound propagation is markedly influenced by the fact that the adiabatic compressibility diverges at the critical point. The divergence can be traced to the fact that the spontaneous density fluctuations, which are normally correlated over a distance of only a few Å, become correlated over distances as large as thousands of Å. Since any compression requires transport over distances of the order of the correlation range to restore local equilibrium, such transport requires an ever longer time as the critical point is approached, and a compression must occur at an ever lower frequency in order to correspond to the static compressibility. This implies substantial dispersion and attenuation in sound waves near the critical point.

Cannell and Sarid[207] have measured the speed and attenuation of sound in SF_6 on the approach to the critical point along the critical isochore. All their data can be modelled through a frequency-dependent viscosity, involving a slightly modified form of the frequency dependence calculated by Kawasaki in the mode-coupling approach of Kadanoff and Swift, together with a single relaxation to account for energy exchange with the vibrational states.

Hohenberg and Halperin[208] have exhaustively discussed the cross-over that occurs in approaching the critical point from the hydrodynamic to the so-called critical region, as the correlation length for density fluctuations becomes much longer than the scattering wavelength. In the region the hydrodynamic value of the relaxation frequency ($\omega_\mathrm{R} = D_T(T,\rho)k^2$) is replaced by a power-law behaviour $\omega \propto k^z$ with a dynamical critical exponent $z \approx 3$.[209]

7.6 Binary Fluids

7.6.1 *Thermodiffusion*

A temperature gradient applied to a binary fluid will also cause a relative flow of the two species in opposite directions. Let us consider a binary mixture in which there are N_α molecules of species α each with molecular mass m_α ($\alpha = 1, 2$). When a concentration gradient is imposed on the isothermal mixture, a temperature gradient develops as interdiffusion occurs. This so-called Dufour effect is rather hard to measure in liquids, but there is more success for the reverse process, known as the Soret or thermodiffusion effect. Here a temperature gradient creates a species concentration gradient as specified by the equation

$$T\nabla x_1 = -k_T \nabla T\,, \tag{7.39}$$

where x_α are the mass fractions, $x_\alpha = N_\alpha m_\alpha/(N_1 m_1 + N_2 m_2)$. The mass fluxes are[210]

$$\mathbf{J}_\alpha = \rho x_\alpha (\mathbf{u}_\alpha - \mathbf{u}) , \tag{7.40}$$

where ρ is the total mass density, $\mathbf{u}_\alpha$ is the centre-of-mass velocity of species α and $\mathbf{u}$ is the centre-of-mass velocity of the mixture. The mass currents of the two species are related by $\mathbf{J}_2 = -\mathbf{J}_1$, so that only one need be considered.[211]

Linearised irreversible thermodynamics relates[212] the flows of concentration and heat to driving forces through Onsager's phenomenological coefficients L_{ij},

$$\begin{cases} \mathbf{J}_1 = L_{11}\mathbf{X}_1 + L_{1Q}\mathbf{X}_Q \\ \mathbf{J}_Q = L_{Q1}\mathbf{X}_1 + L_{QQ}\mathbf{X}_Q , \end{cases} \tag{7.41}$$

where $\mathbf{X}_1 = -T^{-1}\nabla(\mu_1 - \mu_2)$ and $\mathbf{X}_Q = -T^{-2}\nabla T$, the μ's being the chemical potentials of the two species. The Dufour and Soret effects are determined by the cross coefficients in Eq. (7.42), with $L_{1Q} = L_{Q1}$ from Onsager's reciprocity relations.

In the absence of a temperature gradient, Fick's law gives another expression for $\mathbf{J}_1$,

$$\mathbf{J}_1 = -\rho D_1 \nabla x_1 , \tag{7.42}$$

where D_1 is the bulk diffusion coefficient of species 1 relation to the centre of mass. Using the Gibbs–Duhem relation gives

$$D_1 = \frac{L_{11}}{\rho x_2 T}\frac{\partial \mu_1}{\partial x_1} . \tag{7.43}$$

In the experimental fixed volume frame D_1 and D_2 are related to the mutual diffusion coefficient D by $D = \rho \mathrm{v}_2 D_1 = \rho \mathrm{v}_1 D_2$, v_α being the partial specific volumes.

The cross coefficient L_{1Q} characterises the Soret effect. Combination of Eq. (7.39) with Eq. (7.42) gives for the Soret coefficient

$$k_T = \frac{L_{1Q}}{\rho T D_1} . \tag{7.44}$$

Finally, the thermal conductivity of the mixture follows from Fourier's law by setting $\mathbf{J}_1 = 0$,

$$\lambda = T^{-2}\left(L_{QQ} - L_{1Q}^2/L_{11}\right) . \tag{7.45}$$

The Onsager coefficients L_{ij} can be obtained in MD studies[210] either through the Green–Kubo formula expressing them as integrals of the time correlation functions of the currents, or by non-equilibrium methods as described in Sec. 7.2. Although the definitions of the mass and heat fluxes in Eq. (7.41) are not identical to those measured experimentally, the Green–Kubo simulations by MacGowan[211] show that the differences are small and can be ignored in practice.

7.6.2 *Hydrodynamic modes*

Bhatia and Thornton[213] have introduced number-concentration (N-c) structure factors to describe diffraction and inelastic scattering from a binary liquid, with main reference to alloys.

The number–number (or mass–mass) spectrum is closely related to the van Hove function $S(k, \omega)$ for a monatomic liquid. In the hydrodynamic regime[214] the Rayleigh peak in the alloy is the sum of two Lorentzians, the width of one of them being mainly controlled by thermal conductivity and that of the other by mutual diffusion. Sound attenuation in the Brillouin doublet is determined not only by viscosity and thermal conduction, but also by interdiffusion. The N-c spectrum contains again the same thermodynamic modes, whereas the c-c spectrum is determined solely by heat and interspecies diffusion.

Binary liquids with chemical short-range order are reviewed in Chap. 8 below. The c-c static and dynamic structure factors acquire special relevance in the presence of this type of order, including the possibility of collective relative oscillations of the two species.

7.7 Superfluid Helium

The normal boiling point of the bosonic isotope of helium (^{4}He) lies at 4.21 K and, when the temperature is further reduced, the liquid remains stable under the saturated vapour pressure — apparently down to $T = 0$. Rather high pressures are needed to obtain the solid phases, as is shown in the phase

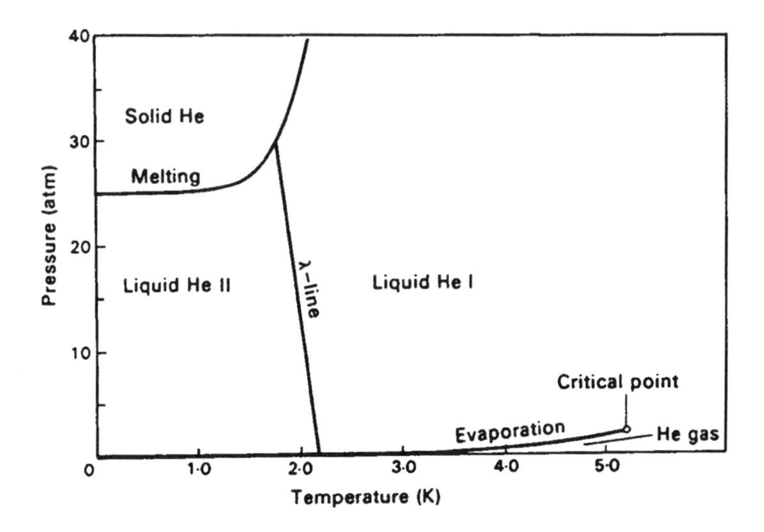

Fig. 7.4. Schematic phase diagram of ^{4}He.

diagram reported in Fig. 7.4. Application of the Clausius–Clapeyron relation to the melting curve in the figure shows that the melting entropy is virtually zero for all temperature below 1 K, so that the liquid cannot loose entropy by freezing.

Just below its boiling point the liquid behaves as an ordinary fluid with low viscosity, but at 2.17 K it undergoes a liquid–liquid transition to a superfluid (He-II) phase. The transition is signalled by a specific heat anomaly, whose characteristic shape has led to the name "λ-line" being given to the coexistence curve of the two liquid phases. The microscopic origin of this behaviour lies in the combination of weak interatomic cohesive forces with a small atomic mass favouring quantum-mechanical delocalisation of the atoms.

Non-Newtonian behaviour and heat-wave propagation are macroscopic manifestations of quantum mechanics in liquid He-II. Chapter 14 in this book is devoted to quantum fluids, but it seems appropriate to conclude the discussion of momentum and energy transport in fluids by reference to some of the phenomena which are observed to occur when Newton's law (6.3) and Fourier's law (7.1) do not apply. In this section a brief review of the main facts on transport in He-II will lead us into the two-fluid model and into the study of excitations out of a collective ground state by means of neutron inelastic scattering experiments.

7.7.1 *Transport properties of superfluid ^4He*

Liquid He-II seems to have virtually zero viscosity in experiments designed to measure viscous resistance to flow.[215] The liquid will flow through fine capillaries in the absence of a driving pressure gradient, at least as long as the flow rate does not exceed a critical value at which viscous resistance appears. Persistent currents can be induced in a torus-shaped vessel packed with porous material to provide very narrow channels. A glass beaker containing He-II will rapidly empty itself by a siphon effect through a very thin film completely wetting the glass surface, the film thickness being typically of order 100 atomic layers under saturated vapour and the driving force being the difference in gravitational potential between the ends of the film.

On the other hand, viscosity appears in experiments which detect the drag on a body moving through liquid He-II. These experiments use oscillating-disk or vibrating-wire viscometers to measure the rate of decay of torsional or vibrational oscillations and demonstrate the existence of viscous drag, the magnitude of the viscosity coefficient of He-II being comparable with that of normal liquid He or of He gas.

These different viscous behaviours can be reconciled by viewing He-II as if it were a "mixture" of two fluids: the normal fluid possessing Newtonian viscosity and the superfluid being capable of frictionless flow through capillaries or past obstacles. In this so-called two-fluid model, going back to the work of Tisza[216] in 1938, the liquid can simultaneously execute two types of motion, having local velocities $\mathbf{v}_n$ for the normal fluid at local density ρ_n and $\mathbf{v}_s$ for the superfluid at local density ρ_s. The total density is

$$\rho = \rho_n + \rho_s \tag{7.46}$$

and the total current density $\mathbf{j}$ is

$$\mathbf{j} = \rho_n \mathbf{v}_n + \rho_s \mathbf{v}_s \,. \tag{7.47}$$

It should be stressed that the two fluids do not correspond to two different classes of ^4He atoms and cannot be physically separated.

The two-fluid model is useful when the velocities are small: at high velocities the superfluid becomes dissipative and the normal fluid becomes turbulent. In a classical experiment carried out by Andronikashvili[217] in 1946, the effective density ρ_n of the normal component was measured as a function of temperature from the period of torsional oscillations (and hence from the

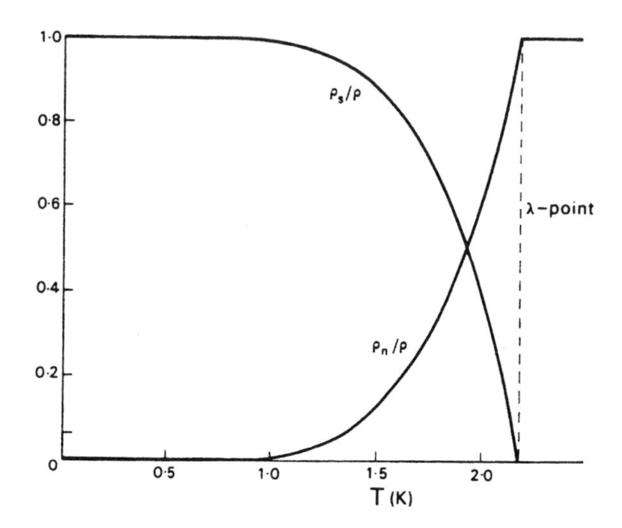

Fig. 7.5. Fractions of normal fluid and superfluid in liquid ^{4}He as functions of temperature, from Andronikashvili's experiment.

moment of inertia) of a pile of thin metal disks, which were closely spaced to ensure that all the normal fluid in the interstices would be dragged along while the superfuid remained stationary. The experiment shows that ρ_n/ρ decreases from unity at 2.17 K towards zero near 1 K (see Fig. 7.5).

The two-fluid model can also explain the main features of heat transport in He-II. Early experiments showed that the thermal conductivity is so high that it is impossible to establish a temperature gradient in the bulk liquid. Indeed, boiling ceases as the liquid is cooled across the λ-line: no large temperature fluctuations can occur in bulk He-II to nucleate gas bubbles, so that evaporation can occur only at free surfaces. A temperature gradient can instead be set up between two volumes of He-II connected by a channel filled with microporous material: since ρ_s/ρ decreases with increasing temperature, the superfluid flows to the hotter volume in order to reduce the "concentration" gradient and in so doing produces a pressure difference.

In fact, heat is not transported in He-II by conduction or convection of the whole liquid. A variety of thermomechanical effects[215] show that the normal component flows from the source to the sink of heat, while the superfluid flows in the opposite direction to maintain the total density constant. Thus, when heat is supplied periodically (e.g. by passing an oscillating electric current in a resistor immersed in a liquid volume) the two components oscillate in antiphase at constant total density. In this way temperature waves can propagate in liquid He-II. The name attributed to these waves is *second sound*, to distinguish them from ordinary longitudinal pressure waves (*first sound*) involving isothermal fluctuations of the total density.

The two-fluid model also describes the behaviour of He-II inside a rotating vessel. As for an ordinary liquid, the normal component undergoes solid-like rotation with the vessel by being dragged from friction against its walls. The superfluid component experiences instead vortex motion at sufficiently high rotational velocities: discrete vortex lines thread the fluid and the superfluid rotates round each line, with an angular momentum which is quantised according to

$$\oint \mathbf{v}_s \cdot d\mathbf{l} = \frac{nh}{m}. \tag{7.48}$$

The integral is taken around any contour surrounding a vortex line. This integral is known as the circulation and from Eq. (7.48) it is an integral multiple ($n = 0, 1, 2, \ldots$) of a quantum of circulation given by h/m, h being Planck's constant and m the mass of the ^{4}He atom.

The hydrodynamic equations governing normal and superfluid flow in He-II were derived by Landau,[218] who also discussed the main features of the microscopic excitation spectrum. An introduction to the hydrodynamic modes is given in Appendix 7.2 and a full discussion can be found in the book of Khalatnikov.[219] Extensions to higher frequencies and to inhomogeneous superfluids have also been given.[220,221] Here we focus on quantised excitations in He-II.

7.7.2 *Inelastic neutron scattering from superfluid ^{4}He*

In the early days of the two-fluid model the appearance of a superfluid fraction in liquid helium was believed to be related to Bose–Einstein condensation in the ideal gas of Bose particles, when a macroscopic number of bosons start occupying the single-particle state of lowest energy (see Chap. 14). Experimentally and under conditions specified immediately below, a condensate in liquid Helium should contribute to its inelastic neutron scattering spectrum a narrow peak superposed on a Doppler-broadened peak due to scattering against atoms out of the condensate.[222] The appropriate conditions are that the dominant contribution to the cross-section comes from the scattering of neutrons against single atoms: this requires measurements at high momentum and energy transfers, so that the role of the interatomic forces in the scattering process may be negligible. These experiments show that a fraction of about 10% of He atoms may be in a Bose–Einstein condensate in superfluid He-II at very low temperature.[223] There is, therefore, a clear difference between the superfluid (the whole liquid at very low temperature) and the condensate (a relatively small, though macroscopic, fraction of the atoms). This "depletion of the condensate" is due to the interatomic forces.

We must thus take the correlated collective ground state of the superfluid as our starting point and study the quantised excitations that it can sustain. The normal component of He-II at low temperature can be viewed as a dilute gas of such excitations, which are thermally created according to Bose statistics out of the ground state. The latter is governed by the interactions between the atoms and has no simple relation to the ground state of an ideal Bose gas.

With regard to the dispersion relation $\omega(k)$ between frequency and wave number for the elementary excitations, at low k they should merge into the first-sound waves ("phonons") with a linear dispersion relation $\omega = c_s k$. From the property of superfluidity Landau[218] argued that with increasing k the dispersion curve should go through a maximum and bend over into a minimum at $k = k_0$ say, with $\omega(k_0) = \Delta$, a finite "excitation gap". The elementary excitations in the region of the minimum are the "rotons", which may be visualized as microscopic quantised motions requiring a minimum amount Δ of energy for their excitation.

The results of inelastic neutron scattering experiments on superfluid ^{4}He confirm this viewpoint.[224] A scattering event involves the creation of excitations through transfer of momentum and energy from the neutron beam to the liquid. Figure 7.6 reports the dispersion curve of elementary excitations as obtained from these experiments. The data clearly show the linear part of the dispersion curve associated with single phonons (with a slope which agrees with the speed of sound as independently measured by ultracoustic techniques) as well as the roton minimum at wave number $k_0 \approx 2$ Å^{-1}, in approximate coincidence with the position of the main peak in the structure factor $S(k)$ of the liquid. However, the decay of an elementary excitation into two becomes possible at $k \approx 2.3$ Å^{-1}. This leads to a broadening of the dispersion curve

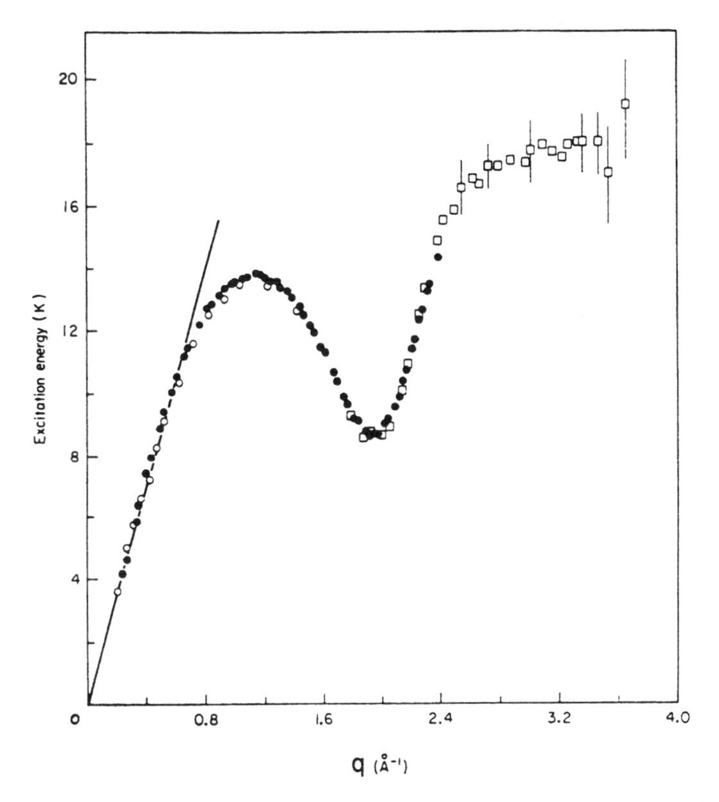

Fig. 7.6. Dispersion relation of elementary excitations in He-II from neutron inelastic scattering experiments. (Redrawn from Cowley and Woods, Ref. 224.)

till it can no longer be seen at $k \approx 3.5$ Å^{-1} and frequency $\omega \approx 2\Delta$, where the decay of a single excitation into two rotons becomes allowed.

While an elementary excitation appears in the spectrum as a narrow peak, a second and much broader peak is seen at higher energies and becomes dominant at $k \approx 2.3$ Å^{-1}. A full representation of the observed spectrum in the momentum and energy transfer plane is reported in Fig. 7.7,[224] where the broad peak is shown by the hatched area. This contribution is due to the creation of two (or more) correlated elementary excitations. The figure also shows that it is this second spectral contribution which ends at high energy and momentum transfer into scattering by single atoms.

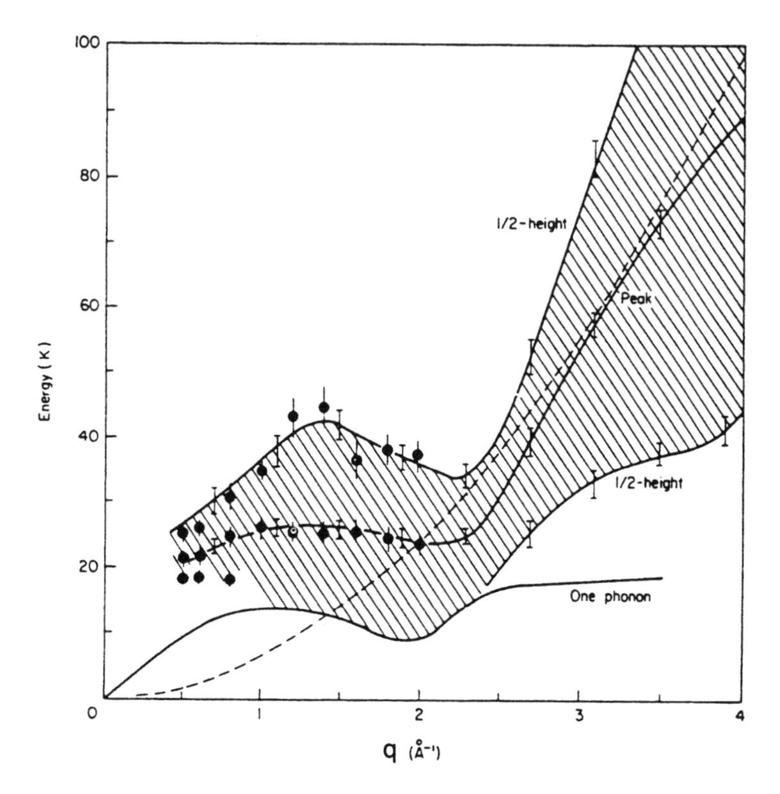

Fig. 7.7. Excitation spectrum of He-II as determined by neutron inelastic scattering. The hatched region describes the broad "multi-excitation" peak within its half-width. The dashed line shows the dispersion relation $\omega(q) = \hbar q^2/2m$ for free particles. (Redrawn from Cowley and Woods, Ref. 224.)

Appendix 7.1 Kinetic Theory of Thermal and Electrical Conductivity

We report in this appendix a kinetic calculation of transport coefficents in an electron gas. Treating first heat conduction in a classical gas, we take the temperature gradient along the z axis and consider the transfer of energy by particles crossing the xy plane. A particle traveling over a distance equal to the mean free path ℓ and striking the plane at an angle θ acquires an energy $(\ell \cos\theta)(\partial E/\partial z)$, the energy gradient being related to the temperature gradient by $n(\partial E/\partial z) = C(\partial T/\partial z)$ where n is the number density and C the heat capacity per unit volume. The number of particles which cross unit area of the plane per unit time, in a direction making an angle between θ and $\theta+\delta\theta$ with the z axis is $nu\cos\theta\sin\theta d\theta/2$, with u the average velocity. Thus, the net flow of energy against the temperature gradient is

$$j_{Qz} = -\frac{1}{2}Cu\ell\frac{\partial T}{\partial z}\int_0^\pi d\theta\cos^2\theta\sin\theta = -\frac{1}{3}Cu\ell\frac{\partial T}{\partial z} \tag{A7.1.1}$$

and Fourier's law gives

$$\lambda = \frac{1}{3}Cu\ell\,, \tag{A7.1.2}$$

for the thermal conductivity. This expression can be used for an electron gas in a semiclassical approximation, in which $C = \pi^2 nk_B^2 T/mu^2$ and u is replaced by the velocity of an electron on the Fermi surface (see Chap. 14).

Turning to electrical conductivity, in an applied electric field $\mathbf{E}$ and in the presence of dissipative scattering we can write the semiclassical equation of motion

$$m\left(\frac{d\mathbf{v}}{dt} + \frac{1}{\tau}\mathbf{v}\right) = e\mathbf{E}\,, \tag{A7.1.3}$$

for the drift velocity $\mathbf{v}$ of the electron gas. This equation has the particular solution $\mathbf{v} = (e\tau/m)\mathbf{E}$, corresponding to a constant electric current density given by

$$\mathbf{j} = ne\mathbf{v} = \left(\frac{ne^2\tau}{m}\right)\mathbf{E}\,. \tag{A7.1.4}$$

This is Ohm's law, with the electrical conductivity being given by $\sigma = ne^2\tau/m$. The inertial term in Eq. (A7.1.3) plays a role in electron dynamics under a time-dependent electric field.

The mean free path in Eq. (A7.1.2) can be written as the product of u times a relaxation time. Setting $\ell = u\tau$ implies the assumption that the scattering processes for transport of energy and charge by electrons are the same. Differences arise in solid metals at very low temperatures.

Appendix 7.2 Hydrodynamics of Superfluid Helium in the Two-Fluid Model

Superfluid Helium was the first system with a broken continuous symmetry (see Sec. 6.1) to be treated in the linearised hydrodynamic regime.[218] The theory attributes to the superfluid a complex order parameter $\psi = |\psi| \exp(i\varphi)$, whose phase determines the superfluid velocity as

$$\mathbf{v}_s(\mathbf{r}, t) = \left(\frac{\hbar}{m}\right) \nabla\varphi(\mathbf{r}, t) . \tag{A7.2.1}$$

The crucial point is that the superfluid velocity field is irrotational, so that transverse shear viscosity resides entirely in the normal component. The function $\psi(\mathbf{r}, t)$ can be viewed as the wave function of the condensate (see Chap. 14).

If $\mathbf{v}_s$ is non-zero, the free energy increases by $\frac{1}{2}m\rho_s \int d\mathbf{r}\mathbf{v}_s^2$ where ρ_s is the superfluid density. We consider next a Galilean transformation in which the whole fluid is taken to move with velocity $\mathbf{v}_n$ relative to the laboratory rest frame. The free energy acquires an extra term $-Nm\mathbf{v}_n^2/2$ (see Eq. (7.15)); in addition, there is a free-energy contribution associated with the phase of the order parameter if $\mathbf{v}_s \neq \mathbf{v}_n$. We can therefore write the free energy as

$$F(T, N, V, \mathbf{v}_n, \mathbf{v}_s) = F_0(T, N, V) - \frac{1}{2}Nm\mathbf{v}_n^2 + \frac{1}{2}m\rho_s \int d\mathbf{r}(\mathbf{v}_s - \mathbf{v}_n)^2 . \tag{A7.2.2}$$

The momentum density $\mathbf{g} = \mathbf{P}/V$ is obtained as for an ordinary fluid (see Eq. (7.16)):

$$\mathbf{g} \equiv -\frac{\partial(F/V)}{\partial \mathbf{v}_n} = m\rho\mathbf{v}_n - m\rho_s(\mathbf{v}_n - \mathbf{v}_s) = m(\rho_n\mathbf{v}_n + \rho_s\mathbf{v}_s) , \tag{A7.2.3}$$

with $\rho_n = \rho - \rho_s$. We have thus recovered Eqs. (7.46) and (7.47) of the two-fluid model. Finally, the field conjugate to the superfluid velocity is

$$\tilde{\mathbf{g}} = \frac{\partial(F/V)}{\partial \mathbf{v}_s} = m\rho_s(\mathbf{v}_s - \mathbf{v}_n) \tag{A7.2.4}$$

and vanishes unless the two fluids are in relative motion.

The treatment of thermodynamic quantities for fluid He-II then follows along the lines given in Sec. 7.4 for an ordinary fluid, except that (i) the velocity $\mathbf{v}$ appearing there is replaced by $\mathbf{v}_n$, and (ii) there is an extra term associated with $\mathbf{v}_s$. In particular, the differential for the entropy density s is

$$Tds = du - \mu d\rho - \mathbf{v}_n \cdot d\mathbf{g} - \tilde{\mathbf{g}} \cdot d\mathbf{v}_s \,. \tag{A7.2.5}$$

This relation will yield the constitutive relations between currents ($\mathbf{j} = \mathbf{g}/\mathrm{m}$ and $\mathbf{v}_s$) and fields ($\mathbf{v}_n$ and $\tilde{\mathbf{g}}$) in the superfluid.

In the linear regime, the constitutive relation for $\mathbf{j}$ evidently involves a Newtonian stress tensor written in terms of $\mathbf{v}_n$ and in addition a dissipative coupling to the divergence of $\tilde{\mathbf{g}}$. We also need a constitutive relation for $\mathbf{v}_s$ and, in view of Eq. (A7.2.1), its time derivative must be the gradient of a scalar quantity. We write

$$m\frac{\partial \mathbf{v}_s}{\partial t} = -\nabla(\mu + X)\,, \tag{A7.2.6}$$

where μ is the chemical potential and X is a dissipative term, determined in the linear regime by the divergence of the fields $\mathbf{v}_n$ and $\tilde{\mathbf{g}}$. The appearance of μ in Eq. (A7.2.6) can be justified by noticing that at equilibrium there must be free exchange of atoms between condensate and non-condensate (see Chap. 14): the condensate must therefore lie at the chemical potential in this case, so that its phase must vary linearly with time according to $\varphi(t) = \varphi(0) + \mu t/\hbar$.

We refer the reader for further details to the book of Khalatnikov[219] and continue the discussion for the superfluid in a non-dissipative linear regime. From the foregoing discussion we see that we can write for the hydrodynamic modes the following equations:

(i) a continuity equation for the total particle density fluctuation,

$$\frac{\partial \delta\rho}{\partial t} = -\nabla \cdot \mathbf{j}\,; \tag{A7.2.7}$$

(ii) a non-dissipative Navier–Stokes equation involving the pressure fluctuation,

$$m\frac{\partial \mathbf{j}}{\partial t} = -\nabla \delta p\,; \tag{A7.2.8}$$

(iii) a continuity equation for the entropy density fluctuation with the entropy current being carried (according to Eq. (A7.2.5)) by the normal component,

$$\frac{\partial \delta s}{\partial t} = -s\nabla \cdot \mathbf{v}_n\,; \tag{A7.2.9}$$

and (iv) a non-dissipative constitutive relation following from Eq. (A7.2.6)),

$$m\frac{\partial \mathbf{v}_s}{\partial t} = -\nabla\delta\mu\,. \tag{A7.2.10}$$

At the low temperatures of present interest we can neglect the coupling between density and heat fluctuations. Equations (A7.2.7) and (A7.2.8) can then be combined to yield

$$\frac{\partial^2\delta\rho}{\partial t^2} = \frac{1}{m}\nabla^2\delta p = \frac{1}{m\rho K_S}\nabla^2\delta\rho\,. \tag{A7.2.11}$$

Taking $\delta\rho(\mathbf{r},t) \propto \exp[i(\mathbf{k}\cdot\mathbf{r}-\omega t)]$ we find two first-sound modes at frequencies given by

$$\omega = \pm(m\rho K_S)^{-1/2}k\,, \tag{A7.2.12}$$

as for an ordinary fluid in adiabatic conditions.

From Eqs. (A7.2.3), (A7.2.8) and (A7.2.10) we get $m\rho_n(\partial\mathbf{v}_n/\partial t) = \rho_s\nabla\delta\mu - \nabla\delta p$, where $\delta\mu = \rho^{-1}(\delta p - s\delta T)$. Upon neglecting the coupling to density fluctuations, Eq. (A7.2.9) yields

$$\frac{\partial^2\delta s}{\partial t^2} = \frac{\rho_s T S^2}{\rho_n m C_V}\nabla^2\delta s\,, \tag{A7.2.13}$$

where S and C_V are the entropy and the isochoric heat capacity per particle. This equation admits propagating entropy-wave solutions, $\delta s(\mathbf{r},t) \propto \exp[i(\mathbf{k}\cdot\mathbf{r}-\omega t)]$ with eigenfrequencies given by

$$\omega = \pm\left(\frac{\rho_s T S^2}{\rho_n m C_V}\right)^{1/2} k\,. \tag{A7.2.14}$$

This is the dispersion relation for second sound in He-II.

Chapter 8

Chemical Short-Range Order: Molten Salts and Some Metal Alloys

Hitherto we have mainly concerned ourselves with monatomic fluids. This is a correct description for liquid argon, for example. But already for liquid sodium it is not correct in all respects — although in many of its properties this liquid behaves as if made of "pseudoatoms", for some others it is necessary to view it as formed of positive ions and conduction electrons. More obviously, one has to start from an ionic picture in describing a sodium chloride melt.[225]

We shall refer to liquid metals as ion-electron liquid in Chap. 14 on quantum fluids, in view of the Fermi degeneracy of the conduction electrons. Here we focus on molten halides and on some alloys of metallic elements in which electronic charge transfer between the components is also taking place. The main new concepts are those of (static and dynamic) screening and of chemical short-range order. Molten chalcogenides will be met in Chap. 10 on glassy materials.

8.1 Classical One-Component Plasma: Static and Dynamic Screening

We introduce ionic fluids by reference to the one-component classical plasma (OCP). This is a model of identical point-like ions, with charge e and mass m, which are embedded in a uniform background of charge and obey the laws of classical statistical mechanics. The charge density of the background is chosen so that the whole system is electrically neutral.[226]

The OCP is, just as the hard sphere fluid, a basic model of liquid state physics. It already exhibits some important features that characterise real Coulomb fluids, such as classical ion-electron plasmas and electrolyte solutions. These are the phenomena of static screening and plasma oscillations, arising from the long-range nature of the Coulomb interactions.

8.1.1 *Debye screening*

Let $\phi(r)$ be the electrical potential created in the OCP by an "average" ion, taken at the origin ($\mathbf{r} = 0$). The Poisson equation relates the Laplacian of $\phi(r)$ to the charge density:

$$\nabla^2 \phi(r) = -4\pi[e\delta(\mathbf{r}) + eng(r) - en]\,. \tag{8.1}$$

The three terms on the RHS of Eq. (8.1) are (i) the ion at the origin, with a point-like density described by a delta-function; (ii) the surrounding ions at density $ng(r)$, where n is the average density of the plasma and $g(r)$ is the pair distribution function; and (iii) the background.

In the self-consistent theory of Debye and Hückel,[227] the probability of finding a second ion at a distance r from the ion at the origin was taken as a Boltzmann factor involving the potential energy $e\phi(r)$. If we further assume that $e\phi(r) \ll k_B T$, we can write in Eq. (8.1)

$$g(r) \approx e^{-e\phi(r)/k_B T} \approx 1 - \frac{e\phi(r)}{k_B T}\,. \tag{8.2}$$

The Poisson equation reduces to $\nabla^2\phi(r) = -4\pi e\delta(\mathbf{r}) + \kappa^2\phi(r)$, where

$$\kappa^2 = \frac{4\pi n e^2}{k_B T} \tag{8.3}$$

defines the Debye length $1/\kappa$. The solution of the Poisson equation is

$$\phi_{\mathrm{DH}}(r) = \left(\frac{e}{r}\right) e^{-\kappa r}\,, \tag{8.4}$$

as can be checked *via* Fourier transform. The length $1/\kappa$ has the meaning of a *screening length*: it is the distance over which the electrical potential due to the ion at the origin and to the surrounding equilibrium distribution of ionic charges is exponentially cut down.

From Eq. (8.2) we find $g_{\mathrm{DH}}(r) = 1 - (e^2/rk_B T)\exp(-\kappa r)$ and hence by Fourier transform, we get the structure factor $S_{\mathrm{DH}}(k)$ of the plasma in the

Debye–Hückel theory as

$$S_{\mathrm{DH}}(k) = 1 + n\int d\mathbf{r}[g_{\mathrm{DH}}(r) - 1]e^{i\mathbf{k}\cdot\mathbf{r}} = \frac{k^2}{(\kappa^2 + k^2)}\,. \tag{8.5}$$

The important point is that $S_{\mathrm{DH}}(k)$ vanishes proportionally to k^2 for $k \to 0$. While in an atomic fluid the structure factor takes in this limit a finite value determined by the compressibility (see Sec. 4.4), macroscopic fluctuations of charge density are suppressed in the plasma.

The simplest route to the thermodynamic functions of the OCP within the Debye–Hückel theory is to notice from Eq. (8.4) that $\phi(r) \to (e/r) - e\kappa$ for $r \to 0$. The first term is due to the ion at the origin and hence the term $-e\kappa$ is the potential created at the origin by the surrounding ionic charges. Thus, the shift in internal energy due to the Coulomb interactions is

$$\Delta U_{\mathrm{DH}} = -\frac{1}{2}Ne^2\kappa\,. \tag{8.6}$$

The corresponding free energy term is easily found to be $\Delta F_{\mathrm{DH}} = 2\Delta U_{\mathrm{DH}}/3$.

In summary, the plasma at equilibrium distributes itself around each ion so as to screen its bare potential within a distance of order $1/\kappa$. The Debye–Hückel results for the thermodynamic functions become correct in the limit $\kappa^3/n \ll 1$, namely at high temperature and low density.[a] This parameter is a measure of the coupling strength in the OCP, since $(\kappa^3/n)^{2/3} = 4\pi e^2 n^{1/3}/k_{\mathrm{B}}T$ is of the order of the ratio (potential energy)/(thermal energy).

The screening length $1/\kappa$ was originally introduced by Gouy and Chapman[228,229] in the theory of how the ions in an ionic solution screen a charged planar electrode. In this problem $1/\kappa$ describes the thickness of the dipole layer generated at the interface between the solution and the electrode, and thus determines the electrical capacitance of the interface per unit area. The extension to a multi-component plasma is immediately effected by redefining the screening length through $\kappa^2 = 4\pi\sum_i(n_i e_i^2)/k_{\mathrm{B}}T$, with the sum running over all charged species.

[a]Equation (8.5) for $S(k)$ in the OCP is valid in general for $k \to 0$. However, the DH form of $g(r)$ is incorrect at small r: $g(r)$ has to vanish for $r \to 0$ in a classical fluid with repulsive interactions.

8.1.2 *Dynamic screening and plasma excitation*

Whereas the Debye–Hückel theory refers to a weakly coupled plasma at equilibrium, a signature of screening also emerges in the dynamics of the plasma. Let us suppose that each ion in the OCP is moved relative to the background by an amount ξ. The induced surface charge per unit area is $en\xi$ and by Gauss law creates an electric field $\mathbf{E} = -4\pi en\xi$. This acts as a restoring force in the equation of motion $d^2\xi/dt^2 = e\mathbf{E}/m$, which describes an oscillator having frequency

$$\omega_{\mathrm{p}} = \left(\frac{4\pi ne^2}{m}\right)^{1/2} . \tag{8.7}$$

Thus, the plasma has an oscillatory mode of motion at the *plasma frequency* ω_{p} in Eq. (8.7). An external electric field oscillating at this frequency will be in resonance with the plasma.

For the classical plasma we may now write the screening length $1/\kappa$ as the ratio between a characteristic velocity v_{c} and the characteristic frequency ω_{p}, $1/\kappa \approx v_{\mathrm{c}}/\omega_{\mathrm{p}}$. From the expressions for κ and ω_{p} we find $v_{\mathrm{c}} \approx (k_{\mathrm{B}}T/m)^{1/2}$, which is the mean speed of thermal agitation. This argument shows the underlying connection which exists between static and dynamic screening.

8.1.3 *Structure and dynamics of the strongly coupled OCP*

Historically, extensive studies of the structure of the OCP as a function of its coupling strength (conventionally denoted by $\Gamma \equiv e^2/k_{\mathrm{B}}Ta$ with $a \equiv (4\pi, n/3)^{-1/3}$) were first carried out by Brush, Sahlin and Teller[230] by the Monte Carlo computer simulation method. Figure 8.1 is taken from their work and shows that, starting from a monotonic shape of $g(r)$ at very low Γ as in the Debye–Hückel theory, the radial distribution function develops with increasing Γ an excluded-volume region and a first-neighbour peak. In fact, the state of short-range order in the strongly coupled OCP is not dissimilar from that of a fluid of neutral hard spheres.[231] With ever increasing coupling strength one observes that at $\Gamma \approx 180$ the OCP freezes into a classical body-centred-cubic crystal, under the sole effect of the Coulomb repulsions. The short-range order in $g(r)$ is combined with complete screening as expressed in the long-wavelength form of the structure factor $S(k)$ given in Eq. (8.5), $S(k) \to k^2/\kappa^2$ for $k \to 0$.

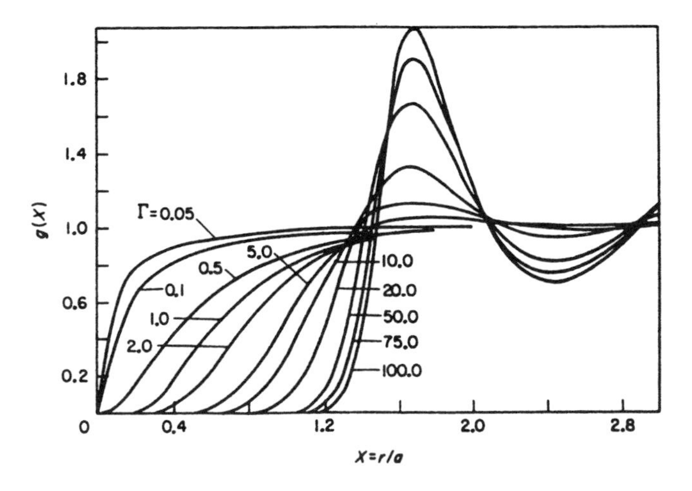

Fig. 8.1. Pair distribution function of the one-component classical plasma as a function of distance (in units of $a = (4\pi n/3)^{1/3}$) for a series of values of the plasma coupling-strength parameter $\Gamma = e^2/(k_B T a)$. (Redrawn from Brush *et al.*, Ref. 230.)

The most important aspect of the dynamics of the OCP concerns the collective oscillations of the charge density. In the limit of long wavelengths ($k \to 0$), the result (8.7) for the plasma frequency remains valid in the fluid state at all values of the coupling strength and also applies to longitudinal vibrations in the crystalline state. The plasma mode appears as a sharp peak in the dynamic structure factor $S(k, \omega)$,[232] which progressively broadens and shifts as k increases. At large k the spectrum takes a Gaussian shape centred at $\omega = 0$, as for an ideal gas.

8.2 Macroscopic Properties of Molten Salts

The crystal structures of halide compounds arise from electronic charge transfer and local compensation of positive and negative ionic charges through chemical order. Nature achieves charge compensation in two qualitatively distinct ways. The first involves halogen sharing and high coordination for the metal ions, as for example in alkali, alkaline-earth and lanthanide metal halides. In the second type charge compensation takes place within well defined molecular units, either monomeric ones as for example in $HgCl_2$ and $SbCl_3$ or dimeric ones as in $AlBr_3$.

Neutron diffraction studies of metal halide melts have shown that melting usually preserves the type of chemical order found in the crystal. For example, the melting of $MgCl_2$ or YCl_3 can be viewed as a transition from an ionic crystal to an ionic liquid (ionic-to-ionic, in short) and that of $SbCl_3$ or $AlBr_3$ as a molecular-to-molecular transition. However, $AlBr_3$ and $FeCl_3$ are known instances of ionic-to-molecular melting (see below). Intermediate-range order (IRO), extending over distances of 5 to 10 Å say, has been revealed in both network-type and molecular-type melts. This type of order is well known in glassy materials (see Chap. 10).

In this section we present macroscopic data (melting parameters and transport coefficients) in selected halides, which along with the microscopic evidence from diffraction experiments allow a broad classification of melting mechanisms and liquid structures. The next sections will give representative cases of structural evidence for various types of ordering. More details and complete references can be found in specialised reviews.[233]

8.2.1 *Selected macroscopic data for chlorides*

Table 8.1 collects, for several chlorides to be discussed in the following sections, (i) the measured values of the melting parameters, (ii) the structure of the hot crystal phase and (iii) transport coefficients of the melt near freezing (ionic conductivity σ and shear viscosity η).

Table 8.1. Macroscopic properties of metal chlorides*.

Salt	χ_M	Crystal	T_m (K)	ΔS_m (e.u.)	$\Delta V_m/V_l$	$\sigma(\Omega^{-1}\cdot cm^{-1})$	η (cp)
NaCl	0.40	NaCl	1074	6.30	0.28	3.6	1.0
CuCl	1.20	Wurtzite	696	2.43	0.16	3.7	4.1
$SrCl_2$	0.55	CaF_2	1146	3.44	0.11	2.0	3.7
$CaCl_2$	0.60	$CaCl_2$	1045	6.44	0.043	2.0	3.4
$MgCl_2$	1.28	$CdCl_2$	980	9.74	0.28	1.0	2.2
$HgCl_2$	1.32	$HgCl_2$	554	9.11	0.21	3×10^{-5}	1.6
$ZnCl_2$	1.44	$ZnCl_2$	570	4.09	0.14	1×10^{-3}	4×10^3
$LaCl_3$	0.705	UCl_3	1131	11.49	0.16	1.3	6.7
YCl_3	0.66	$AlCl_3$	994	7.56	0.0045	0.39	—
$FeCl_3$	0.99	$FeCl_3$	573	17.80	0.39	0.04	—
$AlCl_3$	1.66	$AlCl_3$	466	18.14	0.47	5×10^{-7}	0.36
$GaCl_3$	1.68	$GaCl_3$	351	7.84	0.17	2×10^{-6}	1.8
$SbCl_3$	2.08	$SbCl_3$	346	8.96	0.17	2×10^{-4}	—

Each compound in Table 8.1 is labelled by a parameter χ_M as an indicator of the character of the chemical bond, which has been taken for the metal atom from the chemical scale of the elements proposed by Pettifor.[234] The ordering of the compounds in the table corresponds to increasing covalency against ionicity in moving downwards for each value of the valence.

8.2.2 *Melting parameters*

The entropy change ΔS_m and the (estimated) volume change $\Delta V_m/V$ on melting give useful indications on the melting mechanism. The empirical relation

$$\Delta S_m = \nu R \ln 2 + \gamma C_V \frac{\Delta V_m}{V} \tag{8.8}$$

has been proposed.[235] Here ν is the number of atoms in a formula unit, R the gas constant, γ the Grüneisen parameter and C_V the specific heat. Insofar

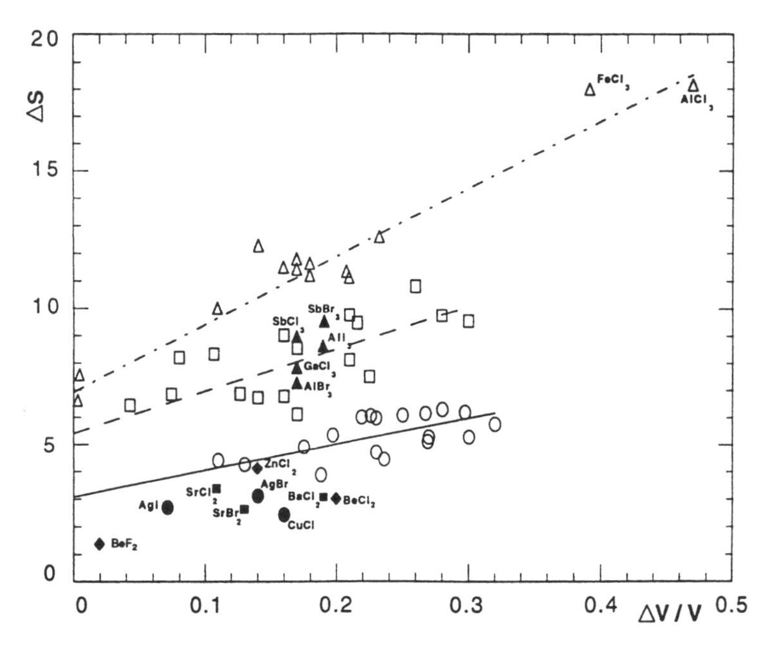

Fig. 8.2. Relation between entropy of melting ΔS (in cal K^{-1} mol^{-1}) and relative volume change $\Delta V/V = (V_\ell - V_s)/V_\ell$ for monohalides (○), dihalides (□) and trihalides (Δ). Filled symbols mark systems showing appreciable deviations from Eq. (8.8). (Redrawn from Akdeniz and Tosi, Ref. 236.)

as γ and C_V take similar values for similar systems in corresponding states, Eq. (8.8) implies an approximate linear relationship between ΔS_{m} and $\Delta V_{\mathrm{m}}/V$, extrapolating for $\Delta V_{\mathrm{m}}/V \to 0$ to a constant value $\Delta S_{\mathrm{m}}/\nu \to R\ln 2$.

Figure 8.2 shows that such a linear relationship is approximately verified by a number of halides of mono-, di- and tri-valent metals[236] (○, □ and △). Some systems, however, show a deficit in ΔS_{m} relative to the "norm" (filled symbols). These exceptions are associated with special melting mechanisms, i.e. from a disordered solid (● and ■) or into a network-forming liquid (◆) and into a molecular liquid (▲). The deficit in ΔS_{m} is associated in the first case with disordering in the hot crystal before melting and in the other cases with residual order in the melt. The points for $AlCl_3$ and $FeCl_3$ at large values of ΔS_{m} and $\Delta V_{\mathrm{m}}/V$ reflect drastic changes in the state of order on melting.

8.2.3 *Alkali halide vapours and critical behaviour of ionic fluids*

Even for alkali halides, the vapour at coexistence with the hot melt is made of molecular monomers and dimers. The same basic ionic model can account for cohesion in these molecules as in the solid and dense liquid states, provided that distortions of the electron shells of the ions from electrical and overlap effects are accounted for.[237]

The issue of critical behaviour in ionic fluids is a topic of high interest (for a short review see Fisher[238]). While fluids with short range or van der Waals interactions tend to exhibit Ising-type critical exponents (see Sec. 4.8), various observations on ionic solutions show either classical criticality or a crossover to Ising behaviour occurring only very close to criticality. A persistence of classical behaviour was attributed by Mott[239] to the presence of long-range Coulomb forces.

It is well established from work on model ionic fluids of charged hard spheres that inclusion of ion pairing and of the associated free-ion depletion is essential at low density. The results of this work still predict only classical criticality, but are in quite satisfactory agreement with Monte Carlo data. However, at higher temperatures all pairing theories violate thermal convexity, owing to problems in implementing a chemical picture of ion pairing and in specifying appropriately the association constant. To overcome these problems an exact thermodynamic formulation of chemical association has been developed.[240]

A number of novel results obtained in this context for density and charge correlations[238] concern (i) the use of the density correlation length to evaluate the cross-over scale for departure from classical/mean field behaviour; and (ii) the role of charge density oscillations as markers of incipient charge ordering and its competition with density fluctuations in the critical region.

8.3 Structural Functions for Multicomponent Fluids

The description of the atomic pair structure in a multicomponent fluid such as a molten salt requires partial radial distribution functions $g_{\alpha\beta}(r)$, which are defined so that $4\pi r^2 n_\beta g_{\alpha\beta}(r)dr$ is the average number of β-type particles lying in a spherical shell of radius r and thickness dr centred on an α-type particle. Here, n_β is the partial number density of β-type particles.

The features of foremost interest in the function $g_{+-}(r)$ for cation-anion correlations in a molten salt are (i) the position of the main peak, giving the average bond length d; (ii) the depth of the minimum after the peak, as a gauge of the stability of the local structure against relaxation and exchange of ions with the rest of the liquid; and (iii) the first-neighbour coordination number. Further information can be obtained from second-neighbour bond lengths and coordination numbers if the whole set of distribution functions $g_{\alpha\beta}(r)$ can be resolved.

The partial structure factors of the molten salt are related to $g_{\alpha\beta}(r)$ by

$$S_{\alpha\beta}(k) = \delta_{\alpha\beta} + 4\pi(n_\alpha n_\beta)^{1/2} \int_0^\infty r^2 dr [g_{\alpha\beta}(r) - 1] \frac{\sin(kr)}{kr} \,. \tag{8.9}$$

With this definition the intensity $I(k)$ of radiation with wavelength λ, coherently scattered through an angle 2θ in a diffraction experiment, is given by

$$I(k) \equiv \langle f^2 \rangle S(k) = \sum_{\alpha,\beta} \left(\frac{n_\alpha n_\beta}{n^2} \right)^{1/2} f_\alpha f_\beta S_{\alpha\beta}(k) \,, \tag{8.10}$$

where $k = (4\pi/\lambda)\sin\theta$ is the scattering wavenumber, f_α are the scattering amplitudes and n is the total number density. As already noted in Sec. 4.3, the scattering amplitudes for X-rays depend on k and increase monotonically with the atomic number. They are independent of k for neutron scattering and vary rather randomly with the nuclear species.

Evidently, a single diffraction experiment does not suffice to resolve the partial structure factors and to yield the full set of distribution functions $g_{\alpha\beta}(r)$. Nevertheless, it does provide an average pair function in which the first few shells of neighbours may be approximately resolved by suitable fitting procedures. Full resolution of pair correlations can be obtained with the method of multipattern determination by isotopic substitution in neutron diffraction, first applied to a liquid alloy by Enderby *et al.*[241] and to a molten salt by Page and Mika.[242] In brief, for a binary liquid this method requires diffraction experiments on three samples in which the contrast has been changed by varying the isotopic composition at a fixed chemical composition. In particular, the use of the ^{35}Cl and ^{37}Cl isotopes, with scattering amplitudes differing by a factor of about five, has led to detailed structural information on many molten chlorides. Useful structural information may also be obtained from X-ray absorption spectroscopy.[243]

8.3.1 *Number-concentration structure factors*

Some linear combinations of the partial structure factors emphasise specific aspects of the short-range order, as was first noted by Bhatia and Thornton[244] for binary alloys. Thus, if in Eq. (8.10) we take the f_α's to be identical for all species, we obtain

$$S_{\mathrm{NN}}(k) = \sum_{\alpha\beta} \left(\frac{n_\alpha n_\beta}{n^2}\right)^{1/2} S_{\alpha\beta}(k)\,. \tag{8.11}$$

This function, and the corresponding distribution function $g_{\mathrm{NN}}(r)$, describe the correlations between fluctuations in the total number density of particles and reflect the topological order coming from excluded-volume effects. In molten alkali halides the main peak in $S_{\mathrm{NN}}(k)$ at $k_{\mathrm{N}} \approx 2\pi/d$ is very broad and $g_{\mathrm{NN}}(r)$ is almost flat beyond the first-neighbour peak.

On the other hand, if we consider a binary liquid with $n_1 f_1 = -n_2 f_2$, we obtain

$$S_{\mathrm{cc}}(k) = \frac{1}{n}[n_2 S_{11}(k) - 2(n_1 n_2)^{1/2} S_{12}(k) + n_1 S_{22}(k)]\,. \tag{8.12}$$

This function and the corresponding distribution function $g_{\mathrm{cc}}(r)$ describe correlations between fluctuations in composition, i.e. the chemical short-range order. In molten alkali halides $g_{\mathrm{cc}}(r)$ shows marked oscillations around zero, which extend over distances of at least 10 Å: starting from a given ion, one meets regions that are alternatively enriched and depleted in ions of the other species. The period r_c of these oscillations is close to the second-neighbour bond length. $S_{cc}(k)$ correspondingly shows a very strong and rather narrow peak at $k_c \approx 0.7k_N$. This is known as the Coulomb prepeak and is commonly taken as a signature of chemical short-range order.

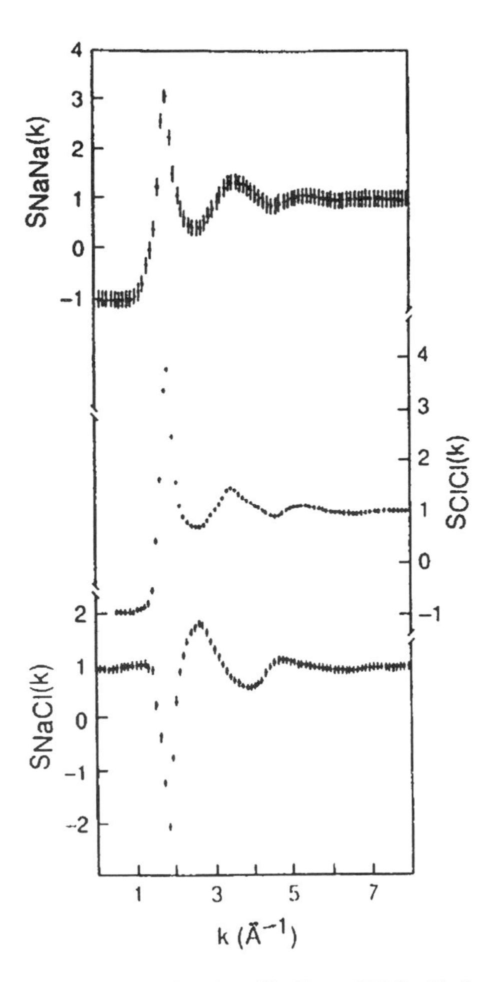

Fig. 8.3. Partial structure factors of molten NaCl at 875°C. (Redrawn from Biggin and Enderby, Ref. 245.) Here and in Fig. 8.5, the definition of partial structure factors is from T. E. Faber and J. M. Ziman, *Phil. Mag.* **11**, 153 (1965).

8.4 Coulomb Ordering in Monohalides and Dihalides

8.4.1 *Alkali halides*

Figure 8.3 shows the partial structure factors of molten NaCl at 875°C as an illustrative example of alkali halides, from neutron diffraction measurements on natural and isotopically enriched samples by Biggin and Enderby.[245] The trademark of Coulomb ordering in a 1:1 melt is the deep valley in $S_{\rm NaCl}(k)$ lying in phase with the main peak in $S_{\rm NaNa}(k)$ and in $S_{\rm ClCl}(k)$. It is immediately seen from Eqs. (8.11) and (8.12) (with $n_1 = n_2 = n/2$) that these features lead to a strong peak in $S_{cc}(k)$ and largely cancel out in $S_{\rm NN}(k)$.

Figure 8.4 compares the experimentally derived $g_{\alpha\beta}(r)$ with those obtained from ionic pair potentials.[246] Coulomb attractions and closed-shell core repulsions between Na^+ and Cl^- ions lead to a first-neighbour shell of unlike ions around any given ion, with a sharply defined region of excluded volume. In spite of the large volume expansion of NaCl on melting (see Table 8.1), the bond length d is shorter in the melt than in the crystal at melting (2.78 Å versus 2.95 Å). Melting thus occurs with reduction in coordination. It is also seen from Fig. 8.4 that the Coulomb repulsions between like ions push them into second-neighbour shells having similar shapes.

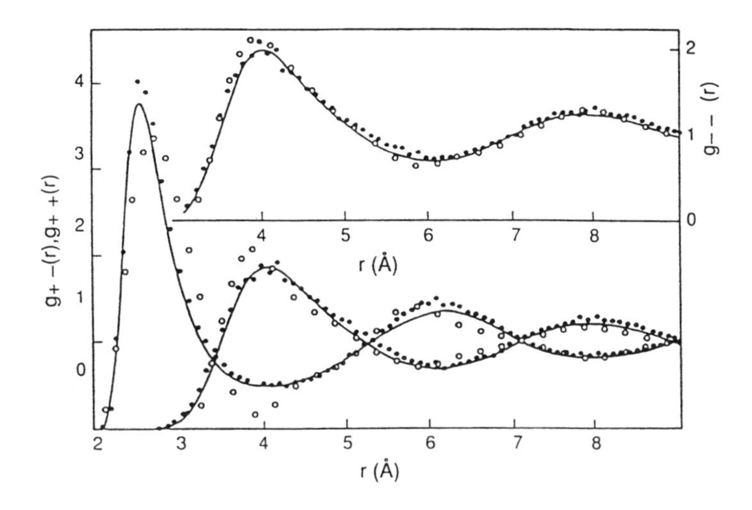

Fig. 8.4. Pair distribution functions of molten NaCl near freezing, from neutron diffraction (∘), computer simulation (•) and theory (curves). (Redrawn from Ballone *et al.*, Ref. 246.)

Thus, the nature of Coulomb ordering in a molten salt is such that the distribution of the screening charge density around any given ion oscillates in space, rather than being a monotonic function of distance as in the Debye–Hückel theory. Nevertheless, a meaningful definition of screening length in a dense ionic fluid can be based on the Debye–Hückel concept of the potential drop across the dipole layer formed by an ion and by the screening charge distribution. The relevant thermodynamic quantity is the Coulombic internal energy U_C of the melt. A simple and physically transparent analytical result is obtained for a 1:1 melt in the primitive model of equisized charged hard spheres as solved in the so-called mean spherical approximation.[247] This yields $U_C = -e^2/(L + d/2)$ where d is the bond length and

$$L = d[(1 + 2\kappa d)^{1/2} - 1]^{-1}, \tag{8.13}$$

with $\kappa = (4\pi n e^2/k_B T)^{1/2}$. The length L can thus be viewed as the screening length in the ionic fluid. More generally, this argument leads to an expression for the screening length as a function of density, temperature, dielectric constant, ionic charges and ionic radii. Contact can then be made with data on the differential capacitance of electrode/molten salt interfaces.[248]

8.4.2 *Noble-metal halides*

The monovalent Cu^+ and Ag^+ ions, with an outer shell of ten *d*-electrons, have small ionic radius and large electronic polarisability in comparison with the corresponding alkali ions. These properties lead to some hybridisation and covalent binding in copper and silver halides, tending to favour low coordination of first neighbours and promoting remarkable transport behaviours.

The ionic conductivity σ of solid CuBr and CuI increases rapidly with temperature, already reaching values of $\approx 0.1\Omega^{-1}\cdot\text{cm}^{-1}$ before attaining, through two structural phase transitions, fast-ion (superionic) behaviour of the Cu^+ ions before melting. A phase transition is also exhibited by AgI at 147°C and is accompanied by a jump in σ to values of $\approx 1\Omega^{-1}\cdot\text{cm}^{-1}$, typical of ionic melts. The Ag^+ ions in the α phase are disordered over many interstitial sites. Solid CuCl, AgCl and AgBr also show premelting phenomena, with σ rising to values of $\approx$ 0.1–0.5 $\Omega^{-1}\cdot\text{cm}^{-1}$.

It can be seen from Table 8.1 and Fig. 8.2 that these materials melt at relatively low temperature with a relatively low entropy change, while the ionic conductivity of the melt is comparable to that of molten alkali halides. Excess entropy has been released in the crystal before melting through the massive disordering of the metal ions. Diffraction data are available for all melts of this family: overall, their liquid structure can be described in term of a random close-packing of halogens, accommodating the metal ions in tetrahedral-like coordinations.

8.4.3 *Fluorite-type superionic conductors*

Fluorite-type materials such as $SrCl_2$ undergo a diffuse transition to a high-conductivity state before melting. The ionic conductivity and the entropy increase rapidly but continuously with temperature across the transition, whereas the heat capacity shows a peak. A high dynamic concentration of anionic Frenkel defects (interstitial-vacancy pairs) is gradually created across the transition, as revealed by neutron diffraction and diffuse quasi-elastic scattering studies on a variety of materials including $SrCl_2$, CaF_2, PbF_2 and UO_2. In other materials, such as $BaCl_2$ and $SrBr_2$, a superionic state is attained through a structural phase transition to the fluorite structure. A deficit of entropy of melting is again apparent from Fig. 8.2.

The liquid structure of $BaCl_2$ and $SrCl_2$ has been determined by neutron diffraction using isotopic substitution.[249] In both melts, within the frame of the divalent cations, the halogen ion component is more weakly ordered. The liquid structure thus shows a remnant of the fast-ion conducting state that the solid attains through an extensive disordering of the anions.

The observed short-range ordering in molten $SrCl_2$ and $BaCl_2$ suggests that freezing may be viewed as a process in which the cationic component is independently crystallising and at the same time modulating the anions into the lattice periodicity.[250] The anionic component in the hot crystal near melting may thus be described as a modulated "lattice liquid". In turn, the diffuse transition from the superionic to the "normal" state on cooling the $SrCl_2$ crystal may be viewed as a continuous process of anionic freezing inside the periodic force field of the metal-ion lattice.

8.4.4 *Tetrahedral-network structure in $ZnCl_2$*

The pair structure is also experimentally known for a number of other dihalide melts. The evolution of the liquid structure with increasing χ_M (in essence, with increasing covalency versus ionicity of the bonding) brings it from a cation-dominated structure to one in which the anions provide a "deformable frame" accommodating the doubly-charged cations. The Cl^-–Cl^- structural correlations are not especially affected: the Cl–Cl bond length stays in the range 3.6 to 3.8 Å.

To illustrate and qualify the above statement, let us contrast $ZnCl_2$ with $SrCl_2$. $ZnCl_2$ melts with relatively low T_m and ΔS_m from a crystal structure formed by corner-sharing tetrahedral units, and in the melt has very low ionic conductivity and very high viscosity (see Fig. 8.2 and Table 8.1). Molten $ZnCl_2$ can be supercooled into a vitreous state, with a glass transition temperature $T_g \approx 115°C$. The deficit in the entropy of melting for $ZnCl_2$ (and also for BeF_2 and $BeCl_2$ in Fig. 8.2) is associated with the presence of IRO in the melt and

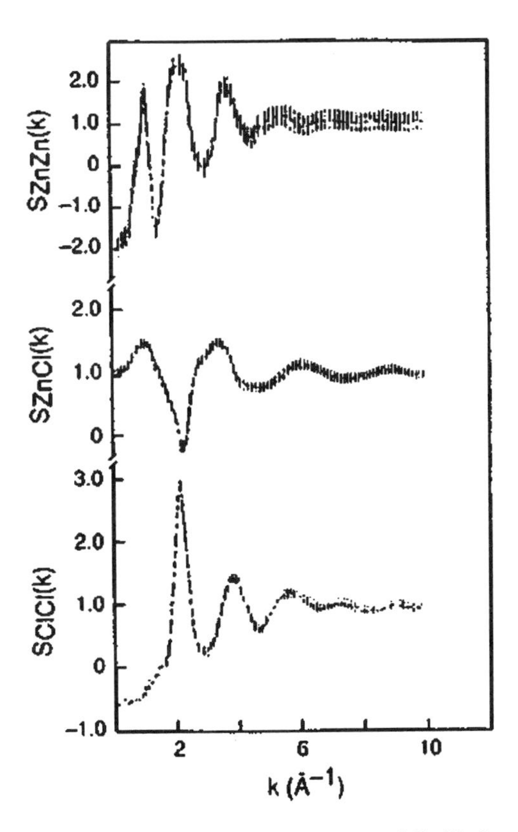

Fig. 8.5. Partial structure factors of molten $ZnCl_2$ at 327°C. (Redrawn from Biggin and Enderby, Ref. 251.)

should be gradually made up on heating the melt. BeF_2 is again a glass-former with a crystal structure built from corner-sharing tetrahedra, while in $BeCl_2$ the local tetrahedral coordination leads through edge-sharing to the chain-like "fibrous SiS_2" structure.

The measured partial structure factors of molten $ZnCl_2$[251] are shown in Fig. 8.5. The peaks associated with topological and chemical short-range order are clearly visible in all three partials. In addition, a prominent first sharp diffraction peak (FSDP) is present at $k \approx 1$ Å^{-1} in the Zn–Zn partial. The FSDP is a marker of the IRO in the liquid.

The state of pronounced IRO in molten $ZnCl_2$ arises from strongly stable local tetrahedral structures through the formation of a network of chlorines. The partial distribution functions can be interpreted as describing a disordered close-packed arrangement of chlorine ions which provides tetrahedral sites for the Zinc ions. Such a structural arrangement is very similar to that of the glassy state of $ZnCl_2$[252]: the Zn–Cl bond length is practically the same in the two states and the average coordination number of Zn is reported as 3.8 in the glass and ≈ 4 in the melt.

8.5 Structure of Trivalent-Metal Halides

Two main trends emerge from liquid structure studies on trichlorides: (i) the trend from cation-dominated Coulomb ordering to loose octahedral-network structures across the series of lanthanide compounds including YCl_3, and (ii) the stabilization of molecular structures with strong intermolecular correlations leading to IRO. The overall structural evolution is governed by the increasing weight of covalency versus ionicity.

The macroscopic properties reported for trichlorides in Table 8.1 reflect melting mechanisms which are consistent with the observed liquid structures. Progressive network formation in the melt from $LaCl_3$ to YCl_3 is signalled by decreasing values of ΔS_m, $\Delta V_m/V_1$ and σ. The melting parameters ΔS_m and $\Delta V_m/V_1$ for $FeCl_3$ and $AlCl_3$ are drastically larger than those of YCl_3: melting occurs from layer-type ionic crystals in these compounds, but brings the former two into a molecular-liquid state with a reduction in coordination number from 6 to about 4. $GaCl_3$ and $SbCl_3$ provide examples of melting from molecular crystal structures into associated molecular liquids with a deficit in ΔS_m (see Fig. 8.2).

8.5.1 *Octahedral-network formation in lanthanide chlorides*

X-ray diffraction data on the series of molten rare-earth trichlorides show similar structural characters.[253] The d_{MCl} bond length lies in the range 2.7–2.9 Å while the second-neighbour bond lengths are $d_{\mathrm{MM}} \approx 5$ Å and $d_{\mathrm{ClCl}} \approx 4$ Å, indicating a Coulomb ordering primarily ruled by the repulsion between the cations as discussed earlier for $SrCl_2$. Ionic conductivity and Raman scattering data suggest that the coordination of the metal ions is becoming more stable through the series, leading to a liquid structure which resembles a loose network of Cl-sharing octahedra.

$DyCl_3$ and YCl_3 crystals are structurally isomorphous and melt with similarly low values of ΔS_{m} and $\Delta V_{\mathrm{m}}/V_{\mathrm{l}}$. In a neutron diffraction experiment on molten YCl_3[254] a well defined FSDP was seen at $k = 0.95$ Å^{-1}, giving unambiguous evidence of IRO. The average coordination number of the metal ions is 5.9, which confirms the Raman scattering findings of a rather long-lived octahedral coordination of the metal ions.[255] The octahedral network must be relatively loose on a time scale longer than the period of the breathing mode of the octahedron at $\nu \approx 260$ cm^{-1}, to be compatible with the value of the ionic conductivity of molten YCl_3 given in Table 8.1.

8.5.2 *Ionic-to-molecular melting in $AlCl_3$ and $FeCl_3$*

YCl_3 is structurally isomorphous to $AlCl_3$ in the crystal phase. Figure 8.6 shows a layer in the $AlCl_3$ and $FeCl_3$ crystal structures, formed by a metal-ion

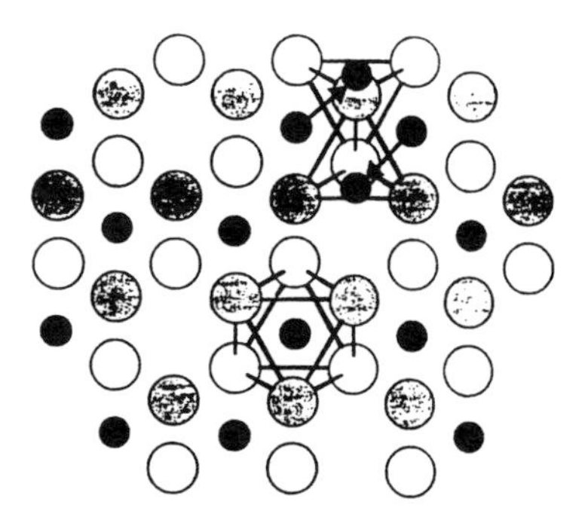

Fig. 8.6. Schematic illustration of a layer in the $AlCl_3$ crystal structure and of melting in YCl_3 and $AlCl_3$.

plane (black spheres) sandwiched between two layers of chlorines (the top plane appears in grey and the bottom plane in white). The lower cluster shows the octahedral coordination of the Y, Al and Fe ions in the crystal, which is basically preserved in YCl_3 on melting. The upper cluster illustrates the cooperative mechanism of metal–ion displacements by which the Al_2Cl_6 and Fe_2Cl_6 molecules can form on melting, each dimer being in the shape of two tetrahedra sharing an edge. In $AlBr_3$ such an arrangement of Al ions in tetrahedral sites already exists in the crystal. The melting of $AlCl_3$ and $FeCl_3$ also involves expansion of the chlorine packing in view of the large value of $\Delta V_m/V_l$.

Joint X-ray and neutron diffraction experiments have been carried out on molten $FeCl_3$.[256] The neutron pattern is very close to the Bhatia–Thornton $S_{NN}(k)$, since the scattering amplitudes of Fe and Cl are almost identical. It exhibits three peaks which, after scaling with an average first-neighbor distance $d = 2.28$ Å, lie at $kd \approx 2.2$, 4.7 and 8.4. These values are typical for the FSDP, the Coulomb peak and the topological (excluded-volume) peak, respectively.[257] The pair function $g(r)$ yields an average first-neighbour coordination number of ≈ 3.6 about the metal ions, definitely ruling out the sixfold coordination found in the crystal.

8.5.3 *Liquid haloaluminates*

In $AlCl_3$ and $AlBr_3$, while the pure melt is a molecular liquid, molten-salt behaviour emerges on mixing with alkali halides. Complex anions are formed with the alkalis playing the role of counterions. Thus, starting from neutral Al_2Cl_6 dimers in the $AlCl_3$ liquid, the $(Al_2Cl_7)^-$ anion in the shape of two tetrahedra sharing a corner has been identified in mixtures with alkali chlorides.[258] This anion is ultimately replaced by $(AlCl_4)^-$ anions at 1:1 stoichiometry.

The fluoroaluminates behave quite differently. The Na_3AlF_4 compound, known as cryolite, presents special interest because of its role in the industrial Hall–Héroult process for the electrodeposition of Al metal from alumina.[259] The Raman spectra of molten $(AlF_3)_c \cdot (NaF)_{1-c}$ and other Al-alkali fluoride mixtures give evidence for a gradual conversion of $(AlF_4)^-$ into $(AlF_5)^{2-}$ and $(AlF_6)^{3-}$ as the solution becomes more basic with c decreasing below 0.5.[260]

8.5.4 *Molecular-to-molecular melting in $GaCl_3$ and $SbCl_3$*

For other trihalides, such as $GaCl_3$ and $SbCl_3$ in the bottom rows of Table 8.1, molecular units can be recognised as constituents in the crystal structure.

Crystalline $GaCl_3$ can be viewed as composed of Ga_2Cl_6 dimers. The crystal structure of $SbCl_3$ is instead built by packing chains of monomers in the shape of trigonal pyramids with metal ions at the apices. The stable molecular units in the vapour phase are the Ga_2Cl_6 dimer and the $SbCl_3$ monomer.

The liquid structure of $SbCl_3$ at 80°C has been studied by a combination of X-ray and neutron diffraction.[261] It can be described as arising from separate monomeric units with strong intermolecular correlations. Each metal ion has three additional chlorine neighbours from other molecules: such a strongly distorted octahedral arrangement could result from stacking the monomers in chains like umbrellas, the dipole axes of molecules within a chain being strongly correlated over at least one or two molecular diameters.

The neutron diffraction patterns measured for molten $AlBr_3$, $GaBr_3$ and GaI_3 show three peaks at approximately 1.0, 1.9 and 3.4 Å^{-1}, which represent the FSDP, the Coulomb peak and the excluded-volume peak.[262] The corresponding pair distribution functions exhibit a very well defined coordination shell of first neighbours, with coordination number 4.0 ± 0.2 for $AlBr_3$ and $GaBr_3$ and 3.75 ± 0.2 for GaI_3. The intermolecular correlations between halogens are quite significant, the corresponding coordination number being in the range typical of a random close-packing in the liquid state.

8.6 Transport and Dynamics in Molten Salts

In a broad sense, the transport properties and the microscopic dynamical behaviour of molten salts are correlated with the types of liquid structure that we have outlined above. Some unambiguous examples of these correlations will be exposed in this section.

8.6.1 *Ionic transport*

We return to Fig. 8.4 and notice that the rather deep main minimum in $g_{\mathrm{NaCl}}(r)$ implies a moderate rate of ionic exchange between the first-neighbour shell and the surrounding liquid.

The persistence times of local structure in molten alkali halides near freezing are of the order of several picoseconds. These correspond to values of ionic conductivity and shear viscosity as given for NaCl in Table 8.1. The self-diffusion coefficients of the two ions in molten NaCl have similar values[263] ($D_{\mathrm{Na}} = 1.7$ and $D_{\mathrm{Cl}} = 1.3 \times 10^{-4}\ \mathrm{cm^2 \cdot s^{-1}}$), in spite of the differences in atomic masses and ionic radii. This is consistent with the similarity of $g_{\mathrm{NaNa}}(r)$ and $g_{\mathrm{ClCl}}(r)$ in Fig. 8.4, which implies similarity in restoring forces and residence times for the two types of ions.

On the other hand, for the melts of noble-metal halides computer simulation studies[264] and theoretical calculations based on the Zwanzig model[265] (see Sec. 5.6.1) indicate a large difference in the mobilities of cations and anions, by up to a factor of order ten in molten AgI (see Table 8.2). This difference has been related to the observed liquid structure as discussed in Sec. 8.4.2, which leads to major differences in the structural back-scattering of cations by cations and of halogens by halogens. On the quantitative side, however, both simulations based on simple pair potentials and theoretical models of transport coefficients are subject to large uncertainties.

Table 8.2. Calculated diffusion coefficients in superionic-conductor melts (in units of 10^{-5} cm$^2 \cdot$ s^{-1}, compared with the results of molecular dynamics (MD).

Salt	T (K)	D_+	$(D_+)_{\mathrm{MD}}$	D_-	$(D_-)_{\mathrm{MD}}$	D_+/D_-	$(D_+/D_-)_{\mathrm{MD}}$
AgI	873	3.6	3.8	0.34	0.3	11.0	12.7
CuCl	773	20.0	10.0	4.3	2.5	4.8	4.0
CuBr	880	13.0	10.0	2.3	2.7	5.6	3.9
CuI	923	18.0	8.8	2.2	1.3	8.1	6.8

A further example of the deep connections which exist between structure and transport in molten salts is provided by the Chemla effect.[266] This was first discovered in experiments on molten (Li, K)Br mixtures, showing that the mobility of K overtakes that of Li with increasing content of KBr. The transport behaviour of the two cations is related to the structural features of the cation-anion pair distribution functions, combined with the volume dilation which accompanies the increase of KBr content. The local liquid structure around Li forces the ion to spend a relatively long residence time in oscillations at a fourfold-type site before being able to diffuse out, and this "trapping" is strengthened with decreasing number density and temperature.

With regard to the relation between ionic diffusivities and ionic conductivity, deviations from the Nernst–Einstein form by up to 20% have been reported for alkali halides both from experiment and from computer simulations.[267] This relation can be written as

$$\sigma = \left(\frac{ne^2}{k_{\mathrm{B}}T}\right)(D_+ + D_-)(1 - \Delta)\,, \tag{8.14}$$

where the factor $(1 - \Delta)$ reflects cross-correlations between the ionic motions. Positive values of Δ, as observed, reflect a tendency of oppositely charged ions to diffuse together.

In molten $ZnCl_2$ near freezing, Sjöblom and Behn[268] report diffusion coefficiencts of about 1.5×10^{-7} cm^2/s for both ions. From the Nernst–Einstein relation and the measured ionic conductivity (see Table 8.1), one would predict diffusivities lower than this by several orders of magnitude. It is evident that motions of neutral units are mainly responsible for diffusion in this melt.

8.6.2 *Viscosity*

As discussed by Hirschfelder *et al.*,[269] dimensional analysis suffices to suggest scaling laws for transport coefficients through microscopic interaction parameters. Examples have been given for shear viscosity and thermal conductivity in Chaps. 6 and 7. In the same spirit Abe and Nagashima[270] have shown that the correlation

$$\frac{1}{\eta^*} = -5.960 + 23.37 V^* T^{*1/2} \tag{8.15}$$

holds for the reduced shear viscosity $\eta^* = \eta\sigma^2/(m\varepsilon)^{1/2}$ in molten alkali halides as a function of $V^* = V/N\sigma^3$ and $T^* = k_B T/\varepsilon$, with σ and ε being parameters

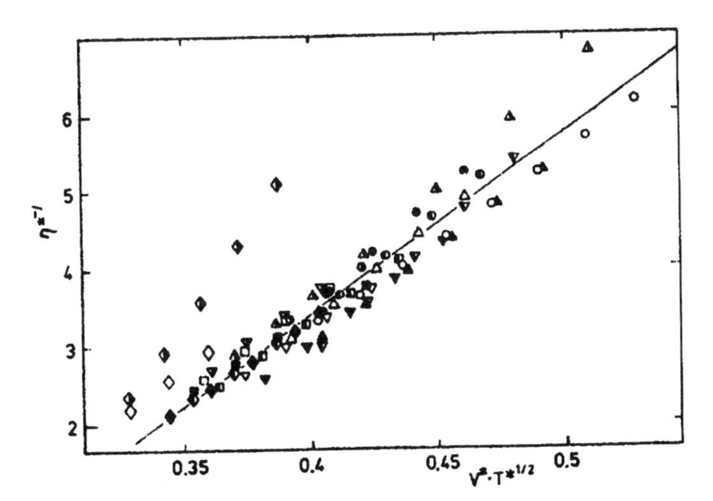

Fig. 8.7. Inverse of scaled shear viscosity against $V^*T^{*1/2}$ in 18 alkali halide melts. (Redrawn from Abe and Nagashima, Ref. 270.)

of the interionic potential and m the molecular weight. The relation (8.15) is shown in Fig. 8.7 together with data on 18 alkali halides, with appreciable deviations being evident for two of them (CsF and CsI). Other systems discussed by the same authors include alkali nitrates, liquid metals, and hydrocarbons.

Precise viscosity measurements on pure and mixed ionic melts (alkali halides and nitrates) including glass-formers have revealed a universal behaviour in the form of a modified Arrhenius law.[271] Figure 8.8 reports the correlation

$$\eta \propto \frac{T}{T_0} \exp\left(\frac{qT_0}{T}\right), \tag{8.16}$$

where $q = 5.9 \pm 0.1$ and T_0 is a suitably chosen characteristic temperature.

Figure 8.9 reports from the work of Voronel *et al.*[272] a fractional-power relation between ionic conductivity and shear viscosity of pure and mixed ionic melts,

$$\sigma T \propto \left(\frac{T}{\eta}\right)^{\alpha}, \tag{8.17}$$

with $\alpha = 0.8 \pm 0.1$. This relation holds over nine orders of magnitude for η in the glass-forming $Ca_2K_3(NO_3)_7$ compound. In a range of temperature where

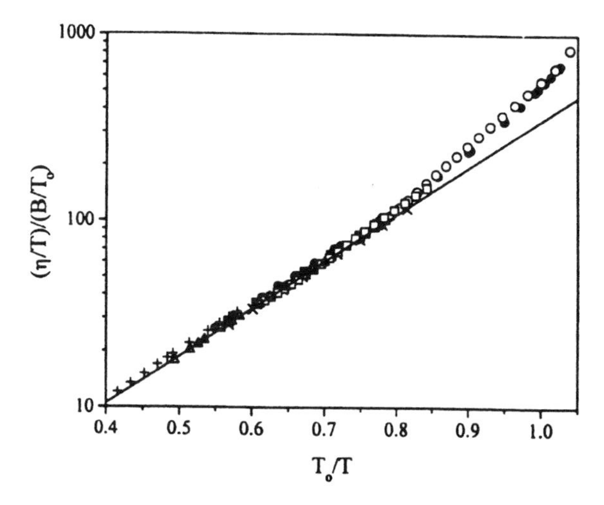

Fig. 8.8. Viscosity of seven alkali halides and nitrates against inverse reduced temperature. (Redrawn from Voronel *et al.*, Ref. 271.)

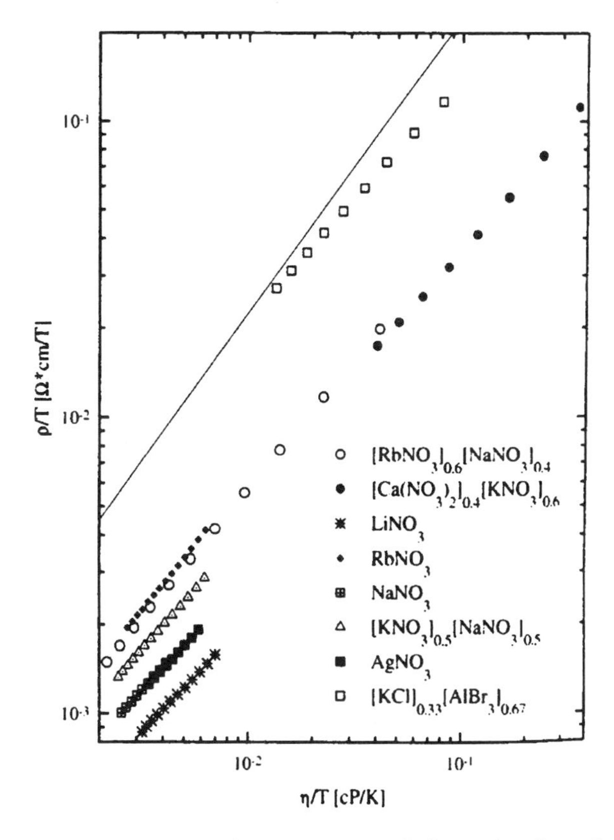

Fig. 8.9. Relation between electrical resistivity ρ and shear viscosity η in eight nitrates and halides. The solid line corresponds to the slope $m = 1$. (Redrawn from Voronel *et al.*, Ref. 272.)

both σ and η^{-1} can be represented by an Arrhenius law, the relation (8.17) implies a systematic difference in the respective activation energies.

8.6.3 *Dynamics of density fluctuations*

In the simplest model of a molten alkali halide, one would expect that the dynamics of density fluctuations be described by a "superposition" of oscillations of the total mass density, leading to sound waves at long wavelengths as described for monatomic fluids in Chap. 6, and of oscillations of the charge density as a counterpart of the propagating plasma mode discussed in

Sec. 8.1.3 for the strongly coupled OCP. This viewpoint, though quantitatively oversimplified, is essentially correct. Thus, Coulomb ordering is responsible for a propagating collective excitation of charge fluctuations. We refer to a specialised review[233] for a discussion of detailed analyses of infrared reflectivity, neutron inelastic scattering and Raman scattering data.

The Raman scattering technique becomes especially useful in network-like melts such as $ZnCl_2$ and in molecular-like liquids such as $AlCl_3$. Experiment shows that the density of vibrational states in these liquids is strongly structured: it mainly reflects the vibrational motions of the basic structural unit and these are only moderately affected by the network-induced (or IRO-induced) coupling. An extensive collection of Raman frequency data for ionic systems has been given by Brooker and Papatheodorou.[273]

8.7 Chemical Short-Range Order in Liquid Alloys

Fully ionised salts with a large band gap, like the alkali halides, remain ionic across melting. At the opposite extreme, melting of covalent semiconductors such as Ge and InSb involves a collapse of the covalent structure, which is directly revealed by an increase of coordination from 4 to values in the range 6–8 and by a sharp increase in electrical conductivity to an essentially metallic type. Between these extremes a number of systems have been identified which show a variety of intermediate electronic behaviour in the liquid phase.

We briefly refer below to alloys from metallic elements with a relatively large difference in electronegativity, which form intermetallic compounds at certain "stoichiometric" compositions in the solid state. Examples are Cs–Au[274] or Li–Pb[275] and other alloys of alkalis with Pb or Sn.[276] The melts of these alloys near certain compositions show a marked preference for unlike first neighbours in the local structure, as well as a maximum in the electrical resistivity and a minimum in its temperature coefficient. We limit ourselves to an outline of the chemical order near "stoichiometry" and refer to specialised reviews[233] for the parallels that can be made between these alloys away from stoichiometry and the solutions of metals in molten salts.

8.7.1 *The CsAu compound*

The stoichiometric CsAu compound crystallises in the CsCl-type structure and is a strongly polar semiconductor with an optical band gap of 2.6 eV at room temperature. Its electrical conductivity drops on melting to a value which is comparable to molten salts. Electromigration experiments give evidence that Cs migrates to the cathode and Au to the anode, one Cs^+ and one Au^- being transported per elementary charge to the electrodes.

A neutron diffraction study of the liquid structure of the Cs–Au alloy[277] shows a structure in the neutron structure factor at $k = 1.2$ Å^{-1}, which is interpreted as the "Coulomb prepeak" characteristic of chemical order. After Fourier transform of these data, the Cs–Au first neighbour distance at 3.6 Å can be followed up to 80% Cs, while the Cs–Cs distance at 5.3 Å characteristic of the pure Cs metal start emerging at 70% Cs.

8.7.2 *Other alkali-based alloys with chemical short-range order*

Interspecies ordering as shown by the Cs–Au system has been reported for a number of other alkali-based alloys, the alloying partners being elements of group III, IV or V. The formation of chemical short-range order at certain compositions is signalled by anomalies in electronic properties such as the electrical resistivity and the magnetic susceptibility, which reflect a minimum in the electron density of states at the Fermi level if not the opening of a gap due to full charge transfer. Three different kinds of compound formation can be identified: (i) compound formation near the electronic octet composition A_4B as in Li–Pb or Li–Sn; (ii) compound formation near the equimolar composition AB, as in K–Pb or Rb–Pb; and (iii) compound formation near both these compositions, as in Li–Si, Li–Ge or Na–Sn. The data show increasing stability of the octet composition through the sequence Si, Ge, Sn and Pb, and decreasing stability through the sequence from Li to Cs.

A neutron diffraction measurement of the Bhatia–Thornton concentration–concentration structure factor in Li_4Pb has shown chemical order extending over a range of about 20 Å in the corresponding $g_{cc}(r)$ distribution function.[275]

With regard to alkali-group IV alloys in the second and third classes mentioned above, such as K–Pb or Na–Sn, Meijer *et al.*[276] have proposed a model for order at equimolar composition which invokes formation of essentially tetrahedral Pb_4 or Sn_4 polyanions. Such tetrahedral "Zintl ions" are seen in the crystal structure of the equiatomic compound. In such a tetrahedral cluster the p-type electron states of Pb, say, would be split into bonding and antibonding states and the former could be filled by electron transfer from the alkali atoms.

The presence of polyanions in Zintl alloys also has dynamical consequences. A striking case is NaSn, in which the Sn_4 polyanions are observed to undergo jump reorientations and thereby to enhance the diffusivity of the Na cations by a paddle-wheel mechanism.[278] These two types of disorder appear simultaneously as the melting point is approached.

Chapter 9

Bonds, Rings and Chains

9.1 Outline

This chapter deals with molecular liquids, including polymers and liquid crystals. We begin with liquid nitrogen, which under normal conditions consists entirely of strongly bonded N_2 molecules. Then, in turn, we shall consider carbon, selenium and sulphur. One unifying theme for these three liquids (and possibly also embracing water) is the concept of liquid–liquid transitions occuring in the phase diagram. Studies of the structure of liquid boron will also be described, as an instance of very unusual covalent bonding which transforms into a metallic state on melting. The precise formulation of liquid structure in assemblies of simple molecular units will then be set out and illustrated, with special focus on the structure of water.

Polymers and liquid crystals follow in turn, with main attention to their structural description as introductory to the overview on their flow properties that will be given in Chap. 11. In these two areas, the interaction between first-principles statistical mechanics and technical applications is of considerable importance.

Following this brief outline, we turn first to summarise some properties, with especial attention to phase boundaries, in a number of elemental molecular assemblies.

9.2 Elemental Molecular Liquids

9.2.1 *Nitrogen*

Let us begin with liquid nitrogen. As already remarked, under normal conditions of temperature and pressure it consists of an assembly of triple-bonded, non-spherical N_2 molecules. The new element of structural order in the condensed phases arises from the orientational correlations between the molecules, and the quantitative structural description of such liquids in terms of these correlations will be given in Sec. 9.3. Liquid O_2 is another example, though this differs from nitrogen in that the ground state of the molecule in free space has two unpaired spins and the molecule carries a magnetic moment. The phase diagram of O_2 has solid-state regions where cooperative magnetism (e.g. order in the molecular spins taking up an antiferromagnetic arrangement) exists.

Homonuclear diatomic molecules might be thought to form some of the simplest condensed molecular systems. Their phase diagrams can, however, be quite complicated. In the case of nitrogen, nine solid phases have been

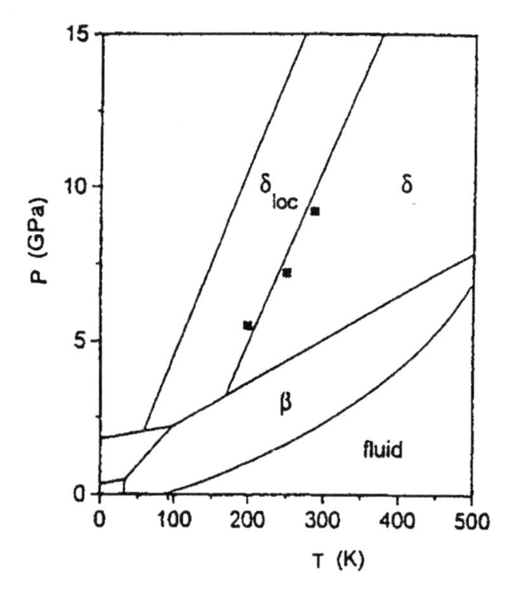

Fig. 9.1. Phase diagram of nitrogen, showing plastic β-crystal phase above melting curve. (Redrawn from Bini *et al.*, Ref. 279.)

reported for temperatures in the range from 4 to 300 K and pressures up to 130 GPa. At standard pressure the α-N_2 phase, where the molecules are translationally and orientationally ordered on the sites of a cubic lattice, is stable up to 35.6 K. In the β-N_2 phase, which is stable from 35.6 K up to the melting point at 63.1 K, the lattice is instead hexagonal-close-packed and there is a high degree of molecular disorder. As is seen from the phase diagram[279] shown in Fig. 9.1, this disordered solid phase is in equilibrium with the liquid phase on the melting curve as a function of pressure. The solid phase in equilibrium with the liquid is a "plastic crystal"[280]: while it has long-range crystalline order, it only has short-range order in the orientational correlations between the molecules.

As discussed by Tozzini *et al.*,[281] the relation between the melting curve and the transition boundary from conventional crystal to plastic crystal can be treated in a model going back to Pople and Karasz.[282] These authors generalized the melting phenomenology of Lennard–Jones and Devonshire[283] to include orientational disordering together with positional melting (see also Appendix 9.1). The new feature is the height of a potential barrier between different allowed orientations of the diatomic molecules on crystal lattics sites. The height of this barrier links the melting curve and the plastic crystal boundary.

9.2.2 *Phase diagram of carbon: Especially liquid–liquid transformation*

Ferraz and March[284] (see also Ghosli and Ree[285]) suggested the existence of a liquid–liquid phase transition (LLPT) in the phase diagram of carbon. The first quantitative treatment of this transition was a semi-empirical equation-of-state modelling of carbon by van Thiel and Ree.[286] However, it was only later that the first experimental evidence, albeit indirect, for the LLPT in carbon became available in work by Togaya.[287] This study on the melting of graphite (see also Fig. 9.2) strongly suggests that the slope of the pressure-temperature melting line has a discontinuity at the temperature maximum and hence at least three stable phases coexist at this point. As stressd by Ghosli and Ree, the most plausible interpretation is that this corresponds to a triple point and the carbon phase diagram exhibits a LLPT.

Support for this conclusion is afforded by the atomistic computer simulations of Ghosli and Ree,[285] which were performed using a bond-order potential

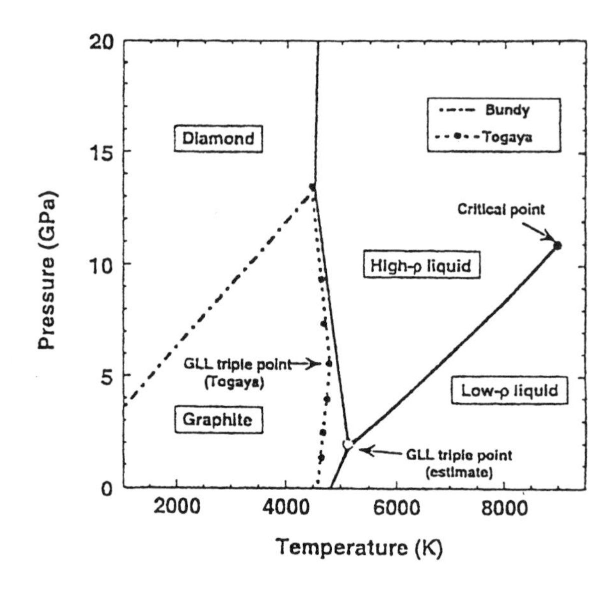

Fig. 9.2. Phase diagram of carbon, showing critical point at end of liquid–liquid "phase boundary". (Redrawn from Ghosli and Ree, Ref. 285.)

developed in earlier work by Brenner. The potential energy function takes the form

$$\Phi = \sum_{\{\text{bonds}\}} [\phi_{\mathrm{R}}(r_{ij}) - b_{ij}\phi_{\mathrm{A}}(r_{ij})] \, . \tag{9.1}$$

In Eq. (9.1), ϕ_{R} and ϕ_{A} consist of exponential functions representing the repulsive and the attractive terms in the bond energy, respectively. The symbol $\{$bonds$\}$ denotes the set of all bonds, but although a sum over bond energies Φ, it is to be stressed, is not built from pair potentials. The bond-order factor b_{ij} is many-body in nature and depends on both bond and torsional angles, bond lengths, and atomic coordination in the neighbourhood of the bond.

While there is semiquantitative accord between the results of Ghosli and Ree from their atomistic simulations and the melting line of graphite as measured by Togaya, the essential point to be emphasised here is the "phase boundary" found between high-density and low-density liquids, ending in a critical point (compare Fig. 9.2). The critical constants from the model are estimated as $T_{\mathrm{c}} = 8800$ K, $p_{\mathrm{c}} = 10.6$ GPa and $V_{\mathrm{c}} = 8.7$ Å^3/atom.

Ferraz and March[284] had already proposed a low-density liquid phase in carbon having *sp* hybridisation. The Ghosli–Ree study leads to considerable insight into the atomic structure and permits them to conclude that the high-density liquid is characterized by sp^3 hybridisation while the low-density liquid has indeed *sp* hybridisation dominantly. Ghosli and Ree stress that refinements of the bond-order potential (9.1) will be required to gain fully quantitative agreement with experiment, and especially note the need to incorporate long-range van der Waals forces.

9.2.3 *Selenium and sulphur: Especially liquid–liquid transitions*

In this section, a brief discussion will be devoted to related behaviour of molten Se and S to the LLPT for carbon discussed in some detail above. Emphasis will be given to the connection between pressure-temperature diagrams of these two elements and experimental observations of volumetric and resistance anomalies in the liquid state. The reader requiring more details is referred to the report of Brazhkin *et al.*[288] and references therein.

Discussing Se first, the short-range order in the molten state is characterised by two-fold coordination, just as for the crystalline phase. Gross modifications in the structure of Se can be effected by changing the temperature, and in particular the mean molecular chain length reduces substantially with increasing temperature. In the vicinity of the liquid–vapour critical point, having critical constants $T_c = 1630°C$, $p_c = 380$ bars and $\rho_c = 8.7$ g/cm^3, the average chain molecule consists of only about ten atoms.[289]

Experimental studies have shown that the liquid structure factor $S(k)$ of Se at pressures of 4.4 and 8.4 GPa differs markedly from that observed at normal pressure. It is thus feasible to produce a major structural transformation of molten Se by compression.[290] Also of significance in the context of a LLPT is that electrical as well as volumetric anomalies have been observed in molten Se in the pressure range from 1 to 10 GPa.[288] These observations are consistent with a structural transformation in liquid Se, as shown in the p-T diagram reported in Fig. 9.3.

Turning to molten sulphur, once again structural changes have been induced by increasing temperature at atmospheric pressure. Near the melting point at 390 K, the short-range order in the melt resembles that in the solid state and is based on the same S_8 rings, which are described as structural units

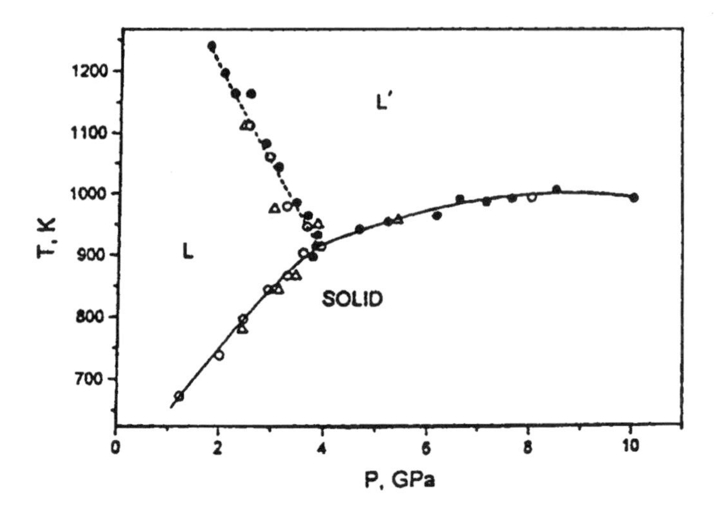

Fig. 9.3. Phase diagram of Se, showing transition between structurally different liquid states L and L'; resistance anomalies are indicated by solid circles. (Redrawn from Brazhkin *et al.*, Ref. 288.)

with an average bond length of 2.05 Å and a bond angle of $\approx 108°$.[291] The melting temperature is about 115°C and sulphur forms a light-yellow liquid of relatively low viscosity. Around 160°C, the viscosity increases markedly with temperature. Viscosity can be (very approximately) related to the liquid structure factor *via* intermolecular forces.

Experiments by Brazhkin *et al.*[292] have clarified the situation regarding various physical properties of molten sulphur in the pressure range between 5 and 12.5 GPa. These workers found, in particular, metallisation in the melt, this taking place near the melting point at 12 GPa. In their report Brazhkin *et al.*[288] give a p-T diagram of solid and liquid sulphur in their Fig. 3, to which the reader interested in further details is referred.

9.2.4 *Structure of liquid boron*

Boron and boron-rich borides have numerous technical applications, especially when a hard and light material is needed. The crystalline structures[293] are dominated by the B_{12} regular icosahedral unit. In the α-rhombohedral form the icosahedra are connected *via* a combination of two-centre and three-centre bonds, giving a mean first-neighbour coordination number of 6.5. At about

1200°C the α-phase transforms into the β-rhombohedral from, in which the icosahedra are surrounded by pentagonal pyramids yielding a unit cell of 105 atoms with mean coordination number of 6.6. These structures are indicative of the unusual character of covalent bonding in pure B, leading to a wide range of physical and chemical properties in boron-rich solids.

Boron melts at about 2360 K through an insulator-to-metal transition accompanied by a volume expansion of about 5%. The liquid is extremely reactive to any container. Therefore, a determination of the liquid structure has required special diffraction techniques,[294] using synchrotron X-rays on levitated samples. The data on $g(r)$ yield a coordination of about 6 and show very little change in bond length on melting, but a broadening of the first coordination shell. They are in reasonable agreement with *ab initio* calculations by molecular dynamics.

Figure 9.4 compares the pair distribution function measured in this study[294] on liquid boron at 2600 K with those of the crystalline phases and of the amorphous state. These data have a bearing on the question of whether the B_{12} iscosahedral units can survive melting and the simultaneous transition to a metallic state. In the α-rhombohedral form the peak at 2.99 Å is associated

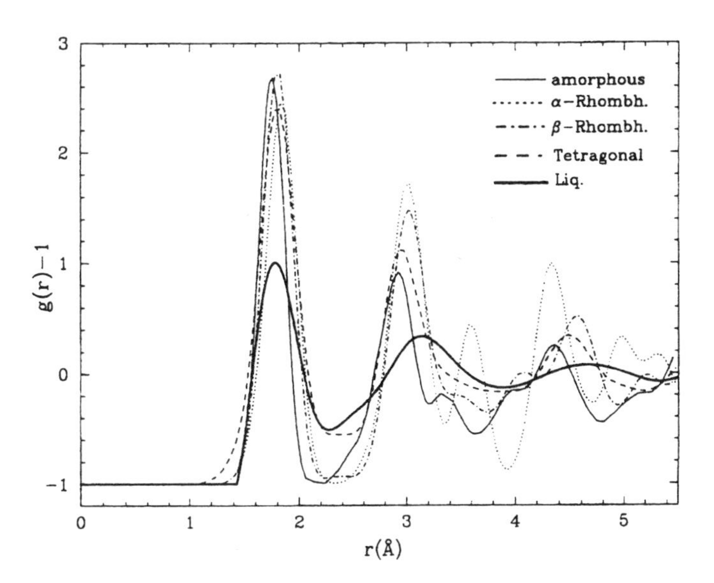

Fig. 9.4. Comparison of the pair distribution function of liquid boron at 2600 K with those of the amorphous and crystalline states. (Redrawn from Krishnan *et al.*, Ref. 294.)

with the icosahedra and that at 3.58 with the bonding between them. The significant shifts and broadening of these structures on melting, which are seen in Fig. 9.4, suggest that in the melt the intermediate-range structure may be rather different.

9.3 Orientational Pair Correlation Function from Diffraction Experiments

The work of Soper[295] has highlighted a number of important points about studies of the structure of simple molecular liquids by diffraction techniques. The structure can be defined by means of the orientational pair correlation function (OPCF) $g(\mathbf{r}, \omega_1, \omega_2)$. This determines the probability of a molecule being found at position $\mathbf{r}$ with orientation ω_2, given that there is a molecule with orientation ω_1 at the origin. Such a description was known, for example, to Blum and Torruella[296] (see also the book of Gray and Gubbins[297]), but has seldom been used.

One reason for this is due to the fact that g as defined above appeared to be a function of nine variables, though only six of them are in fact independent. In the case of two molecules at an arbitrary relative orientation, each molecule has orientation specified by three Euler angles. However, to usefully visualise the orientational correlations one has to orient the laboratory Cartesian axes

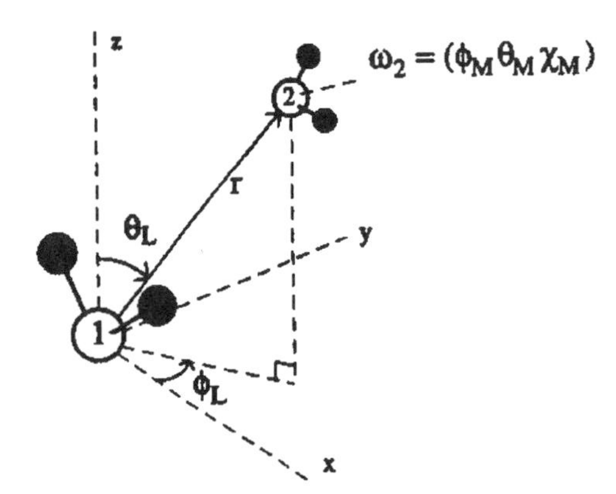

Fig. 9.5. Diagram showing the axes and coordinates of two water molecules. Molecule 1 is held at the origin of the laboratory frame, with the H-O-H plane in the $x-z$ plane and the dipole moment (bisecting the H-O-H angle) along the z axis.

so that the axis of molecule 1 coincides with one of them: i.e. $\omega_1 = 0$. This reduces the OPCF to a function of six independent variables. Such a procedure is correct in an isotropic homogeneous liquid, but would not apply, for instance, in the presence of a surface.

The remaining variables are $\mathbf{r} = (r, \theta_\mathrm{L}, \phi_\mathrm{L})$ for the position of the second molecule relative to the first, and $\omega_2 = (\phi_\mathrm{m}, \theta_\mathrm{m}, \chi_\mathrm{m})$ for the Euler angles used to define the relative orientation of the two molecules (see Fig. 9.5 for the case of two water molecules). If the first molecule is linear there is no dependence on ϕ_L, and if the second molecule is also linear then $\chi_\mathrm{m} = 0$.

9.3.1 *Use of generalised rotation matrices*

In, for example, the book of Gray and Gubbins,[297] the spherical harmonic expansion of the OPCF has been exploited. This expansion may be written as

$$g(\mathbf{r}, \omega_1, \omega_2) = \sum_{(l_1 l_2 l)} \sum_{(m_1 m_2 m)} \sum_{(n_1 n_2)} g(l_1 l_2 l, n_1 n_2; r) C(l_1 l_2 l, m_1 m_2 m) \times D^{l_1}_{m_1 n_1}(\omega_1)^* D^{l_2}_{m_2 n_2}(\omega_2)^* D^{l}_{m0}(\omega) . \tag{9.2}$$

In this equation $D^l_{mn}(\omega)$ are the generalised rotation matrices, $C(l_1 l_2 l, m_1 m_2 m)$ are the Clebsch–Gordan coefficients while $g(l_1 l_2 l, n_1 n_2; r)$ are the coefficients of the series that need to be found.

As Soper points out, for the case of water it turns out that an accurate representation of the orientational structure results with $l_\mathrm{max} = 4$, involving 158 coefficients. With, say 100 values of r for each coefficient, this still amounts to a reduction in storage by more than six orders of magnitude. However, it must be noted that as the orientational correlations become stronger, so the number of coefficients increases. One satisfactory feature of such an expansion is that it can embody any molecular symmetry present: any coefficients not obeying molecular symmetry rules are to be put to zero. Also it is worthy of note that Eq. (9.2) can be applied to any pair of molecules, regardless of whether or not they are of the same type.

Given the expansion (9.2) as the underlying equation to be solved for any particular liquid, the problem of structural refinement reduces to that of finding the coefficients in this expansion. This needs to be carried out in such a manner that available data are embodied as satisfactorily as possible. Obvious data for liquids come from diffraction experiments, but ideally there should be attention given to others: say constraints in the orientational coefficients and on the overlap of molecules, and, for instance in the case of hydrogen-bonded liquids, on which atom in the first molecule is correlated with particular sites on the second molecule.

9.3.2 *Example of orientational structure in water*

Liquid H_2O has been widely studied with diffraction techniques and a complete set of site–site radial distribution functions are available from experiment over a range of thermodynamic states[298] (see Fig. 9.6 for ambient water). Soper[295] has discussed this case in some detail.

Figure 9.7 shows an example of four spherical harmonic coefficients derived in an analysis of diffraction data. Calculations using three different reference potentials lead to similar distributions of molecular centres, as given by $g(000, 00; r)$. It is encouraging that they produce quite similar results for the higher-order coefficients also. Thus, although the OPCF is sensitive to the model interactions, a range of potentials can be found which produce OPCF's

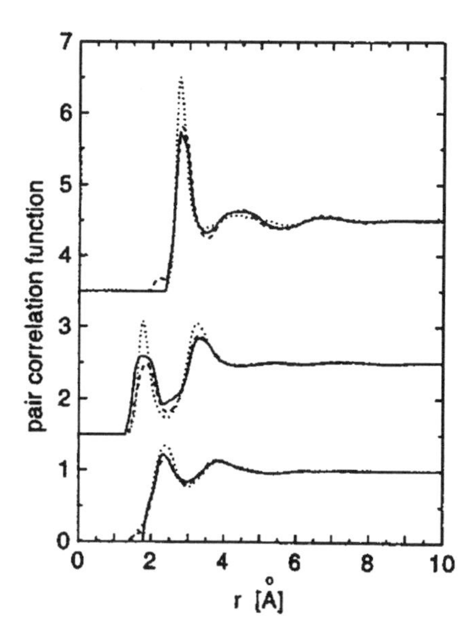

Fig. 9.6. The three site–site radial correlation functions in ambient water (HH, OH and OO from bottom to top). The results from two different analyses of diffraction data are compared with simulation results (•). (Redrawn from Soper *et al.*, Ref. 298.)

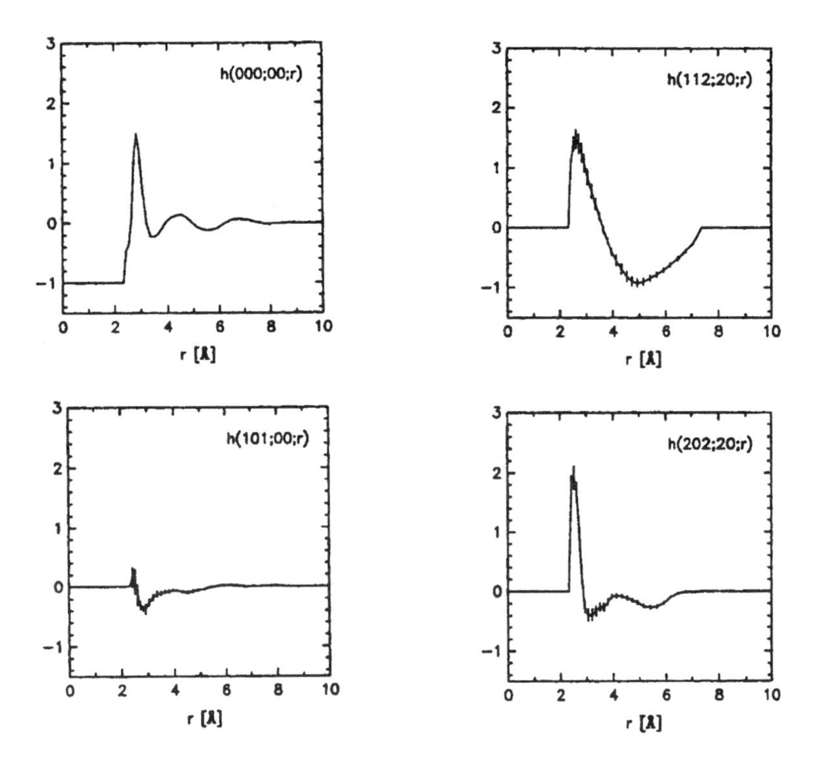

Fig. 9.7. Four spherical harmonic coefficients derived from the analysis of water diffraction data. (Redrawn from Soper, Ref. 295.)

which are sufficiently similar to be considered all consistent with available experimental data.

More generally, there has been continued interest in the structure and properties of water over a wide range of the equation of state, including the deeply supercooled region where the coexistence of high- and low-density fluids has been proposed[299] as well as the supercritical regime which is of great importance in extraction and reaction process technology.[300] Badyal *et al.*[301] have carried out measurements of the X-ray structure factor of water in ambient conditions with synchrotron radiation and compared the results with those predicted from neutron partial structure factors describing the nuclear positions, with different assumptions for the electron distribution. They estimate a charge transfer of about $0.5e$ from each hydrogen to the oxygen on the same molecule, implying an effective dipole moment of 2.9 ± 0.6 Debye.

9.4 Polymers

So far the building blocks of the liquids that we have considered have mostly been single atoms as in argon or small molecules as in water. But when we turn to polymers, the building blocks become huge — a polymer chain being formed by linear repetition of a group of atoms, termed a monomer. For instance, polyethylene is a chain of CH_2 groups, i.e. $[CH_2]_N$ with N up to 10^5, terminated at both ends by a CH_3 group. Two other polymers that are present in everyday life are polystyrene and polyvinylchloride, described by $[CH_2 - \bar{C}H]_N$ where the bar over the second carbon denotes that a benzene side ring or a chlorine atom is attached to it. In fact, the number N of monomer groups does not have a fixed value for a given species and can vary in ways that depend on the conditions under which the material has been prepared.

There are three main physical states for dense assemblies of flexible linear polymers: (i) a liquid of entangled chains at high temperature, (ii) a glass obtained most often by cooling the melt (this is, for instance, the case for the polystyrene that is used in plastic packing), and (iii) under favourable circumstances a crystal — although starting from the melt one usually obtains only partial crystallisation, i.e. a composite formed by crystalline parts inside an amorphous matrix. In addition, the macromolecules may be dispersed in certain liquids to form a solution: at very low concentration they are essentially independent (though in strong interaction with the solvent molecules), but with increasing concentration they soon start to get tangled with each other. Reference to a number of introductory books on these topics may be made.[302]

9.4.1 *The isolated polymer molecule*

A macromolecule does not have a fixed configuration: the majority of linear polymers in solution or in the melt are quite flexible. Returning to polyethylene for an illustration, the whole chain, formed by maybe 25,000 CH_2 groups, can be likened to a long thin thread which twists and turns in irregular fashion, going from pieces where it is more or less straight to parts where it might form a compact ball.

The chain lengths are therefore statistically distributed, the most probable end-to-end distance being $a\sqrt{N}$ for an ideally flexible chain of N monomers each of length a. This result is obtained by assuming that there is no correlation between the directions that different bonds can take and that all directions have the same probability: in this case, the configuration of the polymer chain can be mapped into a random walk on a lattice. Inclusion of short-range interactions between bonds simply replaces a by an effective bond length and preserves the $\sqrt{N}$ dependence on the number of monomers.

The distribution of segments inside the chain can be described by means of the segment pair distribution function, defined as

$$g(r) = \frac{1}{N}\sum_{n=1}^{N} g_n(r) = \frac{1}{N}\sum_{n,m=1}^{N} \langle \delta(\mathbf{r} - (\mathbf{R}_n - \mathbf{R}_m)) \rangle , \tag{9.3}$$

where $\mathbf{R}_n$ are the position vectors of the segments and the brackets denote a statistical average. The Fourier transform of $g(r)$ is accessible to experiments such as light scattering or small-angle X-ray scattering. These experiments yield the radius of gyration R_g defined by

$$R_\mathrm{g}^2 = \frac{1}{2N^2}\sum_{n,m=1}^{N} \langle (\mathbf{R}_n - \mathbf{R}_m)^2 \rangle . \tag{9.4}$$

This quantity provides a precise measure of the average distance between the segments and the centre of mass of the chain. For an ideal chain one finds $\mathbf{R}_\mathrm{g} = a\sqrt{N/6}$. On account of non-ideality, and especially of excluded-volume effects, this law is modified into

$$R_\mathrm{g} = N^\nu a , \tag{9.5}$$

where the exponent ν is close to 0.6.

9.4.2 *Polymer solutions*

Even when quite dilute, polymer solutions flow only with difficulty. It was seen in the discussion of viscosity given for simple liquids in Chap. 6 that when a liquid flows in contact with a solid wall, the various layers parallel to the wall move at different velocities and viscosity results from the interaction between successive layers of molecules sliding over each other. But macromolecules have such great lengths that different pieces of the same chain can be dragged along at varying velocities, thus creating internal tensions inside each macromolecule. Since the different pieces may well find themselves in different liquid layers, this effect interlinks the various layers and can greatly diminish the speed of flow. A cyclist who, long ago, had to repair the inner tube of his tyre would have used a solution of rubber in toluene, where such viscous behaviour would be evident. Measurements of viscosity in dilute solutions can indeed be employed to estimate the chain length of the polymer.

The critical concentration at which different molecules begin to entangle in a polymer solution can be estimated by setting $c^* R_g^3 \approx N$, where c^* is the critical number of segments per unit volume and R_g^3 is an approximate measure of the volume taken by each polymer molecule. From Eq. (9.5) this relation yields $c^* \propto N^{-0.8}$, so that overlap starts at very low concentration ($\approx 1\%$) if N is large ($\approx 10^5$).

Flory and Huggins[303] introduced a simple lattice model for describing the thermodynamic properties of polymer solutions. Let $\phi = n_p N/\bar{N}$ be the volume fraction occupied by n_p macromolecules, each consisting of N segments, over a lattice of $\bar{N}$ lattice points. The remaining points are occupied by a number $n_s = \bar{N} - n_p N$ of solvent molecules. The model yields the Helmholtz free energy of mixing as $F_m = \bar{N} k_B T f_m(\phi)$, where

$$f_m(\phi) = \frac{1}{N}\phi \ln \phi + (1-\phi)\ln(1-\phi) + \phi(1-\phi)w\,. \qquad (9.6)$$

The first two terms come from the ideal entropy of mixing and the third is an interaction term.

The osmotic pressure Π, defined as the extra pressure needed across a semi-permeable membrane to maintain the equilibrium of solvent molecules, follows from the relation $\mu_s(0, p, T) = \mu_s(\phi, p + \Pi, T)$ on the chemical potential of the solvent:

$$\Pi = \frac{\bar{N} k_B T}{V}\left[\frac{\phi}{N} - \ln(1-\phi) - \phi - w\phi^2\right]. \qquad (9.7)$$

This equation yields van't Hoff's law $\Pi = n_p k_B T/V$ at very low polymer concentration ($\phi \ll 1$). On the other hand, if N becomes large the first term can be neglected and Eq. (9.7) yields $\Pi = (\bar{N} k_B T/V)(\frac{1}{2} - w)\phi^2$: the osmotic pressure becomes independent of the molecular weight, as observed. In fact, the observed dependence on polymer concentration c for $c > c^*$ is somewhat steeper than the law $\Pi \propto c^2$ predicted by the Flory–Huggins mean-field theory. A scaling argument showing that $\Pi \propto c^{9/4}$ is given immediately below.

Scaling concepts have proved very useful in understanding the physical behaviour of polymeric systems.[302] On purely dimensional reasons, the osmotic pressure can be written as a function of the segment concentration c and the number N of monomers as

$$\Pi = ck_{\mathrm{B}}Tf(ca^3, N)\,, \tag{9.8}$$

where a is an effective segment length and f denotes some function of the indicated arguments. We consider a scaling transformation defined by $N \to N/\lambda, c \to c/\lambda$ and $a \to a\lambda^\nu$, which corresponds to grouping together λ segments at constant radius of gyration R_g. We take this unit of λ segments as representing a new segment and impose that the osmotic pressure should be unchanged. This condition reduces the form (9.8) to

$$\Pi = \frac{c}{N}k_{\mathrm{B}}Tf\left(\frac{c}{N}(N^\nu a)^3\right)\,. \tag{9.9}$$

This form holds for solutions in both the dilute ($c < c^*$) and the semidilute ($c > c^*$) regime. In the so-called semidilute regime, however, chain entanglement sets in and the osmotic pressure should become independent of the chain length N. This requires $f(x) \propto x^{1/(3\nu-1)}$ in Eq. (9.9) and hence $\Pi \propto c^{3\nu/(3\nu-1)} \approx c^{9/4}$, as anticipated above.

A poor solvent may accommodate only a limited number of macromolecules. The polymer will in this case tend to aggregate and beyond a certain concentration there will appear two phases, a dilute solution and a concentrated solution. In the Flory–Huggins theory the homogeneous solution is stable as long as the free energy $f_{\mathrm{m}}(\phi)$ has only one minimum, but separation into two phases (at volume fractions ϕ_1 and ϕ_2, say) occurs if $f_{\mathrm{m}}(\phi)$ has two local minima, allowing a common-tangent construction to determine the phase equilibrium or equivalently permitting equality of the chemical potentials in the two phases at coexistence. Since the shape of the function $f_{\mathrm{m}}(\phi)$ depends on temperature entering Eq. (9.6) through the interaction parameter w, in the $\phi - T$ plane we obtain a phase diagram comprising a region of homogeneous solution and a phase-separation region culminating in a critical point. The latter is determined by the conditions $\partial^2 f_{\mathrm{m}}/\partial\phi^2 = 0$ and $\partial^3 f_{\mathrm{m}}/\partial\phi^3 = 0$ (see the analogous discussion of the liquid–vapour coexistence curve and critical point in Chap. 4). Equation (9.6) yields the values of the critical parameters as

$$\phi_{\mathrm{c}} = (1 + N^{1/2})^{-1}\,, \qquad w_{\mathrm{c}} = \frac{1}{2}(1 + N^{-1/2})^2\,. \tag{9.10}$$

Therefore, as N increases the critical volume fraction decreases and the critical temperature increases, as is indeed observed experimentally.

9.4.3 *Polymer blends*

The Flory–Huggins model can also be used to evaluate the thermodynamic functions and the fluctuation concentrations in polymer blends. If we mix two different polymers A and B, with respective number of monomers N_{A} and N_{B} and volume fraction ϕ_{A} and ϕ_{B}, the free energy of mixing is determined in the model by

$$f_{\mathrm{m}}(\phi_{\mathrm{A}}, \phi_{\mathrm{B}}) = \frac{1}{N_{\mathrm{A}}}\phi_{\mathrm{A}} \ln \phi_{\mathrm{A}} + \frac{1}{N_{\mathrm{B}}}\phi_{\mathrm{B}} \ln \phi_{\mathrm{B}} + \phi_{\mathrm{A}}\phi_{\mathrm{B}} w\,. \tag{9.11}$$

The two entropic terms in this equation become rapidly irrelevant as both N_{A} and N_{B} increase, so that it is not possible to mix the two polymers unless w becomes very small (of order $1/N$).

With regard to concentration fluctuations in the case of a miscible polymer combination, their correlations are described by a structure factor $S_{\mathrm{cc}}(k)$ defined as

$$S_{\mathrm{cc}}(k) = c_{\mathrm{A}}c_{\mathrm{B}}[c_{\mathrm{B}}S_{\mathrm{AA}}(k) + c_{\mathrm{A}}S_{\mathrm{BB}}(k) - 2(c_{\mathrm{A}}c_{\mathrm{B}})^{1/2}S_{\mathrm{AB}}(k)]\,. \tag{9.12}$$

In the Flory–Huggins model (taking for simplicity the case $N_{\mathrm{A}} = N_{\mathrm{B}} = N$) we have

$$S_{\mathrm{cc}}(k) = \frac{S_{\mathrm{cc}}(0)}{1 + k^2\xi^2}\,, \tag{9.13}$$

where

$$S_{\mathrm{cc}}(0) = \frac{1}{(\partial^2 f_{\mathrm{m}}/\partial\phi^2)} = \frac{N\phi(1-\phi)}{1 - 2Nw\phi(1-\phi)}\,, \tag{9.14}$$

$$\xi^2 = \frac{Na^2}{12[1 - 2Nw\phi(1-\phi)]}\,. \tag{9.15}$$

Analogous relations arise in the Ornstein–Zernike theory of the liquid–vapour critical point (see Chap. 4). In the present case of a polymer blend, the critical point of demixing is given by $\phi_{\mathrm{c}} = 1/2$ and $w_{\mathrm{c}} = 2/N$ so that the denominator in both Eqs. (9.14) and (9.15) diverges there. That is, on the approach to the critical point the scattering intensity at low momentum transfer, which is dominated by $S_{\mathrm{cc}}(0)$, and the range of the correlations in the concentration fluctuations, which is given by the length ξ, become very large. On the other hand, far away from the critical point the above denominator is of order 1, so that the scattering intensity is proportional to N and the correlation length is proportional to $\sqrt{N}$.

9.4.4 *Polymeric materials*

We conclude this section by a brief mention of some materials with great relevance to polymer science and applications. We have already remarked that polymeric crystals prepared from the melt usually consist of ordered (but not perfectly crystalline) domains interspersed inside a disordered matrix. Taking again polyethylene as an example, if the axes of the ordered domains are essentially randomly distributed the material has a cloudy appearance because of its heterogeneous structure and is very flexible because the disordered domains are easily deformed. In oriented polymers, instead, crystalline regions do not have random orientations. These materials include natural fibers such as cotton and synthetic fibers such as nylon. There is indeed a variety of techniques by which a polymer melt can be made to solidify in the form of a continuous filament of uniform diameter and high tensile strength (see also Sec. 11.4).

Solid polyethylene formed by slow evaporation of dilute solutions appears in the form of very thin diamond-shaped lamellae. Electron and X-ray diffraction show that they are small single crystals, with the polymer chains oriented *normally* to the plane of the lamella.[304] Since the lamellar thickness is of order 100 Å, each chain must be folded onto itself, with the parallel strands of equal length being joined at their ends by loops consisting of a few CH_2 groups. Chain folding is due to kinetic reasons[305] — crystallisation in this way is most rapid and the resulting crystals are not in their most stable state but will tend to it e.g. on subsequent heating (for a discussion of the nucleation of metastable phases with lower barriers for nucleation, see Appendix 9.2). Lamellae have also been identified in the bulk melt-crystallised product, where they usually are components of more complex aggregates such as sheaves or spherulites.

Finally, with regard to amorphous polymers we only mention here the process of vulcanisation leading to rubber. Back in 1839, it was noticed by Goodyear that natural rubber obtained from trees such as *Hevea brasiliensis* was considerably improved for practical purposes by incorporating a small amount of sulphur into its heating treatment. Starting from a liquid of linear chains, the vulcanisation process chemically joins together some chains that happen to be close to one another. This yields a random lattice of connected chains which, while locally is still a fluid, at the macroscopic level resists compression with a finite elasticity modulus. Rubber in everyday life is indeed flexible and very elastic. However, if it is cooled to $-100°C$ it becomes hard and brittle like glass. The material has been brought by cooling across its glass transition temperature — a concept that will be central to the discussion of the glassy state in Chap. 10.

9.5 Liquid Crystal Phases

The term "liquid crystal" is used to designate certain phases of macromolecular substances showing some degree of order which is intermediate between that of a molecular crystal and that of a molecular liquid which has no long-range translational or orientational order. Some one hundred years ago, it was first observed that a crystal of the organic material cholesteryl benzoate melted into a viscous and cloudy liquid at about 145°C and that, on further heating to about 179°C, this liquid became fluid and transparent. On cooling, the reverse transformations were seen at the same temperatures. Both transitions were accompanied by volume and heat changes. Hundreds of organic materials have subsequently been found that possess such an intermediate phase or even a succession of different intermediate phases.

Many substances exhibiting such mesophases have a common feature: the molecules which are their building blocks have an elongated shape like a rod, a typical length being about 20–30 Å. Generally the rod is somewhat flattened, with a cross-section of some 4×6 Å, and rigid, at least in its central regions. The liquid crystalline properties arise from the tendency for the molecules to lie with their long axes aligned and the transitions are most easily induced by changing the temperature of the sample (for this reason this class of materials is known as "thermotropic" liquid crystals). The so-called "lyotropic" liquid crystals, instead, are solutions of non-spherical macromolecules in which phase transformations are effected by changing the concentration of the solute or the characteristics of the solvent. Mesophases are also found with molecules shaped like flat discs forming stacks ("columnar" liquid crystals). Reference to a number of textbooks on these systems can be made.[306]

We consider below the most important liquid-crystalline phases. They can be classified in terms of the single-particle distribution function ρ_1, which in the crystalline solid is a function $\rho_1(\mathbf{r}, \Omega)$ of the position vector $\mathbf{r} = (x, y, z)$ of the molecular centre-of-mass and of the vector Ω describing the molecular orientation, while in the isotropic liquid it becomes the constant particle density ρ. In liquid crystals ρ_1 depends on some, but not all coordinates.

9.5.1 *Smectic phase*

Proceeding one step in the direction from crystal to isotropic liquid, in the smectic phase layers of rod-like molecules persist but they can slide over each other, which allows the order between successive layers to be disrupted. Furthermore, the molecules within a layer no longer have long-range translational order, although their areal density remains close to the crystalline value. The one-body density in smectics is periodic in one spatial direction, $\rho_1 = \rho_1(z, \Omega)$. Each layer may therefore be likened to a two-dimensional orientationally ordered liquid.

There are in fact many possible smectic arrangements. The simplest is smectic-A [see Fig. 9.8(a)], in which the orientational distribution of the molecules is axially symmetric and the direction of the symmetry axis (the "director" $\mathbf{n}$) is perpendicular to the plane of the layers. The work smectic comes from the Greek for "soap", the consistency of these phases reminding one of a soft soap-like material.

A thin layer of smectic placed between two parallel glass plates becomes optically anisotropic, i.e. shows birefringence as a crystal platelet. The action of the glass surfaces is to orient the smectic planes making them parallel to the plates throughout the whole layer.

9.5.2 *Nematic phase*

The simplest and best known liquid-crystal phase is the nematic mesophase. Nematics have no long-range positional order, but preserve orientational order. That is, $\rho_1 = \rho f(\Omega)$ where $f(\Omega)$ is a normalised orientational distribution function. In uniaxial nematics [see Fig. 9.8(b)] $f(\Omega)$ is axially symmetric around the director $\mathbf{n}$ and is a function only of the angle ϑ between Ω and $\mathbf{n}$. The centres of the molecules are disordered, however, as in a normal liquid.

A thin nematic layer placed between two parallel glass plates is, again as for smectics, subject to constraints from the interactions of the molecules with the glass surfaces. Depending on the treatment of the glass surface the molecules may take an orientation which is either parallel or normal to the surface. Using two glass plates that have been rubbed in perpendicular directions may produce

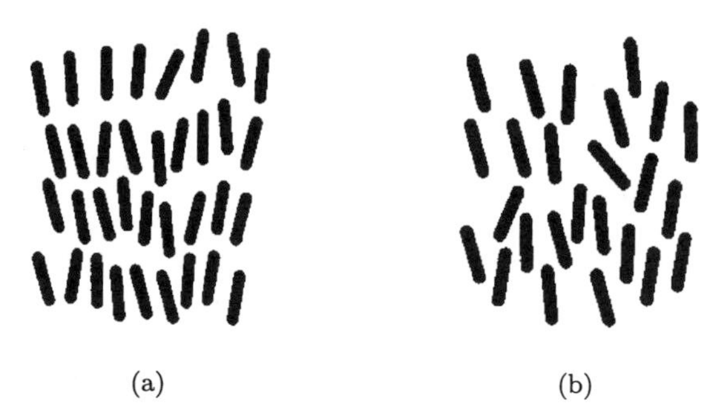

Fig. 9.8. Schematic snapshot of a molecular configuration in a smectic liquid crystal (a) and in a nematic liquid crystal (b).

a twisted structure, with the molecules turning continuously by 90°. The plane of polarisation of light passing through such a nematic specimen follows the orientation of the molecules and is therefore rotated by 90° upon crossing the nematic layer (this behaviour, in combination with the orientational effects of an applied electric field, is exploited in liquid-crystal displays). Molecular orientations can also be controlled by the application of an electric field, as mentioned just above, or by applying a magnetic field.

It should be remarked at this point that some materials go from the crystal to the isotropic liquid through the smectic phase only, and others pass only through the nematic phase. There are also systems, however, that follow the sequence of transformations crystal → smectic → nematic → liquid. Even more complex transitions can occur, with various subspecies possible for the smectic phase, one after the other in the same substance. A variety of the nematic phase, termed a cholesteric, will be briefly referred to below in view of its applications.

In materials displaying such phases as described above, the transition temperatures are frequently in the range between 0° and 150°C, though some are found with higher values.

9.5.3 *Cholesteric phase*

Again, as with smectics and nematics, the molecules providing the building blocks for this type of phase have a rod-like geometry, being elongated and rigid. They have, however, an additional property: chirality. Each molecule is asymmetric in such a way that it cannot be made to superpose on its plane-mirror image by simple movements such as displacement or rotation.

The structure of such a cholesteric phase differs, as a result of chirality, from that of an oriented nematic phase. The asymmetry of the molecules results in the molecules in one layer making a small angle with those in an adjacent layer. This process, being repeated from one layer to another, produces a helical structure: the preferred molecular orientation is a function of position in one direction, i.e. $\rho_1 = \rho f(\Omega(z))$. A cholesteric specimen in contact with a plate takes a twist which is proportional to the distance from the solid surface.

This helical structure endows the cholesteric phases with remarkable optical properties:

(i) Their optical rotatory power is great — typically some hundred times larger than that of "normal" optically active condensed systems.

(ii) The periodic layer arrangement of the sheet arising from the helical arrangement of the molecular axes means that its effect on a beam of light is akin to the way periodic layers of atoms in a crystalline solid act on a beam of X-rays. However, the scales differ by a factor $\approx 10^3$, for the periodicity and the wavelengths. For a given angle of incidence there is a wavelength which is reflected much more than all others, so that the cholesteric sheet displays colour under illumination by white light. A sheet that looks blue under grazing incidence can appear green or yellow when the angle of incidence is closer to normal. Furthermore, with a specified geometry, the colour observed depends on the periodicity and hence on temperature. The colour change is quite noticeable with a variation of temperature as small as 0.1°C. This property is utilised to measure the temperature, say of a surface, by laying down a cholesteric film on it.

9.6 Nematic Liquid Crystals and their Phase Transitions

To make a start on a quantitative description of a uniaxial nematic liquid crystal, the molecular centres have a random liquid-like distribution, subject to geometric constraints, while the local degree of orientational order is more appropriate to a solid. To describe this, we introduce the director $\mathbf{n}(\mathbf{r})$, which is a unit vector parallel to the local preferred orientation and can vary from point to point.

The local degree of orientational order can be quantitatively specified by expanding the distribution function $f(\Omega)$ in the full set of orthogonal Legendre polynomials $P_\ell(\cos\vartheta)$, where ϑ is the angle between the symmetry axis and the director. The coefficients of the expansion play the role of a set of order parameters. In particular, the Zwetkoff parameter S is defined as

$$S = \left\langle \frac{1}{2}(3\cos^2\vartheta - 1) \right\rangle , \tag{9.16}$$

the brackets representing a time or ensemble average over the molecular distribution. The above definition, involving the Legendre polynomial $P_2(\cos\vartheta)$ is convenient since it reduces to $S = 1$ for perfect orientational alignment and to $S = 0$ for ideal isotropic behaviour. Values of S up to 0.8 have been observed, but typically S varies from 0.4, decreasing with increasing temperature and thermal disorder and discontinuously dropping to zero at the nematic–isotropic transition temperature. At this temperature, the optical activity and turbidity disappear.

The definition of nematic order parameter in Eq. (9.16) is not entirely satisfactory, since in the absence of external aligning forces the direction of $\mathbf{n}$ is not known *a priori*. However, it is possible to introduce a tensorial order parameter and define the director $\mathbf{n}$ and the Zwetkoff parameter S in terms of its eigenvalues.[307]

9.6.1 *Landau–de Gennes theory*

A convenient starting point to discuss the transition from the isotropic to the nematic phase (I-N transition) is to consider nematic ordering as a weak perturbation of the isotropic liquid. We follow Landau[308] in assuming that the free energy difference can be expanded in terms of the order parameter S. That is,

$$F = F_0 + a(T - T^*)S^2 - bS^3 + cS^4 . \tag{9.17}$$

We have assumed that in the vicinity of the I-N transition the coefficient of the leading term in the free energy difference between the two phases varies linearly with temperature while the other coefficients are essentially constant. The presence of a cubic term implies that a fluctuation with position S is not equivalent to one with opposite S, and yields a first-order I-N transition.

In order to see what the Landau–de Gennes theory predicts for the I-N transition, it is convenient to rewrite Eq. (9.17) in the form

$$F - F_0 = \left[a(T - T^*) - \frac{b^2}{4c} \right] S^2 + cS^2 \left[S - \frac{b}{2c} \right]^2 . \tag{9.18}$$

It is then easily seen that the I-N transition occurs at the temperature $T_{\rm NI} = T^* + b^2/4ac$, with a discontinuous jump in the order parameter from zero to $S = b/2c$. For $T < T_{\rm NI}$ the temperature dependence of the order parameter is obtained from minimisation of the function $F(S)$ as

$$S(T) = \frac{3b}{8c}\left(1 + \sqrt{1 - \frac{8(T - T^*)}{9(T_{\rm NI} - T^*)}}\right). \tag{9.19}$$

All other thermodynamic quantities in the vicinity of the phase transition can be obtained from Eq. (9.17). Of course, the theory does not provide predictions for the magnitude of the coefficients of the expansion. An example of a molecular theory will be reported below.

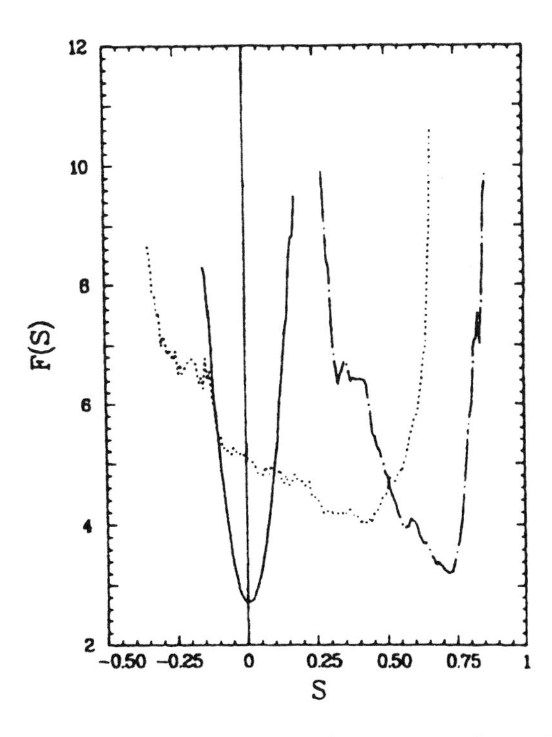

Fig. 9.9. Landau free energy associated with order parameter fluctuations in a nematic liquid crystal, from Monte Carlo simulations. ——: low-density isotropic phase; ···: nematic phase close to I-N transition; — · — · —: high-density nematic phase, where only small fluctuations around a non-zero order parameter S occur. (Redrawn from Frenkel, Ref. 307.)

Spontaneous fluctuations in the order parameter near the phase transition may in principle be observed by light scattering. To treat this the theory needs extending to include account of the gradient of the order parameter and will then lead to an expression similar to Eq. (9.11) with a correlation length of order $\xi = \sqrt{L/a(T-T^*)}$. An effective way to examine order-parameter fluctuations is through computer simulations[309] (see Fig. 9.9).

The Landau–de Gennes approach has been extended by Matsuyama and Kato[310] to treat the I-N transition in a mixture of a liquid crystal and semiflexible polymer chains. Such mixtures have technical applications in non-linear optics and electro-optical devices, and their phase behaviour has attracted the interest of a number of workers.[311] The study of Matsuyama and Kato indicates new phase behaviours in the temperature-concentration plane, such as nematic unstable and metastable regions, a critical solution point in the nematic phase, azeotrope points and triple points. This general area should continue to be of interest.

9.6.2 *Molecular mean-field theory of isotropic-nematic transition*

In a molecular mean-field theory of orientational melting one calculates the potential energy of a representative molecule in the mean field of its neighbours, no account being taken of the reaction of the molecule on its environment. Maier and Saupe[312] proposed an anisotropic pairwise intermolecular potential $\phi_{12} = u(r)(3\cos^2\theta_{12} - 1)/2$, where θ_{12} is the relative orientation of the two molecules and $u(r)$ is attributed to van der Waals dispersion forces. If we average over all spatial and orientational configurations of the second molecule, the potential energy of the first is $\phi_1 = \bar{u}S(3\cos^2\vartheta_1 - 1)/2$ and the molecular orientational distribution is given by the single-particle Boltzmann expression

$$f(\cos\vartheta) = \exp\left[-\frac{\bar{u}S(3\cos^2\vartheta - 1)}{2k_{\mathrm{B}}T}\right]. \tag{9.20}$$

The order parameter S can then be determined self-consistently from its definition in Eq. (9.16).

The excess Helmholtz free energy can also be evaluated from the mean one-body potential energy and is found to vanish at the I-N transition temperature, which is given by

$$k_B T_{NI} \cong -0.22\bar{u}. \tag{9.21}$$

Within the Maier–Saupe model $\bar{u}$ is determined by the anisotropy in the electrical polarisability of the molecule — the same effect that leads to molecular orientation by an applied electric field. An analogous anisotropy in the diamagnetic susceptibility of the molecule leads to molecular orientability in a magnetic field and to the Freedericksz transition that will be discussed in Chap. 11. Some further work on model potentials for molecular liquids and liquid crystals is recorded in Sec. 9.6.4 below.

We conclude this section by quoting the work of Sluckin and Shukla[313] on the I-N transition, that was carried out within a density functional context.

9.6.3 *The isotropic-nematic-smecticA transition*

The phase change from nematic to smecticA is characterised by a relatively small latent heat in many systems and by obvious pre-transition effects. Much more variety of behaviour is observed from one material to another than is seen in the nematic-isotropic transition.[314] In a few materials the transition appears to be second order, while in others it is strongly first order. It offers an area where the effects of molecular structure on ordering can be usefully studied.[315]

In the smecticA phase the density of molecular centres-of-mass becomes periodic in one dimension, parallel to the director (the z axis, say). The distribution function $f(z)$ can be Fourier analysed as $f(z) = [1 + 2\sum_n \sigma_n \cos(2\pi z/d + \varphi_0)]$, where d is the layer thickness and φ_0 fixes the origin with respect to the centre of the layer. For perfect ordering all the order parameters σ_n equal unity, while in the nematic phase they are all zero.

For the purpose of theoretical treatment σ_1, or $\sigma_1 \exp(i\varphi_0)$ may be chosen as the main order parameter of the smecticA phase.[316] Ignoring for the present the orientational order, the Landau expansion for the free energy would contain only even powers of σ_1,

$$F = F_0 + \frac{1}{2}a(T - T^*)\sigma_1^2 + \frac{1}{4}b\sigma_1^4 + \frac{1}{6}c\sigma_1^6 + \cdots, \tag{9.22}$$

and if $b > 0$ a second order phase transition occurs at the temperature T^* into an ordered phase with $\sigma_1(T) = [a(T^* - T)/b]^{1/2}$. On the other hand, the transition turns out to be first order if $b < 0$ and $c > 0$. A tricritical point at $b = 0$ separates these two regimes. Meyer[314] discusses the role of a coupling between σ_1 and σ_2 in lowering the value of b.

McMillan has extended the Maier–Saupe analysis to smecticA systems and given a molecular theory of the whole series of phase transitions. He has calculated order parameters and thermodynamic properties, with results in qualitative agreement with experiment. Both the Kobayashi–McMillan model[316] and the Meyer–Lubensky model[317] can be seen as special cases of density functional theory. For more details the reader may refer to the review by Frenkel.[307]

9.6.4 *Model potentials for molecular liquid and liquid crystals*

Hess[318] has shown that an augmented van der Waals theory can yield acceptable results, over a wide density range, for the equation of state of a Lennard–Jones (LJ) fluid. In his study, the short-range repulsive part of the interaction potential is accounted for by the use of a modified Carnahan–Starling equation of state (see Sec. 2.6).

Subsequently, Hess and Su[319] have extended this approach to model potentials for molecular liquids and liquid crystals. They make estimates in the latter case for the isotropic-nematic transition. Their study is somewhat in the spirit of the studies of Onsager[320] and the later study of Colter.[321] They propose that, because of its relative simplicity, their model is suitable for computational work on the phases of liquid crystals in restricted geometries[322] and also on transport processes.[323]

The model used by Hess and Su is, in essence, a generalised LJ potential where the r^{-6} attractive part depends on the relative orientations of the axes of the interacting molecules and the vector joining their centres. Various types of anisotropy in the interaction are accounted for. The augmented van der Waals results for the free energy and the pressure can then be obtained. In such expressions, an orientation-dependent second virial coefficient and the orientational distribution functions of a pair of particles enter. The short-range part of the interaction is again taken account of by a modified Carnahan–Starling approach.

The model may be tested further by calculating the Frank elasticity coefficients (see Chap. 11) and also interfacial properties, along lines provisionally laid down by Osipov and Hess.[324]

Appendix 9.1 Melting and Orientational Disorder

In their work Lennard–Jones and Devonshire[283] presented a simple model for the melting of positional order as arising from the cooperative generation of point defects in the hot crystal. Consider a situation in which a fraction Q of the atoms are on normal lattice sites and the others are in interstitial sites. Assuming that the energy of interaction between two such neighbouring atoms is W, the partition function acquires a factor $Y(Q)\exp[-NZWQ(1-Q)/k_BT]$ where Z is the number of interstitial sites around a lattice site and $Y(Q)$ is a combinatorial factor counting the number of ways in which the N atoms can be randomly distributed on all sites. The same assumptions enter the Bragg–Williams theory of order–disorder phenomena in alloys and lead to

$$\ln\frac{Q}{1-Q} = \frac{ZW}{2k_BT}(2Q-1), \tag{A9.1.1}$$

for the equilibrium value of the order parameter of the phase transition.

Pople and Karasz[282] extended the above approach to include orientational disordering of the molecules by allowing for two possible molecular orientations on each site, separated by an energy barrier W'. With the further order parameter S giving the fraction of molecules of given orientation on normal lattice sites, their model allows one to write

$$\ln\frac{Q}{1-Q} = L[1-2S(1-S)y](2Q-1), \tag{A9.1.2}$$

$$\ln\frac{S}{1-S} = 2Ly[1-2Q+2Q^2](2S-1), \tag{A9.1.3}$$

where $L = ZW/2k_BT$ and $y = Z'W'/ZW$. These two equations admit a transition from a positionally and orientational ordered solid to an orientationally disordered solid and then to a fully disordered liquid phase, provided y is not too large. The two reduced transition temperatures $t_m = 2k_BT_m/ZW$ and $t_c = 2k_BT_c/ZW$ depend on the parameter y and an explicit relation between them can be found by eliminating y, as discussed by Tozzini *et al.*[281]

The Pople–Karasz phenomenology then seems to capture the qualitative features of a phase diagram such as that shown in Fig. 9.1 for nitrogen, in which it appears that small extrapolations of the melting line and of the $\beta-\delta$ boundary line would result in a meeting of these phase boundaries.

Appendix 9.2 Crystallisation from Solution

Crystallisation from solution is of considerable importance from both technological and purely scientific perspectives. As examples, one may cite wax and polymer crystallisation in hydrocarbon assemblies and protein crystallisation in aqueous systems.

Supercooling (see also Chap. 10) is a central issue where materials processing and the resultant crystal size and morphology are concerned.[529] The supercooling is determined by the temperature at which nucleation takes place. As Sirota[530] has emphasised, such nucleation does not necessarily imply homogeneous nucleation of the stable phase, but could involve nucleation of a metastable phase with a lower barrier for nucleation (cf. the historically important study of Ostwald[531]). There is a substantial degree of interest in proposals that transient metastable phases, of higher density than that of the liquid, but where the symetry of the stable phase is not yet developed, may be responsible for mediating crystallisation. Such systems could embrace polymers, colloids, proteins and alkanes.

The interest in *n*-alkanes in particular arises since they are the simplest organic homologous series and are the main building blocks of lipids, surfactants and a good many polymers. Equilibrium properties are already well characterised and are known to strongly influence the properties of derivative molecules. Also, the equilibrium solubilities of alkanes in a variety of solvents have been extensively studied.

Of special interst in the present context is the knowledge that dilute alkane solutions supercool[532] whereas the undiluted melts do not.[533] To elucidate such crossover behaviour, Sirota[534] has studied experimentally the supercooling exhibited by bulk solutions of tricosane (C_{23}) as a function of dilution with dodecane (C_{12}). He notes that C_{23} undergoes a solid–solid transition a few degrees below its melting point and demonstrates that supercooling becomes observable at dilutions such that the low-T solid phase is stable at the dissolution temperature, and the observed precipitation coincides with the temperature at which the high-T solid phase becomes metastable. Sirota concludes that the supercooling in bulk *n*-alkane solutions is determined by the metastable phase diagram and proposes that such a mechanism where nucleation is induced by a transient metastable phase may possibly be of frequent occurrence.

Chapter 10

Supercooling and the Glassy State

10.1 Macroscopic Characteristics of a Glass and the Glass Transition

As was emphasised in Chap. 1, a substance can normally be found in one of three equilibrium states: gas, liquid, or ordered crystalline solid. Here, we shall turn to a subject of continuing technical importance, namely solids which are in a disordered non-equilibrium state (the terms amorphous solids or glasses are also used to designate them). Materials of this kind are usually formed by rapid cooling from the normal liquid state and have two main properties that characterise them: (i) their viscosity is so high that they are for all practical purposes rigid, and (ii) they show no crystalline long-range order under examination by X-ray diffraction. A structurally amorphous state can also be created by massive cold working in a number of solid alloys — even in alloys where rapid quenching of the melt has not been found to produce a glassy state.

Taking window glass as an example, it is hard, difficult to deform and easy to break: it is brittle. When it breaks the new surface may have a complex aspect, but is smooth and shows no crystalline grains. Glass is isotropic, i.e. no direction can be singled out in its structure which differs from any other. Finally, transparency in a glassy insulator results from the absence of textural irregularities which may deviate the rays of light as they pass through it.

Supposing that a liquid has been cooled fast enough to prevent crystallisation, the subsequent behaviour on further cooling may be described with reference to a first derivative of the free energy (e.g. the volume or the entropy) or to a dynamical property such as the shear viscosity. The former viewpoint emphasises the thermodynamic aspects of the glass transition, the latter the kinetic aspects. A first-order thermodynamic quantity varies continuously with temperature (and pressure) in the supercooled liquid, but shows a rapid change of slope in a narrow range whose location depends on the rate of cooling and on the thermal history. This temperature *range* is used to define the glass transition temperature T_g, across which the second-order quantities (thermal expansion coefficient, specific heat, compressibility) undergo very rapid changes. These behaviours are illustrated in Fig. 10.1 for the instance of the V–T relationship in the formation of vitreous selenium.[325] On a glass branch the system is thermodynamically unstable and if held at a temperature below T_g it will tend to approach the metastable state of the supercooled liquid at that temperature, on a time scale increasing from minutes near T_g

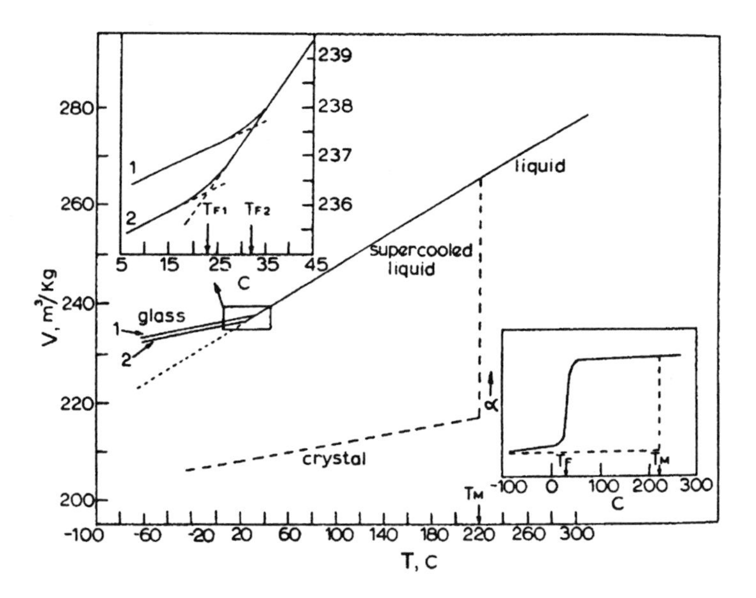

Fig. 10.1. The volume–temperature relation for the formation of vitreous Se. The inset on top left shows the role of the cooling rate (1: fast cooling; 2: slow cooling). The inset at bottom right shows the expansion coefficient in the glass-transition region. (Redrawn from Owen, Ref. 325.)

to centuries well below T_g. On reheating from the glassy state anomalies will occur unless the rate of heating is the same as the original rate of cooling.

The kinetic aspects of the glass transition are reflected in the thermal behaviour of the shear viscosity η and of the characteristic times associated with the primary (so-called α) relaxation of various kinds of structural and dielectric observables (see Sec. 10.5 below). Viscosity is a measure of the response of the liquid to a suddenly applied shear stress and is related to the corresponding relaxation time by the Maxwell formula

$$\eta = G_\infty \tau \,, \tag{10.1}$$

where G_∞ is the high-frequency elastic shear modulus and τ is the average response time of the system to the applied stress. In glass-forming liquids η increases smoothly over many orders of magnitude on cooling towards and beyond the glass transition temperature, the value of T_g corresponding in this context to the point at which η reaches a value of order 10^{13} poise. At this fluidity the configurational relaxation times are in the range of a few minutes to a few hours — that is, in the range of typical experimental times. Thus, in a kinetic viewpoint what is observed reflects the crossing between an internal relaxation time and the experimental time scale.[326]

In many systems the viscosity satisfies over a wide temperature range the empirical Vogel–Fulcher relation,[327]

$$\eta = \eta_0 \exp\left[\frac{E_\eta}{k_B(T - T_0)}\right] , \tag{10.2}$$

where T_0 is a characteristic temperature lying below T_g and E_η may be viewed as an activation energy for viscous flow at $T \gg T_0$. In comparison with an Arrhenius behaviour as discussed in Sec. 6.3 for water, Eq. (10.2) implies that an "excess viscosity" emerges with decreasing temperature in the glass-forming liquid. An alternative way of representing the phenomenon is to plot the effective temperature-dependent activation free energy for α-relaxation in the formula $\tau_\alpha = \tau_{\alpha,\infty} \exp[E(T)/k_B T]$, thus emphasising the crossover from Arrhenius to super-Arrhenius behaviour which is typical of most supercooled liquids. An early hypothesis for this behaviour was that part of the liquid volume would be blocked for flow by the formation of "clusters", as may be suggested by the theory of the viscosity of suspensions briefly reported in Sec. 6.4. More generally, the strong viscous slowing down of a supercooled liquid with decreasing temperature can be viewed as being the result of a collective jamming process.

Many years ago Simon obtained the excess (or configurational) entropy for some low-temperature organic glasses from calorimetric data. In vitreous glycerol it has an approximately constant value of about 5 e.u., extrapolating with little change down towards $T = 0$. This essentially constant value is the entropy frozen in the glass because the relaxation times for configurational rearrangements exceed experimental times. In fact, the anomaly at T_g in the second-order thermodynamic quantities occurs as the entropy of the supercooled liquid is approaching that of the corresponding crystalline phase. The Kauzmann temperature T_K is defined from the heat capacities $C_p(T)$ by the condition

$$\int_{T_K}^{T_F} [C_p(\text{liq}) - C_p(\text{cryst})] \frac{dT}{T} = \Delta S_F \,, \tag{10.3}$$

T_F being the ordinary melting temperature and ΔS_F the corresponding entropy change[328] (see Fig. 10.2). Thus, T_K marks a point of "thermodynamic crisis" at which the entropy of a disordered system (the supercooled liquid) would become equal to that of the ordered phase (the crystal). Kauzmann[329] stressed that it is paradoxical that a purely thermodynamic crisis at T_g is being avoided *via* the purely kinetic crossing between the time scales of the system and of the experimental apparatus. There is here a strong suggestion that the kinetic and thermodynamic phenomena of the glass transition are closely related.[328]

Below we shall go in some more detail into the many issues raised in this introduction. We shall start by giving some attention to the kinetics of

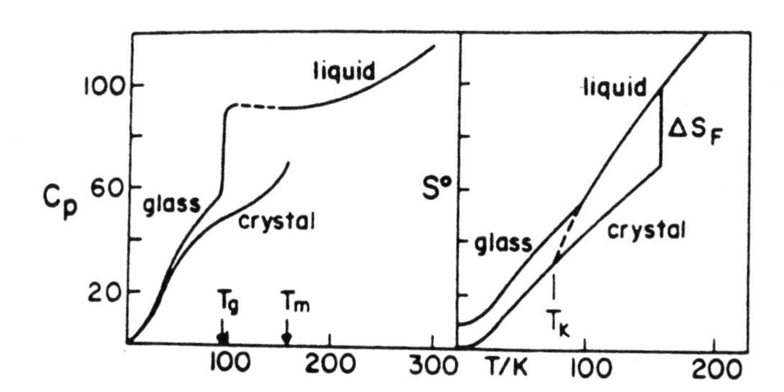

Fig. 10.2. Behaviour of the heat capacity and entropy of ethanol as an example of glass-forming liquids. T_K is the Kauzmann temperature. (Redrawn from Angell, Ref. 328.)

nucleation and crystal growth — the first requirement for glass formation being that crystallisation be avoided as the liquid is supercooled. A review of models of the glass transition can be found in an article of Jäckle.[330]

10.2 Kinetics of Nucleation and Phase Changes

The radial distribution function $g(r)$ that is used to describe the short-range order in a liquid (see Chap. 4) represents a space and time average. The distribution of atoms around any given atom changes rapidly with time, and at some times the local order may be more pronounced and bear resemblance to a crystalline configuration. Above freezing such a local arrangement would be only a transient, thermal motions tending always to disrupt it. However, if the liquid can be brought slightly below freezing, such "seeds" tend to be larger and more frequent and become potential nuclei for the start of crystal formation. Large supercoolings can be achieved if the formation of seeds is somehow prevented, as for instance in water dispersed into very fine droplets, some μm in diameter, or in a viscous liquid like glycerol.

10.2.1 *Homogeneous nucleation and crystal growth*

The free energy change ΔG for the isothermal formation of such an ordered nucleus, assumed to be spherical and of radius R, is the sum of a volume term proportional to R^3 and a surface term proportional to R^2. The coefficients of these two terms are determined by the difference ΔG_{F} in Gibbs free energy between the solid and the liquid (per unit volume) and by the surface free energy (per unit area), respectively. ΔG_{F} vanishes at freezing and becomes negative in the supercooled liquid: at each temperature below equilibrium freezing, therefore, $\Delta G(R)$ as a function of R shows a maximum. Writing $\Delta G(R) = aR^3 + bR^2$ we find that the critical size at which a nucleus could grow spontaneously is $R_{\mathrm{c}} = 2b/3|a|$. The rate of homogeneous nucleation is the rate at which critical-size nuclei can form and the probability of their formation is proportional to $\exp(-\Delta G_{\mathrm{c}}/k_{\mathrm{B}}T)$ with $\Delta G_{\mathrm{c}} \equiv \Delta G(R_{\mathrm{c}})$.

However, for a nucleus to grow atoms must be added to it and this process will also involve an activation energy (ΔG_{d}, say) for diffusion or re-orientation. The rate J of formation and growth of critically sized nuclei is then given by transition state theory[331] as

$$J = \left(\frac{nk_{\mathrm{B}}T}{h}\right) \exp\left[-\frac{\Delta G_{\mathrm{c}} + \Delta G_{\mathrm{d}}}{k_{\mathrm{B}}T}\right], \tag{10.4}$$

where n is the number of molecules per unit volume and h is Planck's constant. This yields[325]

$$J = \left(\frac{nk_{\mathrm{B}}T}{h}\right) \exp\left(-\frac{\Delta G_{\mathrm{d}}}{k_{\mathrm{B}}T}\right) \exp\left[-\frac{16\pi\gamma^3 T_{\mathrm{F}}^2}{3\lambda^2 \Delta T^2 k_{\mathrm{B}}T}\right], \tag{10.5}$$

where γ is the surface tension, $\Delta T = T_{\mathrm{F}} - T$ is the supercooling and $\lambda = \Delta H_{\mathrm{F}}/V_{\mathrm{m}}$, with ΔH_{F} the latent heat of fusion and V_{m} the molar volume.

From Eq. (10.5), J rises very rapidly as T decreases, passes through a maximum and then decreases. The maximum lies at $T_{\mathrm{F}}/3$ if $\Delta G_{\mathrm{d}} = 0$ and moves towards T_{F} as ΔG_{d} increases. Thus, if a liquid has a sufficiently large ΔG_{d} that it can be cooled below the temperature of the maximum in J without crystallising, then crystallisation is less likely to occur on further cooling. Taking $T_{\mathrm{g}} \approx 2T_{\mathrm{F}}/3$ as observed for many glasses formed by normal quenching from the melt, one estimates[325] that the maximum of J would lie at this temperature if $\Delta G_{\mathrm{d}} \approx 40\ k_{\mathrm{B}}T \approx 24$ kcal/mol, roughly of the same order as E_η in Eq. (10.2).

In reality, crystallisation in a supercooled liquid can be fostered by the presence of foreign particles acting as initiators for crystal growth. Such heterogeneous nucleation processes have been probed in great detail for the crystallisation of water into ice,[332] some main issues being the properties of foreign particles that make them effective as heteronuclei and the role of dispersion of the liquid into droplets or in small pores and capillaries. For a modern account of crystal nucleation and growth in viscous liquids, the reader is referred to a review by Turnbull.[333]

Direct knowledge of the critical nucleus is necessarily limited, since its formation is a fleeting event. Experiments can only probe the rate at which crystallites form in a supersaturated solution by observing them after they are formed and grown. Numerical simulations allow one to probe the early stages of nucleation[334] and suggest that the structure and the free energy of a critical nucleus may deviate significantly from the predictions of the classical theory.

10.2.2 *The critical cooling rate for glass formation*

It is possible to make crude estimates of the critical cooling rate R_c that must be exceeded to permit glass formation in a particular liquid. Sarjeant and Roy[335] proposed that R_c is related to the shear viscosity of the melt by the empirical formula

$$R_c = 2 \times 10^{-6} \frac{RT_F^2}{V_m \eta}, \tag{10.6}$$

where R is the gas constant. This relation embodies the fact that the measured R_c decreases as the viscosity of the melt increases and its specific form is roughly in accord with experiment.

The notion of a critical cooling rate implies that it should be possible in principle to quench *any* liquid into a glass. Glasses form from liquids with various bonding types (covalent, ionic, metallic, hydrogen or van der Waals bonded) and the failure to vitrify some materials (pure metals, most metal halides) may just be due to practical limitations in cooling techniques.

10.2.3 *Superheating and vapour condensation*

A nucleation process is also involved in the heating of a liquid until it boils. Steady boiling at the normal boiling temperature is assisted by the presence of nuclei, which often are dissolved gases or pockets of air at solid surfaces. In the absence of foreign nuclei, however, nucleation of bubbles is needed and these can grow only if the internal vapour pressure exceeds $2\gamma/R$. For a bubble of radius $R \approx 0.5$ nm in a liquid such as water, this pressure is ≈ 3000 atm. Carefully purified water can be superheated to several 100 K above the normal boiling point, and once bubbles begin to form and grow the excess pressure is rapidly released by explosion.

The condensation of vapour into liquid droplets is yet another example of a nucleation phenomenon. Droplets of radius R exert a vapour pressure of $2\gamma\rho_v/\rho_1 R$ and, instead of growing by further condensation will evaporate if this exceeds the saturated vapour pressure. Condensation can either be assisted by foreign nuclei such as dust particles or be induced by producing supersaturation of the vapour as through a sudden cooling. It is this second method which was used to reveal trajectories of charged particles in the Wilson cloud chamber.

The microphysics of clouds and the condensation of vapour on solid surfaces are subjects of great practical importance. An introduction to homogeneous nucleation in this context can be found in the book of Abraham.[336]

10.3 The Structure of Amorphous Solids

We have presented in Chaps. 4 and 8 the notions of topological and chemical short-range order (SRO) in liquids. The former is typical of monatomic liquids, with a prototype in the hard-sphere fluid, and is dominated by an atomic diameter σ resulting in a main peak in the liquid structure factor $S(k)$ at $k_m \approx 7.5/\sigma$. Topological SRO is supplemented in an ionic liquid such as molten NaCl by chemical SRO from alternation of the component species, which results in a main peak in the concentration–concentration structure factor $S_{cc}(k)$ at $k_c \approx$

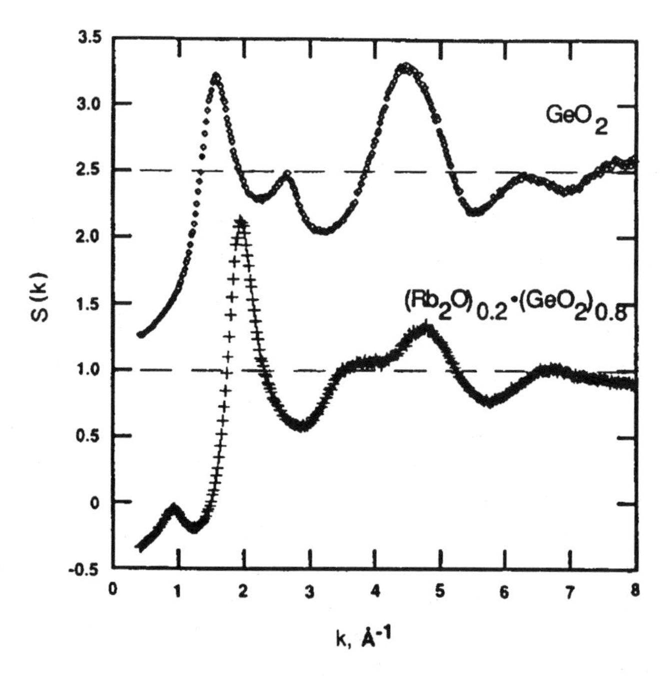

Fig. 10.3. X-ray structure factors of germania and of a rubidium germanate glass. The IRO corresponds to the peak at 1.54 Å^{-1} in GeO_2 and at 2.0 Å^{-1} in $Rb_{0.2}(GeO_2)_{0.8}$. The small peak at 0.95 Å^{-1} in the germanate glass marks the extended range order. (Redrawn from Price and Saboungi, Ref. 338.)

$7.5/\sigma_2$ where σ_2 is a characteristic nearest distance of ions of like charge. In connection with molten salts such as $ZnCl_2$, $AlCl_3$ and $FeCl_3$, however, we have also discussed in Chap. 8 the presence of a first sharp diffraction peak (FSDP) in the diffraction patterns as a marker of intermediate-range order (IRO). IRO is the characteristic feature of structural correlations in glasses, extending over distances of ≈ 1 nm — larger than atomic bond lengths (≈ 0.3 nm) but smaller than the scale of textural inhomogeneities (> 5 nm).[337] Two examples[338] are reported in Fig. 10.3.

10.3.1 *Network and modified-network glasses*

Many common glasses are characterised by directional interatomic bonds forming local structural units with a well defined SRO. In the crystal these units are connected into an ordered network. For instance, the crystal structures of the IV–VI compounds are built from chemically unsaturated tetrahedra which may be connected in two alternative ways, i.e. by corner sharing or by edge sharing. Pure corner sharing in SiO_2, mixed corner and edge sharing in $GeSe_2$ and pure edge sharing in $SiSe_2$ give rise to networks having dimensionality $D =$ 3, 2 and 1, respectively. Among the elements, threefold coordination in P and As yields P_4 and As_4 tetrahedra as basic units (one may describe the crystal in this case as an ordered network having dimensionality $D = 0$). Again, twofold coordination in the group-VI elements yields a variety of crystalline allotropes with structures formed from chains or molecular rings. Covalent bonding in such networks appears to be generally stable across melting, although long range order is lost.

In the continuous random network model proposed a long time ago by Zachariasen,[339] the structure of a glass is viewed as an essentially random assembly of such strong structural units. In fact, the IRO is believed to reflect the way in which the local groups of atoms are preferentially connected into a mesoscopic disordered structure.[340] Alternative viewpoints attribute the IRO to correlations of voids or rings, or even to layer-like correlations. What is certain is that the FSDP implies a remarkably long range in structural correlations inside a disordered system.

The FSDP has a number of observed anomalous properties. Its amplitude increases with temperature and drops with increasing pressure — quite the opposite of what is commonly observed for the other peaks in the diffraction pattern of a disordered material. Compaction of vitreous silica under pressure is known to attenuate the FSDP, as a result of increased frustration of the ordering caused by the decrease in volume available to the network.[341]

IRO in network glasses also has dynamic manifestations.[337] In the vibrational spectra, so-called defect modes and companion modes have been associated with local arrangements of the tetrahedral units in oxide and chalcogenide glasses. Low-frequency excitations (at 10–100 cm^{-1}) leading to anomalies in thermal properties have also been associated with IRO.[342] These excitations are observed in Raman spectra[343] (where they are usually called "boson peaks") and in inelastic neutron scattering.[344]

The IRO is substantially affected by the addition of modifiers to the network. An example is shown in Fig. 10.3[338]: in rubidium germanate glasses an "extended range order" emerges at a wave number below that of the FSDP, while the FSDP moves to higher wave number. It has been suggested[345] that the alkali metal ions enter the larger cages of the network, pushing out the oxygen atoms on the boundary and compressing the smaller cages. The structural data on several classes of modified network glasses are consistent[337] with a model in which the network is built from various structural units that occur in the corresponding crystalline compounds.

10.3.2 *Molten and amorphous semiconductors*

Semiconducting group-IV elements and polar III–V compounds crystallise in tetrahedrally coordinated open structures. They melt into metallic liquids having higher density and coordination number close to seven. The elements can also be prepared in a network-like amorphous state, having fourfold local coordination and semiconducting properties. In a chemical picture the melting of these solids is accompanied by a release of valence electrons from interatomic bonds into conducting states, and the bonds are rebuilt on formation of a disordered network in the amorphous state.

Figure 10.4 shows a comparison between the liquid and amorphous structures of Ge, the positions of Bragg diffraction spots of the crystal being shown at the top.[346] The structure factor of the melt shows a main peak followed by a shoulder. While the position of the main peak does not correspond to any structural features in the crystalline or amorphous states, the shoulder corresponds to the (220) and (311) Bragg reflections and to be main peak in the amorphous structure factor. The FSDP is present in the latter and is in correspondence with the (111) Bragg spots of the crystal. Evidently, the

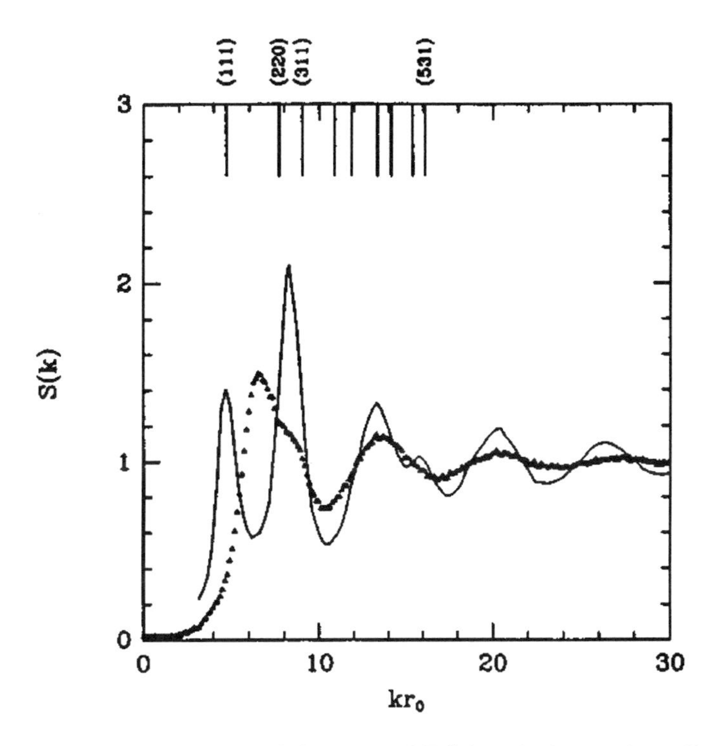

Fig. 10.4. Structure factor of liquid Ge at 1000°C (▲) and of amorphous Ge at room temperature (——). The vertical bars at the top show the location of the allowed Bragg reflections from the diamond crystal structure. For each state of the material, the wave number k has been scaled with the appropriate value of the first neighbour distance r_0.

reconstruction of the interatomic bonds in amorphous Ge has led to IRO. The wave vectors of the (111) star are clearly related to the formation of connectivity between the local tetrahedral units, which would be frustrated if the system did not open up on freezing.

The significance to be attached to the order parameters associated with the (220) and (311) Bragg reflections emerges from quantum-chemical considerations by Stenhouse *et al.*[347] on the correlations between atoms and bond centres in amorphous Si. They show that the discrepancies between the results of a continuous random network model and the observed diffraction pattern can be explained by accounting for the diffraction intensity coming from the distribution of valence electrons among the ionic cores. Their results imply that the shoulder in the liquid structure factor reflects a remnance of bonds and that the emergence of the corresponding structures in the crystalline and amorphous states can be associated to "freezing of the bonds".

10.4 Thermodynamic Aspects and Free Energy Landscape

The inherent non-equilibrium nature of the glassy state implies that thermostatics does not work for glasses. The proper theoretical frame is provided by irreversible thermodynamics.[348] We present below the approach proposed by Nieuwenhuizen[349] in dealing with systems for which the non-equilibrium state involves two well separated time scales.

Such a system is described by three thermodynamic parameters, i.e. the temperature T, the pressure p and an effective temperature T_e reflecting the dependence on cooling rate and thermal history. The heat change $đQ$ in an infinitesimal process is written as

$$đQ = TdS_{eq} + T_e dS_e \, , \tag{10.7}$$

where S_{eq} is the entropy associated with the fast β-processes, having time scales shorter than the observation time, while S_e is the excess (or configurational) entropy of the slow α-processes. In its standard definition[350] the configurational entropy is the entropy of the glass minus that of the vibrational modes of the crystal. From Eq. (10.7) the first law for a glass-forming liquid reads $dU = TdS_{eq} + T_e dS_e - pdV$. The total entropy is $S = S_{eq} + S_e$ and the second law ($đQ \leq TdS$) leads to $(T_e - T)dS_e \leq 0$.

As we have seen in Sec. 10.1, in the glass transition region a glass-forming liquid exhibits a *smeared* jump $\Delta\alpha \equiv \alpha_{\rm liquid} - \alpha_{\rm glass}$ in the thermal expansion coefficient and similarly *smeared* jumps ΔC_p and ΔK in specific heat and compressibility. *Discontinuous* jumps in the same quantities occur across equilibrium second-order phase transitions, where they obey the so-called Ehrenfest relations[351] (see also Appendix 3.1). Similar relations can be obtained by similar methods for the smeared jumps in the present context.[349] Thus, one gets

$$\frac{dp_g}{dT} = \frac{\Delta\alpha}{\Delta K} \tag{10.8}$$

by differentiating the relation $\Delta V(T, p_g(T)) = 0$ and

$$\frac{dp_{\rm g}}{dT} = \frac{\Delta C_{\rm p}}{T_{\rm g} V \Delta\alpha} - \frac{1}{V\Delta\alpha}\left[1 - \left(\frac{\partial T_{\rm e}}{\partial T}\right)_{\rm p}\right]\frac{dS_{\rm e}}{dT} \tag{10.9}$$

by differentiating the relation $\Delta U(T, p_{\rm g}(T)) = 0$. The Prigogine–Defay ratio Π is defined by

$$\Pi \equiv \frac{\Delta C_{\rm p} \Delta K}{TV(\Delta\alpha)^2} \, . \tag{10.10}$$

In equilibrium transitions the second term on the RHS of Eq. (10.9) vanishes and one has $\Pi = 1$. Experimentally, glass formers usually have[352] $\Pi > 1$: e.g. in Se the Prigogine–Defay ratio lies in the range 1.2–2.4.

The implications of the non-equilibrium value of the Prigogine–Defay ratio in the glass transition were emphasised in the early work of Goldstein.[353] He pointed out that in this case on the energy surface $G(p, T)$ of a liquid there is an infinite set of points, each of which defines a *distinct* glass. In this view there is no "ideal" glassy state having a unique structure: rather, the structure of a glass is whatever the liquid structure happens to be when the liquid solidifies at $T_{\rm g}$. This leads us in turn to the vivid picture of a disordered free energy landscape for the glassy state, having its conceptual precursors in the work of Goldstein[353] and Anderson.[354]

10.4.1 *A topographic view of supercooled liquids*

As emphasised in a review article by Stillinger,[355] in order to understand the phenomena involved in supercooling and glass formation it is useful to adopt a topographic view of the potential energy function $\Phi(\mathbf{x}_1, \ldots, \mathbf{x}_N)$. We can imagine a $3N$-dimensional map showing the "elevation" Φ at any "location" $\mathbf{R} = (\mathbf{x}_1, \ldots, \mathbf{x}_N)$ in the configuration space of the N-particle system. Such a Φ-scape presents maxima ("mountain tops") and minima ("valley bottoms"), as well as saddle points ("mountain passes"). A minimum corresponds to a mechanically stable arrangement of the N particles in space and is enclosed in its own "basin of attraction", containing all configurations that are connected to it by strictly downhill motions. The lowest-lying minima are those that the system would select if it were cooled to absolute zero slowly enough to maintain thermal equilibrium. Higher-lying minima represent amorphous packings which may be sampled by the stable liquid above melting. Transition states correspond to saddle points through which the system may pass in migrating

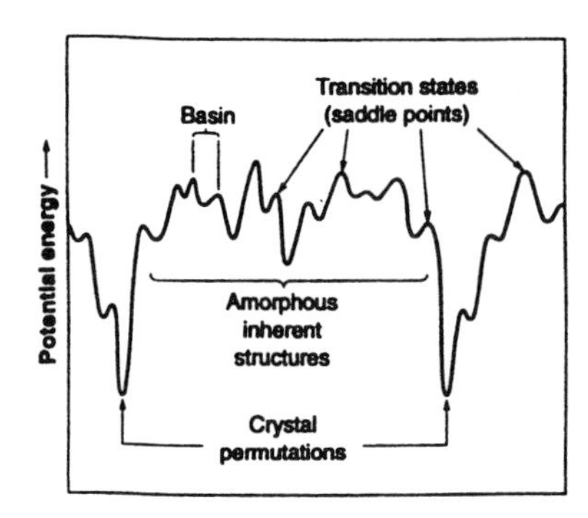

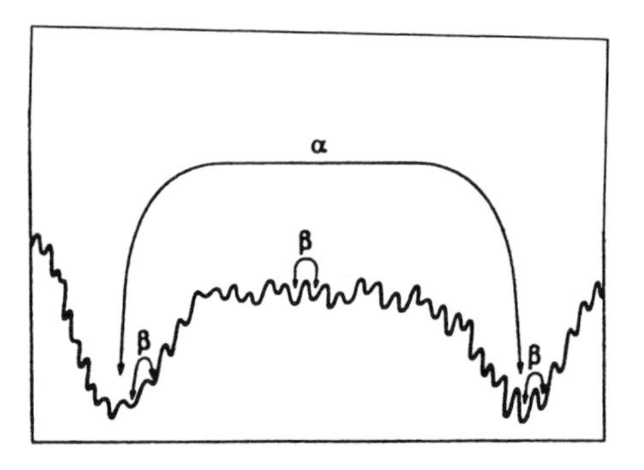

Fig. 10.5. Left: schematic map of the potential energy hypersurface in the configuration space of the many-particle system. Right: illustrating interbasin transitions corresponding to β-type and α-type relaxations in a fragile glass. (Redrawn from Stillinger, Ref. 355.)

from a minimum to another. A schematic view is shown in Fig. 10.5 (left panel).

The Φ-scape picture lends itself to a description in terms of a free-energy landscape on account of the role of temperature. The equilibrium state at any temperature T corresponds to preferential occupation of basins having an optimum depth $\phi^*(T) = \Phi^*(T)/N$. The first-order transition at freezing corresponds to a discontinuous change in ϕ^* as the system switches from liquid to crystalline basins. The supercooled liquid remains instead in basins which refer to higher-lying amorphous structures. However, as long as the configuration point $R(T)$ can move more or less freely among such amorphous structures while avoiding crystal nucleation, the system is in a reproducible quasi-equilibrium state. If the system visits many such minima during its evolution over a certain observation time, it behaves like a liquid over such time scales.

The glass transition occurs when the time scale for jumps among amorphous minima becomes long relative to the experimentally accessible time scales. The supercooled liquid can move among basins whose depths are clustered around $\phi^*_{\text{liquid}}(T)$ only as long as all structural relaxation times are substantially shorter than the time available for measurement. But as temperature drops the mean relaxation times increase according to the Vogel–Fulcher law and cross the experimental time scale at T_{g}. Further cooling fails to lower the depth of the inhabited basins below $\phi^*_{\text{liquid}}(T_{\text{g}})$ and the supercooled liquid has fallen out of quasi-equilibrium.

We shall dwell on the dynamical consequences of this picture in the next section. Let us here briefly comment on its implications with regard to the concept of an ideal glass state.[355] We recall from Sec. 10.1 that at the Kauzmann temperature T_K the crystal and the extrapolated supercooled liquid would attain equal entropies. The configuration entropy of the fully relaxed glass at T_K would then vanish. This realisation, combined with the empirical fact that T_K is closely similar to the relaxation-time divergence temperature T_0 in the Vogel–Fulcher law, has suggested that an "ideal glass state" could be experimentally attained if sufficiently slow cooling rates were available.[330] Such a state, if it exists, would correspond to the inherent structure with the lowest potential energy which is devoid of substantial regions with local crystalline order. There is, however, some ambiguity in qualifying inherent structures with regard to the size and degree of perfection allowed for crystalline inclusions in an otherwise amorphous structure.

10.5 Atomic Motions in the Glassy State

10.5.1 *Relaxation processes*

A visual illustration of the emergence of diffusive slowdown at the level of atomic motions can be given with reference to the discussion of the velocity autocorrelation function in Sec. 5.5. We noted there that in the liquid the particles mostly execute rattling motions from first-neighbour collisions causing trajectory inversions, but occasionally a glancing collision allows a particle to diffuse into a transient void which becomes a new centre of rattling. As the volume per particle decreases with decreasing temperature, each particle spends an increasing fraction of time in rattling motions, since higher cooperation between first neighbours is needed to allow it to diffuse out of its cage. While the time between trajectory reversals remains approximately constant, the diffusion time (which may be viewed as a structural relaxation time) increases. Eventually a wide separation of time scales, of up to 10 orders of magnitude in the temperature range from $1.1T_g$ to T_g, opens up between the rattling time and the structural relaxation time.

More generally, time-dependent response functions to any of a variety of weak external perturbations (mechanical, electrical, thermal, optical and so on) reflect the kinetics of restructuring which results from transitions in the free energy landscape. Denoting such a generic relaxation function as $\phi_n(t)$

with normalisation $\phi_n(0) = 1$ at the initial time, the area under the $\phi_n(t)$ curve defines a mean relaxation time $\tau_n(T)$. Although these times depend to some extent on the property that is being studied, in supercooled liquids they generally increase rapidly with decreasing temperature T and can often be fitted to a Vogel–Fulcher relation as in Eq. (10.2). The so-called CKN melt (the $Ca_2K_3(NO_3)_7$ compound already mentioned in Subsec. 8.6.2, i.e. a 3:2 mixture of the salts KNO_3 and $Ca(NO_3)_2$) has become a model system for experimental test of the predictions made by the mode-coupling theory[356,357] (see below) and has been the object of many physical measurements aimed at understanding glassy dynamics.

In fact, a careful examination of the relaxation functions above the glass transition temperature T_g reveals the presence of distinct processes. Intrabasin relaxation is dominant over time scales of the order of vibrational periods. This domain is followed by an extended time regime in which interbasin structural relaxations take place, and in the long-time limit the relaxation function tends to display a Kohlraus–Williams–Watts "stretched exponential" decay:

$$f_n(t) \approx e^{-(t/t_n)^\gamma} . \tag{10.11}$$

Here, the exponent γ is in the range $0 < \gamma \leq 1$ and the characteristic time t_n is comparable to the mean relaxation time τ_n when $T \cong T_g$. The $\gamma = 1$ limit in Eq. (10.11) corresponds to simple Debye relaxation with a single relaxation time. Smaller values of γ lead to a more or less broad distribution of relaxation times. This becomes evident after transformation to the frequency domain, where peaks appear corresponding approximately to the main relaxation times.

The observed shape of the relaxation spectrum actually depends on temperature as the liquid is cooled towards the glass transition. At equilibrium or moderate supercooling there is a single absorption frequency maximum. On approaching T_g this peak splits into a pair of maxima, the slow primary α-relaxations and the faster secondary β-relaxation (see Fig. 10.6). The former are non-Arrhenius and kinetically frozen out at T_g: they entail escape of the system from one deep basin into another deep one in the free energy landscape. The latter are more nearly Arrhenius and remain operative across the glass transition: they correspond to elementary relaxations between neighbouring basins, through transition processes that require only local rearrangements of a limited number of particles. The nature of the α and β relaxation processes in relation to the free energy landscape is illustrated[355] in Fig. 10.5 (right panel).

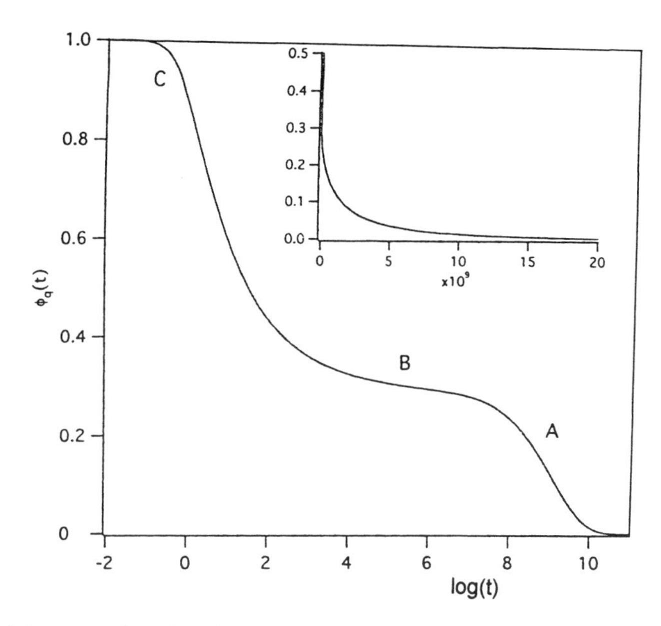

Fig. 10.6. Schematic plot of a relaxation function *versus* $\ln(t)$ in a moderately supercooled liquid, illustrating (A) long-time α-relaxations, (B) intermediate region of two-step relaxation, and (C) microscopic region. On a linear time scale only the α-relaxations are visible (see inset). With decreasing T the α-relaxations move to longer time and eventually freeze, so that the plateau around region B extends and results in elastic scattering of neutrons or light. (Redrawn from Cummins, Ref. 357.)

In mode-coupling theory[356,357] the dynamics of a liquid is described by a Langevin equation for density fluctuations, including a memory term arising from coupling between different modes of motion. Below a critical temperature T_c, say, relaxations are arrested and a spontaneous breaking of ergodicity occurs (that is, the system no longer explores phase space uniformly). Above T_c density relaxation takes place in a two-step form involving a fast β-relaxation and a structural α-relaxation, whereas below T_c only the β-relaxation persists. In further refinements the inclusion of hopping processes smears out T_c and maintains the α-relaxation and transport even at lower temperature.

10.5.2 *Strong and fragile liquids*

The scaling of viscosity data implied by the Vogel–Fulcher representation allows a useful classification of glass-forming liquids between "strong" and "fragile" extremes,[326] depending on the value of the ratio T_0/T_g (with T_g defined in a reproducible way from either relaxation data or from scanning calorimetry at a standard rate). This classification is shown in the Arrhenius plot reported in Fig. 10.7.

The strong limit is realized in open-network liquids such as SiO_2 and GeO_2, which display Arrhenius behaviour corresponding to $T_0 \ll T_g$. The other extreme is realised in liquids characterised by Coulomb interactions such as $ZnCl_2$ or by van der Waals interactions as in aromatic substances with many π electrons: in such fragile liquids the viscosity varies in a strongly non-Arrhenius

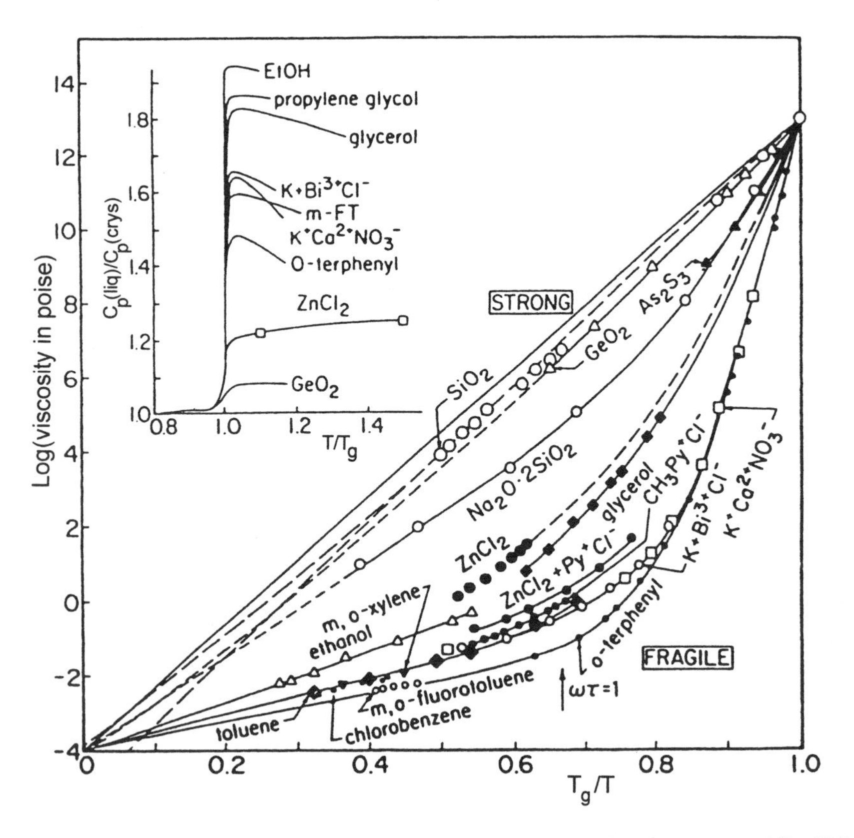

Fig. 10.7. Arrhenius plot of viscosity data showing the strong-fragile pattern of liquid behaviour. The inset shows that the heat capacity jump is generally large for fragile liquids and small for strong liquids. (Redrawn from Angell, Ref. 326.)

fashion. Strong liquids typically display a very small jump in the specific heat at T_g whereas fragile liquids show large jumps (see the inset in Fig. 10.7).

Fragile liquids have glassy structures that may easily reorganise through fluctuations over a variety of particle orientations and coordination states, without much assistance from thermal excitations. Strong liquids, on the other hand, intrinsically resist structural changes and their radial distribution functions and vibrational spectra show little reorganisation even over wide temperature ranges. These differences in behaviour can be traced back to topographic differences in the Φ-scape. In the strong glass-formers only the β-type transitions between contiguous basins are relevant: little reorganisation of the individual basins into deep troughs takes place and the (α, β) bifurcation is weak or absent. In contrast, the most fragile glass-formers exhibit distinctive (α, β) bifurcations corresponding to the onset of strong α-relaxation processes between different deep basins in the free energy landscape.

Fragile glass-formers display near T_g a striking breakdown of the Stokes–Einstein relation between the self-diffusion coefficient D and the shear viscosity η[358]: $D(T)$ may become two orders of magnitude larger than expected from the measured $\eta(T)$ (a similar breakdown of the Nernst–Einstein relation in molten $ZnCl_2$ near freezing was noted in Subsec. 8.6.1). This feature may again be related to the Φ-scape cratering.[355] An α-relaxation process will involve a sequence of elementary interbasin transitions, which may be viewed as a local structural excitation fluidising a mesoscopic domain. Thus, translational diffusion is disproportionately enhanced relative to rotational diffusion, only the latter remaining linked to shear viscosity.

The work of Yamamoto and Onuki[359] demonstrates that the diffusivity of tagged particles is heterogeneous on time scales comparable with, or less than, the stress relaxation time in a highly supercooled model liquid. The particle motions in the relatively active regions dominantly contribute to the mean square displacement, resulting in a diffusion constant larger than the Stokes–Einstein relation would predict.

10.5.3 *Annealing and aging*

We have emphasised above two main features of the viscous liquid state, that is stretched-exponential relaxation and non-Arrhenius behaviour. A third canonical feature of relaxing complex liquids is the so-called non-linearity of relaxation.[360] That is, near and below T_g relaxation takes place in systems which are non-ergodic and are evolving on very long time scales towards an amorphous state in metastable equilibrium. This process of physical aging is of great relevance in the glass industry and in the design, manufacture and use of glassy polymeric materials. The more fragile the liquid, the greater is the dependence of the relaxation time on the departure from equilibrium.[361]

10.5.4 *Anharmonicity and boson peaks*

A sudden increase of slope in the temperature dependence of the Debye–Waller factor, measuring the mean square displacement of the particles of the system, has been observed to occur across the glass transition temperature in a variety of systems (open-network and fragile ionic liquids, organic and inorganic polymers, proteins) by a variety of techniques (Mössbauer and neutron inelastic scattering, computer simulation).

Angell[326] has emphasised that the origin of such breaks is associated with the onset of severe anharmonicity in the molecular motions. He remarks that this view can be reconciled with a view emphasising inelastic processes by recognising that the overdamping of the harmonic motions resulting in the boson peak appears as the inelastic processes detected by neutron scattering and is largely responsible for the increased mean square displacement above T_g.

10.6 Supercooled and Glassy Materials

We conclude this chapter by giving specific reference to some systems that are particularly relevant in the present context, ranging from hard spheres in the amorphous state to supercooled water and to special glassy materials.

10.6.1 *Hard sphere statistics on the amorphous branch*

We have presented in Sec. 2.4 the equilibrium phase diagram of the hard sphere system as a basic model displaying a transition from a fluid state to a crystalline state. The fluid state is stable for values of the packing fraction ϕ on the branch from $\phi = 0$ up to $\phi \cong 0.494$, while the crystalline state is stable on the branch from $\phi \cong 0.545$ to $\phi \cong 0.74$, the latter corresponding to close-packing. The disordered dense hard-sphere state is not represented by any of these lines, but as shown in Fig. 10.8 is a metastable extension of the fluid branch, ending into a state of random close-packing (RCP) at $\phi \cong 0.644$. The RCP can be precisely defined[362] as having the largest packing fraction over

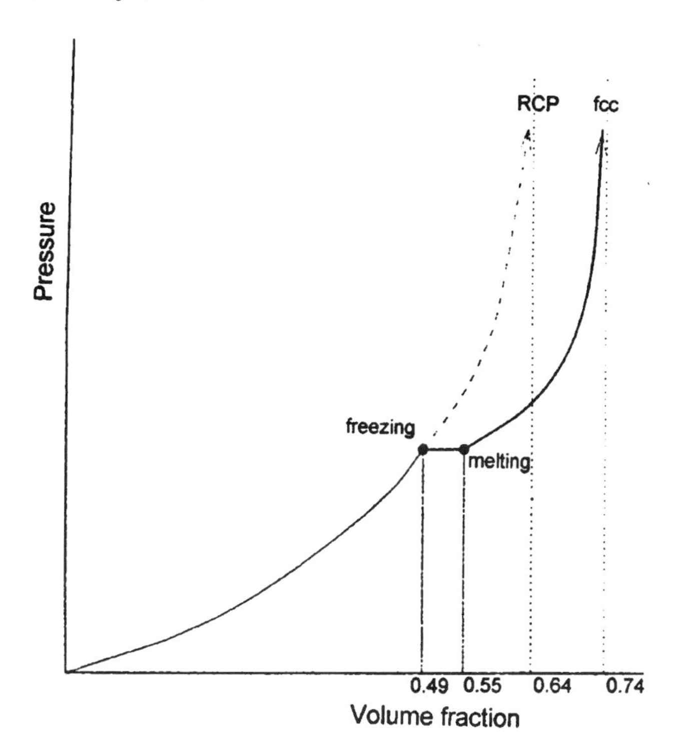

Fig. 10.8. Phase diagram for the hard-sphere system, including the metastable amorphous branch (— — —).

all ergodic isotropic ensembles at which the first neighbour distance equals the hard sphere diameter.

In the context of the hard-sphere phase diagram, the fundamental problem in creating metastable dense systems is to make sure that they are truly random. While there is no perfect measure of order or disorder, Rintoul and Torquato[363] quantify the degree of local order in their computer realisations of hard-sphere assemblies through an order parameter which is a rotationally invariant average over all bonds and is nonzero in the presence of any type of crystallisation. This affords a precise quantitative means of creating dense random systems that lie on the metastable amorphous branch of the phase diagram. On these systems Rintoul and Torquato determine key statistical properties, with special attention to the statistics of voids and to mean pore sizes. The pore length scale determines transport properties, such as the mean survival time of Brownian particles diffusing in a system of traps.

With reference to Sec. 4.9, a random assembly of (non-overlapping, or even overlapping) hard spheres provides a model matrix inside which to study the equilibrium and transport properties of fluids adsorbed in disordered porous materials.[107] Some important questions concern the phase transitions that may occur in such heterogeneous fluids and how they are related to those in the bulk fluid. A great deal of attention has been given to the phenomenon of capillary condensation, i.e. the shift in the bulk liquid–gas transition due to confinement, and to surface phase transitions (layering and prewetting) from interactions with a solid matrix.

10.6.2 *Supercooled water*

The lowest temperatures down to which water remains a liquid in ordinary time-scale observations have been determined in cloud chamber experiments,

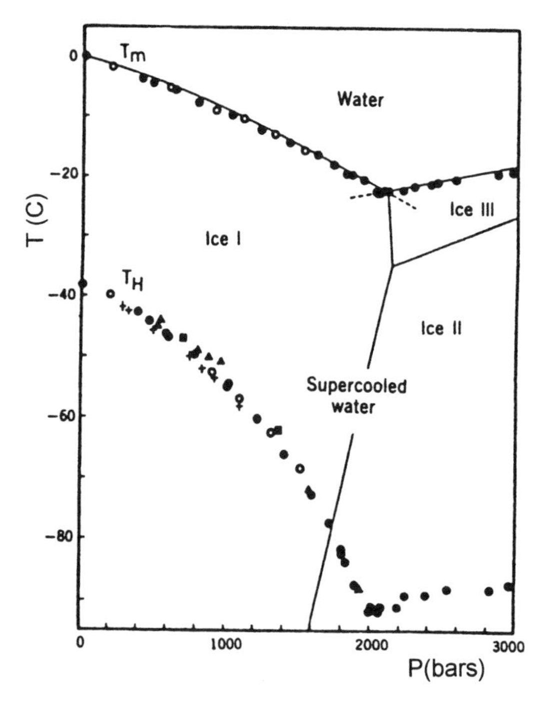

Fig. 10.9. Homogeneous nucleation and equilibrium melting temperatures for water in emulsion form as functions of pressure. (Redrawn from Angell, Ref. 119.)

in which the behaviour of large numbers of droplets can be studied. At standard pressure such small droplets freeze by homogeneous nucleation at $T_H \cong -40°C$. Figure 10.9 reports from the review by Angell[119] the accelerating decrease observed for temperature of homogeneous solidification with increasing pressure, superposed on the equilibrium phase diagram. The vitrification temperature at standard pressure is estimated to be about $-135°C$ by using extrapolations from data on glass-forming aqueous solutions.

Many properties of supercooled water have been measured in great detail.[119] These include thermodynamic properties (density and expansivity, vapour pressure, heat capacity, compressibility and sound velocity), transport properties (diffusivity, viscosity, electric conductivity, dielectric relaxation, sound absorption, nuclear and electron spin relaxations) and spectroscopic properties (Raman and infrared spectra, proton magnetic resonance). An example of such measurements has already been reported in Fig. 5.2, showing an Arrhenius plot of the self-diffusion coefficient from tracer diffusion and nuclear magnetic resonance measurements. Interesting correlations between the temperature dependence of a number of these properties have been brought to light by these studies. We must refer the interested reader to Angell's review article[119] for further details.

10.6.3 *Metallic glasses*

Amorphous metallic alloys were first discovered in 1960 by P. Duwez, who showed that the $Au_{75}Si_{25}$ alloy could be frozen into an amorphous state by rapid cooling from the liquid. Metallic glasses have acquired considerable commercial importance, e.g. in the production of high-strength and wear-resistant materials or of soft magnetic materials. Applications thus range from protective coatings to tape recorder heads and other magnetic devices. Slight or even substantial devitrification, and also surface devitrification, may be used in tailoring material properties.

The realisation of metallic glass requires ultra-rapid cooling. A common laboratory technique is by melt-spinning, in which a fine stream of molten metal alloy is allowed to fall onto a copper wheel in fast rotation and solidifies into a ribbon a few millimeters wide. This method gives cooling rates of the order of 10^5–10^6 degrees per second. A variety of other techniques can be used and a large range of alloy types and compositions can be brought to a glassy state, as listed for instance in the review article by Greer.[364] Examples of glass-forming alloys include (i) transition metals in combination with metalloids, (ii) early and late transition metals, and (iii) aluminium-based, alkaline-earth-based, lanthanide-based and actinide-based systems.

Rapid cooling from the liquidus temperature T_m down to the glass transition temperature T_g is evidently easier when the interval between these two temperatures is small. T_m can be a strong function of composition while T_g usually is not: therefore, glass formation can be expected to be favoured near deep depressions at eutectics in the liquidus curve. A general rule is that the glass forming ability is promoted by stabilising the liquid relative to the solid: this has been exploited by adding solutes and by increasing the number of alloy components. The presence of a multiplicity of components, especially when their atomic sizes are quite different, inhibits crystallisation. An alloy such as $Zr_{41.2}Ti_{13.8}Cu_{12.5}Ni_{10.0}Be_{22.5}$ has a glass-forming ability which approaches that of oxide glasses. However, metallic glasses tend to be much less stable: they crystallise on heating or, in the case of aluminium-based alloys, devitrify into a quasicrystalline phase of icosahedral symmetry.

There have been many attempts at modelling the structure of amorphous metallic alloys. Average coordination numbers are in the range 11–14, consistently with extrapolations from data on liquid metals, and go together with strong chemical ordering. Models are often based on dense random packing and supplemented by the inclusion of local chemical order.

10.6.4 *Superionic glasses*

Ionic glasses are usually realised from a network former such as SiO_2 or Al_2O_3 and a network modifier such as Na_2O or Li_2O. The modifiers lower the glass transition temperature and open up the network structure by introducing non-bridging oxygens, while the alkali ions are mobile and can diffuse through the glassy network. The decoupling of conduction modes from viscous modes which is involved in the generation of ion-conducting glasses below T_g is apparently quite distinct from the (α, β) bifurcation.[365] The modes of small low-charged cations become anharmonic at temperatures far below those of other species and such cations can easily escape from their initial sites and wander through the glassy structure.

A characteristic feature of singly modified ionic glasses is that the ionic conductivity σ rises sharply with the content x of the mobile alkali species,[366] with an effective power law $\sigma \propto x^n$ where, on account of Arrhenius behaviour, the exponent n is proportional to $1/T$. Doping with an alkali halide expands the network and drastically increases the conductivity. Deviations from Arrhenius behaviour are observed in glasses with very high conductivity, which typically contain AgI mixed with silver phosphates, arsenates or borates. Such deviations are most pronounced in the fastest ionic conductors. In mixed-alkali glasses, on the other hand, the conductivity shows a sharp minimum as the relative concentration of the two alkali oxide components is varied, and at the same time the diffusion coefficients of the two mobile cations vary with composition and intersect near the minimum-conductivity composition.

With regard to the a.c. conductivity of singly modified glasses, Roling *et al.*[367] have reported that as a function of composition, temperature and frequency it scales according to

$$\sigma(x, T; \omega) = \sigma_0 f\left(\omega \frac{x}{\sigma_0 T}\right), \tag{10.12}$$

where σ_0 is the d.c. conductivity. This relation holds for the $(Na_2O)_x \cdot (B_2O_3)_{1-x}$ glass up to MHz frequencies. The same scaling function seems to apply to dopant modified glasses and to glass-forming supercooled melts. The temperature dependence of the conductivity changes over from Vogel–Fulcher to Arrhenius behaviour as frequency is increased into the dispersive regime.

10.6.5 *Glassy polymers*

Solid polymers are generally amorphous, that is, they are microscopically disordered and crystallisation is rare. The disorder remains essentially unchanged as the polymer transforms from the amorphous solid state to the melt or liquid state, while elastic behaviour is changing into viscoelastic and ultimately viscous behaviour.

Studies of the microscopic dynamics of polymers below and above the glass transition temperature have given evidence for various processes commonly appearing in glasses: local motions, vibrations, and relaxation processes such as α and β relaxation.[368] Within the glassy state rotations about single bonds of the main chain, which allow diffusion of the polymer chain at high temperatures, are restricted — but this is not the case for local side motions such as the rotations of methyl groups around their threefold axis. These motions are revealed by a drop in the elastic scattering intensity, as the relaxation times for methyl group rotations enter the dynamical window of the neutron spectrometer. With increasing temperature the relaxation times get shorter and the elastic scattering intensity levels off, to enter a further decay step as T_g is approached.

These experiments also show that the inelastic neutron scattering intensity at temperatures still well below T_g contains a prominent boson peak in the region of 1–2 meV (see Fig. 10.10). With increasing temperature the dynamic

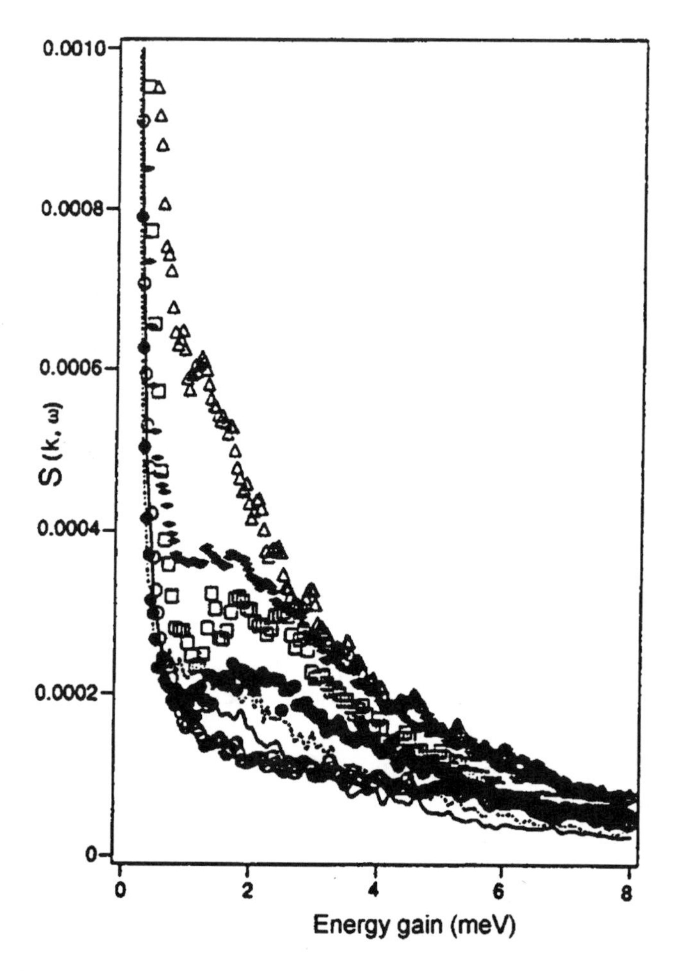

Fig. 10.10. Low-temperature neutron scattering spectra from various amorphous polymers, showing the boson peak as compared to crystalline trans-PB-h6 (o). (Redrawn from Frick and Richter, Ref. 368.)

scattering function starts to convert from the inelastic boson peak spectrum towards a quasi-elastic spectrum. This evolution is also shown in Fig. 10.10.

Fast processes on the picosecond time-scale are revealed by the neutron scattering spectra as the temperature is increased near T_g. Slower processes are observed above T_g, corresponding to α-relaxation of stretched-exponential form. The neutron scattering technique allows one to investigate the wave number dependence of the characteristic relaxation time and of the prefactor, that is the fraction that relaxes *via* the α-relaxation process. This fraction is minimal in correspondence to the main peak position in the structure factor — apparently an instance of the phenomenon of de Gennes narrowing and in agreement with the predictions of mode-coupling theory. An Arrhenius plot of the α-relaxation rate at this wave number $q = 1.88$ Å^{-1}, from neutron scattering data on amorphous polymers,[368] shows that at low temperatures the observed relaxation times follow an Arrhenius behaviour rather than a Vogel–Fulcher law.

Chapter 11

Non-Newtonian Fluids

11.1 Introduction to Non-Newtonian Flow Behaviour

In Chap. 6 we have been concerned with Newtonian liquids in laminar or turbulent-free flow, where according to Newton's law the shear viscosity η is a constant at given temperature and pressure and is independent of the shear rate. In Eq. (6.1) the shear stress σ_{12} has been related to the shear rate $\dot{\gamma} = (\partial v_1/\partial x_2)$ as

$$\sigma_{12} = \eta\dot{\gamma}\,. \tag{11.1}$$

Examples of Newtonian fluids are many pure single-phase liquids of low molecular weight, e.g. water, in which viscous dissipation can be regarded as associated with collisions between small molecules. More properly we might in such fluids think of the shear stress as being represented by a power series expansion in the shear rate,

$$\sigma_{12}(\dot{\gamma}) = \sigma_{12}(0) + \eta\dot{\gamma} + O(\dot{\gamma}^2)\,, \tag{11.2}$$

where $\eta = (\partial\sigma_{12}/\partial\dot{\gamma})|_{\dot{\gamma}=0}$. This expression reduces to Newtonian behaviour in the linear regime of low shear rate, if the time average of the off-diagonal stress tensor is zero.

The time scale τ for stress relaxation can be taken as the time for an average molecule to diffuse a distance of the order of its mean diameter. It is useful to attach a time scale $\tau_s = \dot{\gamma}^{-1}$ to the shear rate, which can be regarded as a "disruption" time scale. If $\tau/\tau_s \ll 1$ the fluid is in the small-perturbation

Newtonian limit: water is Newtonian because $\tau \approx 10^{-12}$ s and, as it is rare to meet shear rates in excess of 10^6 s^{-1}, then $\tau/\tau_s \ll 1$. However, if $\tau/\tau_s \gg 1$ then the molecules are unable to remain close to equilibrium during flow, and there are deviations from Eq. (11.1) and higher-order terms in Eq. (11.2) have to be considered. This is called non-Newtonian flow and its study is termed rheology.[a]

Non-Newtonian fluids typically are made of large and slowly moving molecules, such as polymer melts and colloidal liquids containing solid particles in excess of 0.1 μm in diameter.[b] There are many non-Newtonian fluids in everyday life: for instance yoghurt, tomato ketchup, toothpaste, and household cleaning fluids. Examples to be met in industrial plants, in addition to polymer melts and polymers in solution, are slurries and foams. All of them owe their non-linearity to the fact that their structure changes under shear: for instance, in a suspension of anisotropic particles these tend to align under shear, or in a polymer melt the long-chain molecules are entangled with each other and tend to stretch in the direction of shear. A further point to be noted is that the history of the sample is often an important parameter in determining its non-Newtonian behaviour, especially for polymeric fluids.

Stress generally depends on a number of parameters. An analytic relation between stress and these parameters is called the constitutive equation. Rheology as widely practiced relies on empirical constitutive equations of the form

$$\sigma_{12} = \sigma_y + \eta_p \dot{\gamma}^n \,, \tag{11.3}$$

where σ_y is the yield stress, i.e. the minimum stress that must be applied to the fluid before it will flow. The exponent n allows for a range of non-Newtonian behaviours: in particular $n = 1$ is pseudo-Newtonian with an apparent viscosity η_p describing plasticity. Equation (11.3) with $n = 1$ is known as the Bingham fluid.

[a]In fact the whole stress tensor has to be considered, since the off-diagonal stresses couple to the diagonal components of the stress tensor and these change under shear.

[b]An entirely different type of non-Newtonian behaviour is presented by superfluid liquid Helium, as we have already seen in Sec. 7.7. A further instance that is not included in the present discussion is a plasma of charged particles, which becomes anisotropic when threaded by magnetic lines of force: transverse shear waves with an acoustic dispersion relation (the so-called Alfvén waves) can propagate along the direction of the magnetic field (see e.g. Ichimaru[369]).

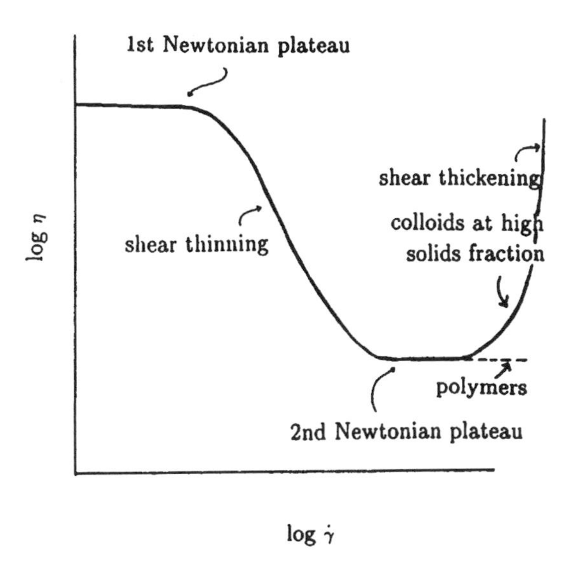

Fig. 11.1. A schematic diagram showing a rheological flow curve for a liquid.

A schematic diagram of a typical rheological flow curve is shown in Fig. 11.1. For $n < 1$ the apparent viscosity decreases with increasing rate of shear — the fluid is said to be shear thinning. If $n > 1$ the fluid is said to be shear thickening or dilatant, usually a situation met with colloidal liquids at high shear rates. A levelling of viscosity in a second plateau is observed for polymer solutions, melts and colloidal suspensions: the latter at high solid fractions (about 50%) show the additional dilatant behaviour. Shear thinning and shear thickening are associated with some form of internal re-ordering of the molecules in the bulk of the fluid, in a manner that affects flow under the applied shear stress or strain rate. The viscosity of all non-Newtonian fluids will approach a limiting value if one waits long enough to take the measurement: this is called Newtonian viscosity η_0 and Eq. (11.1) then holds.

11.1.1 *Linear visco-elasticity*

Time or equivalently "history" is often an essential parameter to include in the description of viscous fluid behaviour. This can arise in two ways. The apparent viscosity, entering for instance Eq. (11.3), could be a function of time (the shear thinning form of this is known as thixotropy). However, time can still be an important parameter in the Newtonian strain rate regime, where there is only a minor perturbation in the internal structure of the fluid. The application of an oscillatory shear strain can excite a visco-elastic response, in which the sample behaves simultaneously (i) as a viscous fluid without memory of its past history, and (ii) as an elastic medium which can store elastic energy of deformation to be recovered on release from the deforming element. We met visco-elastic behaviour in Sec. 6.7.3 in connection with the dispersion of sound velocity ("fast sound") in water and in glass-forming liquids. We return to it here in connection with the response to shears.

Writing the strain applied to the sample as $\gamma(t) = \gamma_0 \cos(\omega t)$, the structure of the liquid is not significantly disrupted provided that the strain amplitude is low ($\gamma_0 < 0.03$, say). Its response in the visco-elastic regime can be described by means of a complex shear modulus,

$$G(\omega) = G'(\omega) - iG''(\omega)\,, \tag{11.4}$$

where $G'(\omega)$ is the elastic storage modulus and $G''(\omega)$ is the viscous loss modulus. The Maxwell model for a classical visco-elastic system describes the stress relaxation following the application of a step of unit strain at time $t = 0$ through the function

$$C_s(t) = G_\infty e^{-t/\tau}\,, \tag{11.5}$$

where G_∞ is the high-frequency elastic modulus and $\tau = \eta/G_\infty$. In terms of the so-called stress relaxation function we have

$$G(\omega) = -i\omega \int_0^\infty dt C_s(t) e^{i\omega t} = -\frac{i\omega\tau}{1 - i\omega\tau} G_\infty \tag{11.6}$$

or

$$\begin{cases} G'(\omega) = \left[\dfrac{\omega^2\tau^2}{1+\omega^2\tau^2}\right] G_\infty \\ \\ G''(\omega) = \left[\dfrac{\omega\tau}{1+\omega^2\tau^2}\right] G_\infty\,. \end{cases} \tag{11.7}$$

The phenomenon may also be described through a complex frequency-dependent shear viscosity, $\eta(\omega) = \eta/(1 - i\omega\tau)$ becoming purely imaginary at high frequency.

We conclude by remarking that stress relaxation can also occur in a solid which may creep e.g. by vacancy diffusion at high temperature. In this case the stress falls in time as the atoms adjust their positions, but there is a limit to the amount of creep that may take place over the observational time scale.

11.2 Viscosity in Uniaxial Liquid

Consider a liquid which is anisotropic when at rest and whose anisotropy is unaffected by flow. We shall in particular be concerned with nematic liquid crystals: these are uniaxial and the axis about which rotational symmetry exists is the director, which is defined from the mean orientation of all molecules (see Chap. 9). The director may, in principle, be anchored in any chosen orientation by applying a suitable field. Magnetic fields of moderate intensity can anchor the director very effectively in a nematic bulk sample. Notice that when the director is anchored the argument used in Sec. 6.2 to show that the stress tensor is symmetric is no longer valid. The viscous stresses can exert a finite torque on any fluid element because, following infinitesimal rotations of the molecules within the element, such viscous torque is balanced by a countertorque exerted by the anchoring field. Below we base our summary of the topic of viscosity in a uniaxial liquid in the linear regime on the account given by Faber.[4]

Whereas the constitutive equations for an isotropic incompressible fluid involve only one viscosity coefficient (see Sec. 6.2), those for a fluid with uniaxial anisotropy require five such coefficients, each of them independent of the others. An even larger number of coefficients are needed for biaxial anisotropy. The flow regimes of nematic liquid crystals are therefore more complex, essentially because the translational motions are coupled to orientational motions of the molecules and flow disturbs the molecular alignment in the absence of orientational anchoring (see Sec. 11.3). Conversely, a change in the alignment, e.g. by application of an external field, will induce flow in a nematic. Optical observations of these effects are impeded by the high turbidity of nematic samples — typically higher than the light scattering by conventional isotropic fluids by a factor of order 10^6, from spontaneous fluctuations in the molecular alignment. Measurements of viscosity in nematics require either imposing an alignment by an external field or using sophisticated probes such as attenuation of acoustic shear waves or inelastic scattering of light.

Suppose that the director of a uniaxial fluid is anchored in the x_3 direction. Then the linear equations for shear stress that replace Eq. (6.3) read[370]

$$\sigma_{12} = \sigma_{21} = \eta_3 \left(\frac{\partial v_1}{\partial x_2} + \frac{\partial v_2}{\partial x_1} \right) , \tag{11.8}$$

$$\sigma_{23} = \eta_4 \frac{\partial v_2}{\partial x_3} + \eta_1 \frac{\partial v_3}{\partial x_2} , \quad \sigma_{32} = \eta_2 \frac{\partial v_2}{\partial x_3} + \eta_4 \frac{\partial v_3}{\partial x_2} , \tag{11.9}$$

$$\sigma_{31} = \eta_4 \frac{\partial v_3}{\partial x_1} + \eta_2 \frac{\partial v_1}{\partial x_3} , \quad \sigma_{13} = \eta_1 \frac{\partial v_3}{\partial x_1} + \eta_4 \frac{\partial v_1}{\partial x_3} . \tag{11.10}$$

Here η_i ($i = 1$–4) constitute four of the five linear viscosity coefficients of the uniaxial fluid. Noteworthy is the fact that in Eq. (11.9) the coefficient of $(\partial v_2/\partial x_3)$ referred to σ_{23} and that of $(\partial v_3/\partial x_2)$ for σ_{32} are the same. An analogous situation arises in Eq. (11.10). These are instances of Onsager's relations which obtain in irreversible thermodynamics (see e.g. Sec. 7.6).

The linear equations for the diagonal stress components, which replace Eq. (6.3) for an uniaxial fluid when the director is anchored along x_3, involve a fifth viscosity coefficient:

$$\sigma_{11} = -p + \frac{2}{3} \left(2\eta_3 \frac{\partial v_1}{\partial x_1} - \eta_3 \frac{\partial v_2}{\partial x_2} - \eta_5 \frac{\partial v_3}{\partial x_3} \right) , \tag{11.11}$$

$$\sigma_{22} = -p + \frac{2}{3} \left(-\eta_3 \frac{\partial v_1}{\partial x_1} + 2\eta_3 \frac{\partial v_2}{\partial x_2} - \eta_5 \frac{\partial v_3}{\partial x_3} \right) , \tag{11.12}$$

$$\sigma_{33} = -p + \frac{2}{3} \left(-\eta_3 \frac{\partial v_1}{\partial x_1} - \eta_3 \frac{\partial v_2}{\partial x_2} + 2\eta_5 \frac{\partial v_3}{\partial x_3} \right) . \tag{11.13}$$

In the case of incompressible flow, i.e. $\nabla \cdot \mathbf{v} = 0$, we can write

$$\sigma_{33} - \frac{1}{2}(\sigma_{11} + \sigma_{22}) = (\eta_3 + 2\eta_5) \frac{\partial v_3}{\partial x_3} . \tag{11.14}$$

Permutation of indices yields the equations that apply when the director is along x_1 or x_2.

Figure 11.2 shows four cases of steady planar shear flow of an uniaxial fluid along the x_1 direction between two plates perpendicular to the x_2 axis, so that v_1 is the only non-zero velocity component and increases linearly with x_2 but is independent of x_1 and x_3. In such a set-up one may in principle measure the ratio between the stress σ_{12} on the plates and the velocity gradient (dv_1/dx_2) to determine (a) η_1 if the director is along x_1, (b) η_2 if the director is along x_2, and (c) η_3 if the director is along x_3. According to Eq. (11.14) one may in

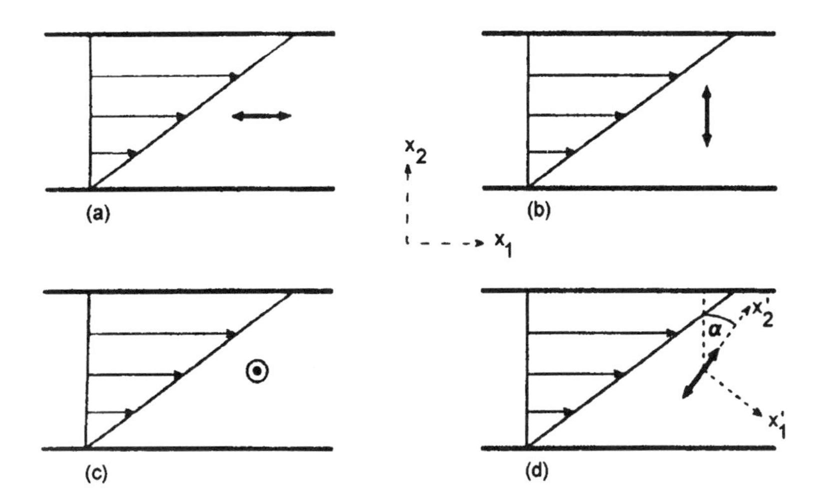

Fig. 11.2. Four cases of planar shear in a uniaxial fluid. The orientation of the director is shown by a double-headed arrow in (a), (b) and (d) and by a point-and-circle in (c). (Redrawn from Faber, Ref. 4.)

principle determine η_5 from the force needed to extend an element of the fluid of known cross-sectional area in which the director is oriented longitudinally.

In Fig. 11.2(d) the director is anchored along the x'_2 axis of a reference frame S' which differs from the original frame S by a left-handed rotation about the x_3 axis through an angle α. As Faber[4] shows in detail, the effective viscosity $\eta^*(\alpha)$ as a function of the angle α is

$$\eta^*(\alpha) = \eta_2 \cos^4 \alpha + 2(\eta_3 + \eta_5 - \eta_4) \sin^2 \alpha \cos^2 \alpha + \eta_1 \sin^4 \alpha \,. \tag{11.15}$$

The normal stress component in the frame S can also be found. Writing σ_{33} in the form

$$\sigma_{33} = -p + \frac{4}{3}\xi_{12}(\eta_3 - \eta_5) \sin \alpha \cos \alpha \,, \tag{11.16}$$

where $\xi_{12} = \frac{1}{2}(dv_1/dx_2 + dv_2/dx_1)$, this is valid even when the fluid undergoes non-planar shear about the x_3 axis (i.e. when dv_2/dx_1 is non-zero as well as dv_1/dx_2).

Whereas the diagonal stresses remain isotropic in an isotropic Newtonian fluid undergoing planar shear, they become anisotropic in an anisotropic fluid under shear. We shall see consequences of this fact in later sections.

11.3 Flow Birefringence and Flow Alignment

It has been known since the time of Maxwell that simple liquids which are completely isotropic when at rest, but are composed of non-spherical molecules, tend to become birefringent when subjected to shear. The phenomenon is called flow birefringence and is also exhibited by solutions made from non-spherical solute molecules and by suspensions of solid rods, fibres, or platelets. We therefore proceed to extend the preceding discussion to cases where no external field is anchoring the director, but preferred axial alignment is induced by flow. We follow again the account of Faber[4] for the situation in which the flow combines a shear component $\xi_{12} = \frac{1}{2}[(dv_1/dx_2) + (dv_2/dx_1)]$ with a vorticity $\omega_{12} = \frac{1}{2}[(dv_1/dx_2) - (dv_2/dx_1)]$.

Consider first a dilute suspension of essentially independent solid particles in the shape of prolate spheroids, immersed in a Newtonian fluid of viscosity η. In the case of planar shear in the (x_1, x_2) plane we have $\xi_{12} = \omega_{12}$: the spheroidal inclusions then precess continuously about the x_3 axis and at the same time undergo orientational fluctuations from rotational diffusion induced by thermal agitation. The angular distribution function $f(\alpha)$, defined so that $f(\alpha)d\alpha$ is the fraction of spheroids whose major axis lies at any given time in the angular range between α and $\alpha + d\alpha$ from the x_2 axis, has a maximum at $\alpha = \pi/4$ when rotational diffusion is dominant, as is often observed experimentally. In the opposite limit one finds instead that the maximum lies at $\alpha = \pi/2$, corresponding to preferred alignment along the x_1 axis.

Another important case leading to flow alignment is that of a concentrated suspension of rod-like particles. If the rods tend to become almost parallel during flow, then the fluid becomes markedly uniaxial and needs to be described by five viscosity coefficients. In this situation the director can go into preferred orientations and remain there in the absence of any anchoring field, precession of the individual molecules being effectively suppressed. In vorticity-free shear ($\omega_{12} = 0$) the preferred orientation is at $\alpha = \pi/4$, i.e. the rods tend to line up in the direction in which the suspension is being stretched. In planar shear, on the other hand, the stable solution is

$$\alpha \cong \frac{\pi}{2} - \sqrt{\frac{\eta_4 - \eta_1}{\eta_2}}, \tag{11.17}$$

assuming $\eta_2 \gg \eta_4 \gg \eta_1$. From measurements of the viscosities of the MBBA nematic liquid crystal[370] Eq. (11.17) predicts that the director sets at $\alpha \cong 82°$ if no external field constrains it.

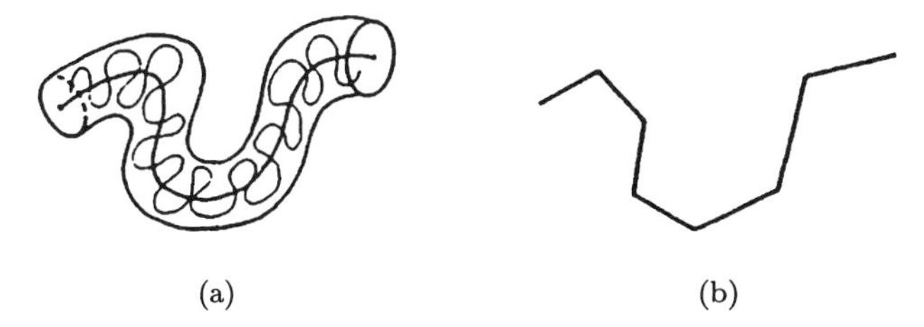

Fig. 11.3. The tube model, showing (a) the polymeric chain together with its tube of constraints and (b) the primitive chain in discretised form.

11.4 Non-Newtonian Behaviour in Polymeric Liquids

The dynamical properties of polymer melts can frequently be traced back to the topological hindrance exerted on each chain by the surrounding chains. These constraints have a rather complex many-body structure and the result has been that, in modelling the melt, their effects have often been represented by the action of a confining tube on the dynamics of a single chain (see Fig. 11.3). Further reduction is sometimes accomplished by replacing the tube by a set of fixed interaction sites selected at random from points that lie along its surface. Such a set of random scattering centres can be considered as among the simplest possible realisations of the microscopic entanglements occurring in a polymer melt.[371,372]

It seems to be rather generally agreed that a polymer molecule finding itself in such an arrangement of scattering centres would move by sliding down its own contour, to be likened to the way that defects travel down the length of a rope. *Reptation* (a word which implies snake-like motion) is the term used to describe segmental motion of this type, which is biased along directions parallel to the backbone of the chain. As to the actual behaviour to be expected of the polymer/scatterer assembly, a transition from free motion to reptation is believed to be exhibited at a particular concentration of scattering centres, just as in the melt an analogous type of transition is thought to occur at a definite chain length of the polymer[371] (see also Loring[373] and earlier references there).

Not only is the flow of a polymeric liquid strongly dependent on its molecular weight, but its flow properties are also sensitive to cross-linking. When the polymer is dispersed in an inert solvent, flow behaviour naturally depends on the concentration. Here we are concerned mainly with long-chain polymers which are not cross-linked, and which are either in molten form or dispersed at concentrations that are not so low for the molecules to behave independently of one another. The situations envisaged here are such that each molecule is constrained by entanglements with other molecules and can extricate itself only by reptation.

11.4.1 *Reptation in concentrated polymer systems*

Flow processes in polymer melts or concentrated polymer solutions are an essential step in the processing of polymeric materials into manufactured items.[374] In treating the motion of a laterally constrained chain along its length de Gennes[371] showed that rearrangements of the chain within the existing constraints obey the mathematics of a model previously developed by Rouse[375] for a single chain[c] and occur in a time proportional to M^2, M being the molecular weight. The time needed for the chain to reptate out of the existing constraints is thus proportional to M^3. As the polymeric chain is deformed, so are the constraints and, if the only way for the chain to return to complete equilibrium is to diffuse out of the deformed constraints, the longest relaxation time will become proportional to M^3. This result can be transferred to the viscosity coefficient, which also is found to be proportional to M^3. The experiments indicate an $M^{3.4}$ law.

In the work of Doi and Edwards[372] and later authors[374] the model is further developed into a constitutive equation for the stress tensor. Throughout the analysis the stress is calculated by considering the entropic forces in the segments of the tube centre-line (the "primitive chain") as given by a Gaussian form and by assuming that the segments of the primitive chain deform affinely with the continuum. In the particular case of a stress relaxation experiment, just after an impulsive deformation is imposed on the polymeric liquid the initial stress is identical to that of an ideal rubber network. The subsequent stress relaxation results from the combination of two processes, which have a different time scale and thus essentially occur one after the other.

The fast relaxation process is a redistribution of monomers along the primitive chain, corresponding to the monomer density going back to its equilibrium

[c]In brief, the model envisages each chain as a multiplicity of friction points ("beads") connected by springs. Stress relaxation is evaluated by introducing a spectrum of relaxation times $\tau_n = \tau/n^2$, where the maximum relaxation time τ is proportional to a friction constant ζ. Hydrodynamic interactions among the beads were later introduced by Zimm. On account of the fact that a real polymeric material is usually polydisperse (i.e. contains a distribution of molecular weights), the Rouse–Zimm model leads to visco-elastic behaviour.

value, and takes place while the primitive chain remains fixed in its deformed configuration. The change of stress in time follows a Rouse-like behaviour with the largest relaxation time being proportional to M^2. The second relaxation process is much slower and corresponds to a longitudinal diffusion of the chain along the tube, bringing the chain back to equilibrium through a reptation out of the deformed constraints. During this process the deformed tube can be considered as fixed, since — although its "walls" are made up of other chains which are also diffusing — its relaxation time results from the cooperative motions of many chains and is much longer than that of a single chain. The deformed tube progressively disappears starting from its ends and is replaced by a new tube having an equilibrium random conformation. The final result of the model is a constitutive equation for the stress tensor, having again the form of a visco-elastic behaviour with a memory function determined by a relaxation spectrum.

11.4.2 *Macroscopic flow phenomena in polymeric liquids*

As noted in Faber's book,[4] an aqueous solution of high-molecular-weight polyacrylamide at a concentration in the range 1–2% by weight will allow a demonstration of most of the behaviours outlined below.

A striking phenomenon, for which the reptation model provides adequate explanation, is that of tension-thickening: that is, the extensional viscosity [$\eta_{\text{ext}} = \eta_3 + 2\eta_5$ in Eq. (11.14)] increases as the rate of extension increases. At large rates of extension the ratio between η_{ext} and the apparent shear viscosity in a polymeric liquid may exceed that of a Newtonian liquid by several orders of magnitude. In such cases η_{ext} is apparently much larger than the other viscosities (η_i for i = 1–4) individually. This property is responsible for the fact that many polymeric liquids may readily be drawn out into fine threads. Whereas a water jet forced under pressure through an orifice will soon break up by a dynamic instability, tension thickening stabilises polymer threads against formation and development of necks in their structure.

Polymer solutions not only have a large extensional viscosity, but are also visco-elastic. Thus, if a falling stream of polymeric solution is suddenly cut, the two parts of it contract in much the same way as a stretched rubber band contracts when it is cut.

A further manifestation of non-Newtonian behaviour in polymeric liquids is the die-swell effect. This is observed when a jet of liquid emerges from a capillary tube through which it has been moving in a laminar fashion under a longitudinal pressure gradient. Except at very small values of the Reynolds number, a jet of a Newtonian viscous fluid would narrow as it emerges from the capillary, but jets of polymeric liquids normally increase markedly in radius. Inside the capillary the liquid feels a diagonal stress given by Eq. (11.16), which is compressive insofar as η_{ext} exceeds $3\eta_3$ and therefore tends to shorten ring-shaped elements of fluid. When the fluid emerges the rate of shear vanishes and the stress disappears, but because of its visco-elasticity the fluid carries memory and responds to the removal of the compressive stress by expanding.

A related effect was demonstrated by Weissenberg. Consider a vertical rod dipped into a dish of liquid and spinning about its axis. The liquid is rotating too, with an angular velocity which decreases with increasing radius. The effect of such rotation on a dishful of a Newtonian viscous liquid is to displace the liquid away from the rod and to lower the level of the liquid surface there. The effect which is observed on a polymeric liquid is quite the reverse and is again related to the large value of its extensional viscosity.

A striking manifestation of shear thinning in polymeric liquids is a phenomenon referred to as the spurt effect, which is often observed when such a liquid is extruded through a cylindrical capillary tube. The discharge rate as a function of the pressure gradient displays a hysteretic transition from a slow to a fast regime.

The rate of flow of water through a capillary may be strongly affected by small additions of polymeric molecules. Such additives not only enable water at large values of the Reynolds number to flow more freely through tubes, but also reduce the drag force exerted on obstacles.

11.5 Flow in Nematic Liquid Crystals

As introduced in Chap. 9, liquid crystals are locally anisotropic mesophases intermediate between a crystalline solid at moderately low temperatures and an isotropic liquid at moderately high temperatures. Liquid-crystal behaviour has been observed in pure compounds of both rod-like and disk-like molecules. In all mesophases the local structural anisotropy is usually uniaxial in the case of rod-like molecules, as determined by the local relative alignment of the molecules. The nematic mesophase has a high degree of such orientational order, but no long-range translational order. The preferred axis of orientation

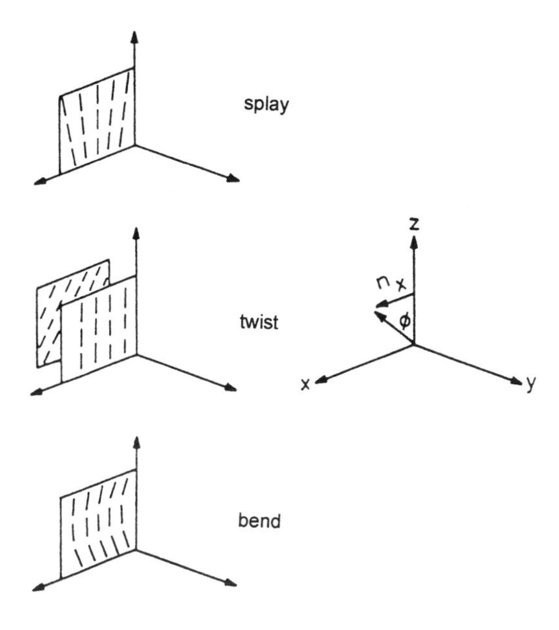

Fig. 11.4. The three main types of deformation in a nematic liquid crystal.

usually varies from point to point in the absence of anchoring fields, but a monodomain sample is optically uniaxial and strongly birefringent. The anisotropy is a function of the orientational order and decreases with rise of temperature, dropping abruptly to zero at a threshold temperature through a weakly first-order transition into the isotropic phase. Such materials commonly feature in liquid-crystal displays.

The orientation of the director in a nematic in state of flow is determined by four competing factors. In addition to anchoring by external fields if present (see Sec. 11.2) and to the emergence of flow alignment (see Sec. 11.3), the containing solid surfaces affect the flow of thin fluid specimens and can be treated in various ways so as to locally anchor the director. The fourth influence is that of the so-called curvature elasticity of the nematic sample, that we introduce immediately below.

11.5.1 *Curvature elasticity and the Freedericksz transition*

Curvature elasticity is associated with stiffness of the nematic sample against orientational deformation of the director.[376] The three main types of deformation are splay, twist and bend, as illustrated schematically in Fig. 11.4. Considering first the case of pure splay, the elastic free energy per unit volume is $\frac{1}{2}K_1(d\phi/dx)^2$, where ϕ is the tilt of the director and K_1 is the splay elastic constant. Taking the director to be of unit magnitude and describing it by a unit vector $\hat{n}$, we have $\phi = n_x$ and the free energy density is $\frac{1}{2}K_1(dn_x/dx)^2$. Extension of the argument to twist and bend is immediate and leads to the expression

$$F = \frac{1}{2}K_1(\nabla \cdot \hat{n})^2 + \frac{1}{2}K_2(\hat{n} \cdot \nabla \times \hat{n})^2 + \frac{1}{2}K_3(\hat{n} \times \nabla \times \hat{n})^2 , \qquad (11.18)$$

for the elastic free energy density. The elastic coefficients K_i are known as Frank's elastic constants.

The most direct way of measuring the elastic constants of a nematic sample is by studying the distortions induced by an external magnetic field. The geometry has to be chosen so that the orienting effect of the field works against the orientations imposed by the surfaces confining the liquid crystal. The three principal experimental geometries are illustrated in Fig. 11.5. Above a critical

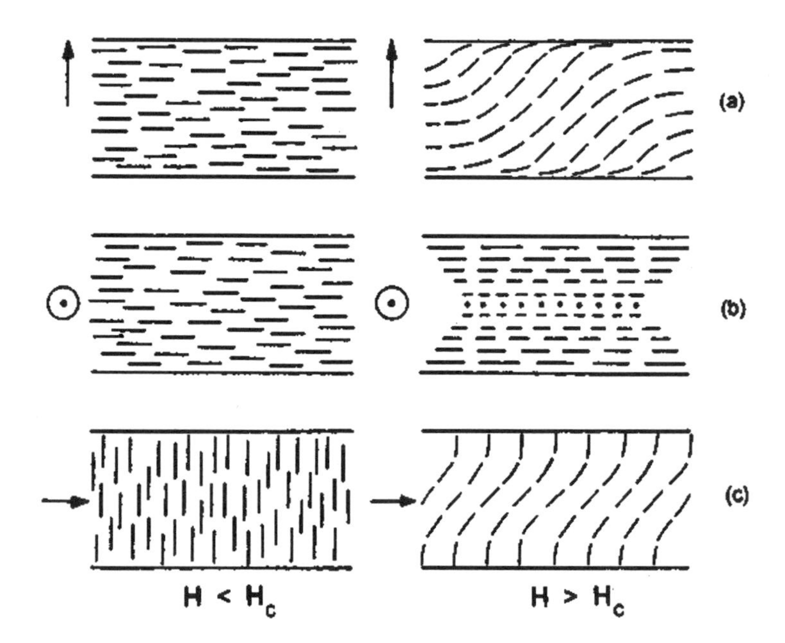

Fig. 11.5. Geometries for measuring the three elastic constants of a nematic liquid crystal.

value H_c of the field, a distortion sets in which can be detected optically. The threshold field is determined by the appropriate Frank constant and by the difference χ_a between the principal diamagnetic susceptibilities along and perpendicular to the axis of the director.

To explain the mechanism of this so-called Freedericksz transition, let us consider the twist geometry shown in Fig. 11.5(b). The free energy density in the presence of the field is

$$F = \frac{1}{2}K_2\left(\frac{d\vartheta}{dz}\right)^2 - \frac{1}{2}\chi_a H^2 \sin^2\vartheta \tag{11.19}$$

and the equilibrium value of the twist angle $\vartheta(z)$ is determined by minimisation of F. For small deformations we can set $\vartheta = \vartheta_m \cos(qz)$, neglecting higher harmonics and with $q = \pi/d$ where d is the sample thickness. Minimisation of the free energy density yields

$$H_c = \left(\frac{\pi}{d}\right)\left(\frac{K_2}{\chi_a}\right)^{1/2}. \tag{11.20}$$

The shape of the energy per unit area as a function of ϑ_m is reminiscent of the free energy curve for a second-order phase transition: its minimum lies at $\vartheta_m = 0$ for $H < H_c$, but the distorted state becomes the stable one above the threshold field.

11.5.2 *Macroscopic flow and disclinations in nematics*

We consider a layer of nematic liquid crystal which is undergoing planar shear flow in the x_1 direction between two plates located at $x_2 = \pm d/2$ and moving in opposite directions. No external field is applied, but the plate surfaces have been treated so that at contact with them the director is anchored along x_3. On account of the shear the director tends to twist away from the x_3 axis into an orientation almost parallel to the x_1 axis, which is favoured by the flow (see Fig. 11.6). There is in this situation a competition between the torque responsible for flow alignment and the counter-torque exerted by curvature elasticity until, at some critical shear rate, the orientation of the director along x_3 becomes unstable against a perturbation twisting it about the x_2 axis by an angle $\beta(x_2)$. The result is that above the critical shear rate the fluid forms rolls, which are parallel to the x_1 axis and have thickness

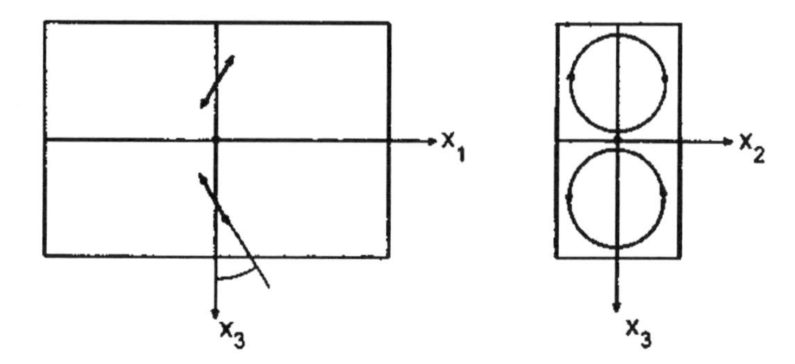

Fig. 11.6. Twist of the director in a nematic under shear (left), leading to the formation of liquid rolls (right).

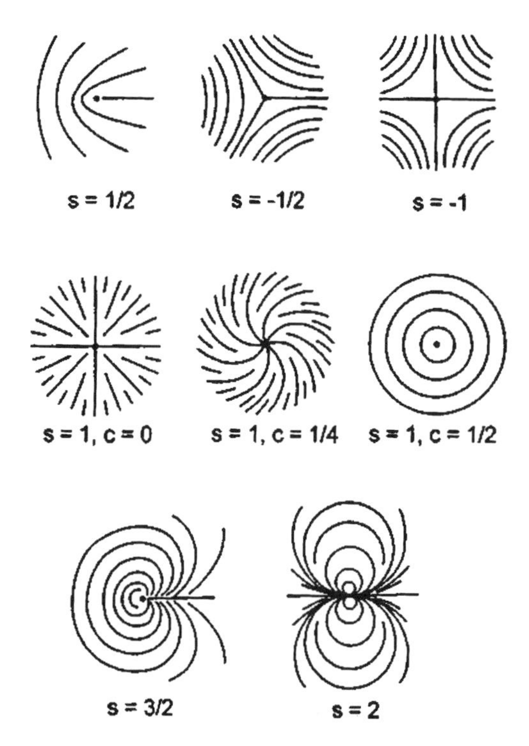

Fig. 11.7. Director field in the neighbourhood of a disclination (after Frank, Ref. 377.)

comparable to d in the x_3 direction. The sense of circulation alternates from one roll to the next.

Rapid circulation of a nematic liquid crystal, induced by increasing the rate of shear or the temperature difference between top and bottom surface, leads to the generation of disclinations. A disclination is a line of singularity around which there is a marked distortion in the director field (splay, twist, bend or their combinations, as the case may be). Figure 11.7 reports from the work of Frank[377] examples of the director field $\psi(\alpha)$ as a function of the angle α around the disclination line, for several values of the parameters s and c entering the relation

$$\psi(\alpha) = s\alpha + \pi c\,. \tag{11.21}$$

With reference to Fig. 11.8, if incident light is polarised at an angle ψ with respect to the x axis its polarisation will remain unchanged at all points on the polar line α and hence will not be transmitted by a crossed analyser, resulting in a dark brush at an angle α. A similar situation arises when ψ changes by $\pi/2$. Thus disclinations become visible as dark brushes when viewed through a microscope between crossed polarisers, the number of dark brushes per singularity being $4|s|$. Neighbouring disclinations connected by brushes are of opposite signs and the sum of the strength of all disclinations in the sample is zero.

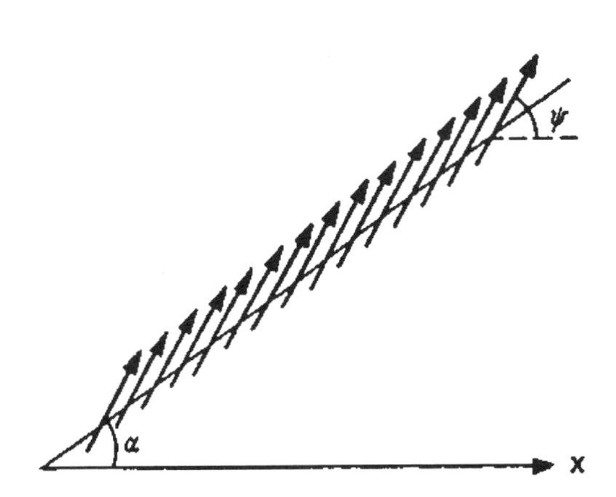

Fig. 11.8. The arrows mark the orientation of the director along a polar line at an angle α. Incident light which is linearly polarised at angle ψ or $\psi \pm \pi/2$ will be extinguished by a crossed analyser and will give a dark brush at angle α.

The deformation energy associated with an isolated disclination in a circular layer of radius R and unit thickness is obtained from elasticity theory as[370]

$$F = \frac{1}{2} K \int_{r_c}^{R} r dr \int_{0}^{2\pi} d\alpha \frac{1}{r^2} \left(\frac{d\psi}{d\alpha} \right)^2 , \tag{11.22}$$

where K is a suitable elasticity constant and r_c is an inner cut-off radius inside which elasticity theory breaks down. From Eq. (11.21) we get

$$F = \pi K s^2 \ln \left(\frac{R}{r_c} \right) . \tag{11.23}$$

For a pair of disclinations at a relative distance r, using $\psi = \psi_1 + \psi_2$ we similarly get

$$F = \pi K (s_1 + s_2)^2 \ln \left(\frac{R}{r_c} \right) - 2\pi K s_1 s_2 \ln \left(\frac{r}{r_c} \right) . \tag{11.24}$$

In the case of singularities of opposite sign ($s_1 s_2 < 0$), F increases with increasing r: therefore, such singularities attract each other with a force which is inversely proportional to their separation, as would be the case for two current-carrying conductors.

For reviews of optical effects and optical applications in liquid crystals the reader may refer to a reprint collection by Jànossy[378] and to an article by Durand.[379]

11.6 Colloidal Dispersions and Suspensions

Colloidal dispersions are two-phase systems involving mesoscopic solid or liquid particles suspended in a liquid. Examples to be met in everyday life are paints, ink, lubricants, cosmetics, and milk. The rheological properties of suspensions are important in many industrial applications. An introductory account may be found in an article by Hansen and Pusey.[380]

The sizes of colloidal particles are typically in the range 10 to 10^3 nanometers — much larger than single molecules but small enough that Brownian motion usually dominates gravitational settling, allowing thermodynamic equilibrium to be attained. Solid colloidal particles may be mineral crystallites, synthetic polymeric particles such as polystyrene spheres suspended in water, or amorphous polymethylmetacrylate (PMMA) particles dispersed in hydrocarbons. These mesoscopic particles are impenetrable and usually attract each other via strong van der Waals forces, which may lead to flocculation or coagulation of the colloids into gel-like structures. Flocculation may be precluded by either steric or electrostatic stabilisation. Steric stabilisation is achieved by grafting polymer brushes on the surface of the colloidal particles, providing an elastic repulsion when two particles come so close together that their brushes are compressed. Colloidal particles in water generally acquire a charge by dissociation of surface groups, forming with ions in solution electrical double layers which repel strongly whenever neighbouring surfaces approach closer than the Debye screening length. In fact, the interactions between colloidal particles can be tuned, e.g. by the addition of salt to a dispersion of charged colloids, leading to a reduction of the screening length, or by the addition of free polymers.

These tunable interactions lead to a rich variety of phase behaviours which have been thoroughly investigated, both experimentally and theoretically.[381] Depending on colloid concentration and on the concentration of added ions or polymers, the suspensions exhibit gaseous, liquid, crystalline, and glassy phases. A schematic phase diagram of the colloid-polymer system[380] is reported in Fig 11.9. Colloidal crystals in suspension are readily detected by Bragg reflection of visible light, whose wavelength is comparable to the spacing of the colloidal lattice. Similarly the dynamics of colloidal systems is slow on the laboratory time scale, permitting detailed studies of metastability and of the kinetics of phase transitions. Statistical mechanical approaches have been developed to account for polydispersity, the inevitable distribution of sizes of colloidal particles.

11.6.1 *Flow properties of colloidal dispersions*

We base the present summary of the rheology of colloidal dispersions on the account given by Pusey,[381] with main attention to simple shear flows in suspensions of essentially hard spheres. Figure 11.10 reproduces from the work of Choi and Krieger[382] the relationship between viscosity and shear stress for suspensions of PMMA spheres of various sizes in a variety of liquid media. The spheres are stabilised by polymer coating. The relative viscosity η_R on the vertical axis is defined as the ratio $\eta_R = \eta/\eta_0$, where η is obtained from the shear stress σ needed to induce flow at strain rate $\dot{\gamma}$ as $\eta = \sigma/\dot{\gamma}$ and η_0 is the viscosity of the pure suspension medium. The reduced stress s_R^* on the horizontal axis, on the other hand, is defined as $s_R^* = R^3\sigma/k_BT$ where R is the average particle radius. With the characteristic time scale of the flow-induced structural rearrangement defined as $\tau_s = \dot{\gamma}^{-1}$ (see Sec. 11.1) and estimating the structural relaxation time as $\tau \approx R^2/D$ with the diffusion coefficient given by $D = k_B T/6\pi\eta R$, we find $\tau/\tau_s = 6\pi s_R$. That is, the reduced stress s_R^* used as the abscissa in Fig. 11.10 is a measure of the ratio τ/τ_s between the two time scales that we have introduced in Sec. 11.1.

The data reported in Fig. 11.10 show a number of noteworthy features. Firstly, plotting relative viscosity against reduced stress leads to superposition of the data for suspensions of particles of different sizes, but preserves a strong dependence on colloid concentration as measured by the volume fraction ϕ. For $\phi \leq 0.2$ the viscosity is independent of shear rate, i.e. the suspensions effectively show Newtonian behaviour. However, for $\phi \geq 0.3$ significant shear thinning is observed after a first Newtonian region at low shear rate, and shear thinning is followed by a second Newtonian plateau at higher shear rates. Shear thinning is observed for values of s_R^* of order unity; that is, when the time scales for Brownian motions and for shear-induced motions of the particles become comparable so that the microstructure of the suspension can be significantly distorted during flow. The second Newtonian regime is believed to result from the persistence of a relatively stable non-equilibrium microstructure over a range of shear rate. There also are indications of shear thickening occurring at still higher shear rates in the data for the more concentrated suspensions.

In his discussion of these data Pusey[381] appeals to various other observations such as (i) computer simulations of shear thinning in simple fluids, showing that it may be associated with the ordering of atoms along streamlines, and (ii) light scattering studies of the distortion of the structure factor of colloidal fluids in a state of weak flow, suggesting string-like ordering of the particles at volume fractions near that for hard sphere freezing. Even more pronounced structures are observed under an oscillatory shear. It is clear, therefore, that suspensions of model colloids under controlled conditions of flow can exhibit a variety of non-equilibrium structures. Pusey[381] also remarks that for their PMMA samples Choi and Krieger found that the low-shear viscosity appears to approach a divergence at $\phi \approx 0.58$ and the high-shear viscosity at $\phi \approx 0.63$. These values of the volume fraction lie on the metastable amorphous branch of the hard sphere phase diagram (see Sec. 10.6.1), suggesting that

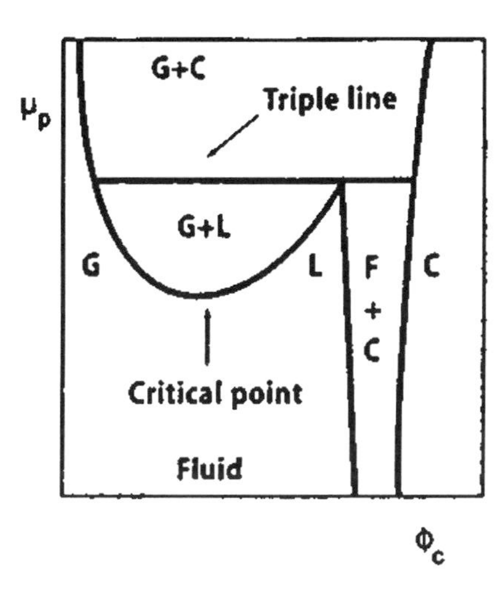

Fig. 11.9. Schematic colloid-polymer phase diagram, plotting the polymer chemical potential against the volume fraction of colloid.

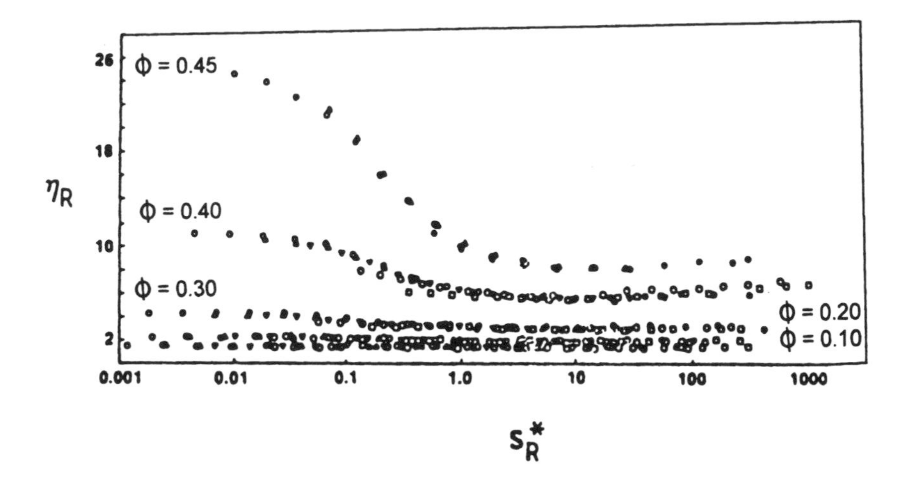

Fig. 11.10. Relative viscosity versus reduced stress for suspensions of PMMA spheres of various sizes in a variety of suspension media. Curves are labelled by the volume fraction. (Redrawn from Pusey, Ref. 381.)

these viscosity divergences may be associated with the glass transition and with the achievement of random close packing.

11.6.2 *The rheology of field-responsive suspensions*

The term "field-responsive liquid matter" embraces a class of soft-condensed-matter systems which undergo significant changes in their properties upon application of an external field. Electro- and magneto-rheological fluids as well as ferrofluids belong to this class.[383,384]

Of particular interest is the family of two-phase systems made from a responsive phase in an inactive fluid medium. The active phase may consist of particles carrying either a permanent or an induced dipole, which are randomly dispersed in the medium at low concentration and tend to aggregate at higher concentration. The present summary is based on a short review by Rubì and Vilar.[385] This is mainly addressed to ferrofluids in which the active phase is made of single-domain magnetic particles with permanent dipole moments.

As in the case of dispersions of more conventional colloidal particles or of polymer solutions, the suspended particles interact with the fluid medium and modify its flow behaviour. These interactions are assisted by the external field and are influenced by the dipolar interactions between the particles, so that the static and dynamic properties of the suspension, such as its magnetisation and its viscosity, may be substantially affected.

In dilute suspension the viscosity increases linearly with the field strength in a constant field, whereas in time-dependent fields it may decrease with increasing field even in the linear regime.[386] The magnetic particles inside a ferrofluid which is in a state of rotation inside an alternating magnetic field may behave as nanomotors or nanogenerators.[387] The essential point is the interplay between the magnetic and rotational degrees of freedom of the particles: in the simplest model the dynamics of the dipole moment is determined by the rotations of the particle to which it is rigidly attached, and these reflect a balance between magnetic, hydrodynamic, and Brownian torques. Additional dissipation arises form the torque exerted on the fluid medium, thus modifying its transport coefficients. Non-linear effects emerge when the magnetic dipole is not rigidly attached to the particle but may rapidly orient itself in the direction of the field.

Coating of the particles at low concentration with a surfactant prevents aggregation and stabilises the suspension. With increasing concentration the main new feature is the formation of chains or more complex branched structures. Chains appear at low temperatures, where Brownian effects are negligible, or in systems with strong dipolar interactions. More compact aggregates may be formed in other cases, including the case of ferrofluids containing nanosized ferromagnetic particles. These structures may evolve in time by growth or fragmentation, under the influence of the external field, of the velocity field, and of fluctuations. The rheological behaviour of the fluid in this regime is very sensitive to the underlying structures.

11.7 Surfactant Systems

In the final section of this chapter, an area which is at the time of writing the focus of a great deal of interest in the general area of soft-condensed-matter physics will be briefly presented. This concerns surface-active agents, or surfactants.

The possibility of forming a monomolecular layer of an insoluble substance on the surface of water was first discussed scientifically by Franklin in 1774. Quantitative studies of the properties of these films were undertaken in the early part of the twentieth century — in particular, it was soon established that the films behave as a two-dimensional ideal gas in the limit of a large available area per molecule and that a variety of phases can be evidenced from measurements of surface pressure as a function of available area.[388] Precise measurements of surface pressure as a function of density and temperature on films of pentadecylic-acid molecules on water substrates have led to the determination of the liquid–vapour coexistence curve and demonstrated the existence of a critical point for such a system of insoluble aqueous surfactant molecules.[389] These measurements show resemblance, in a general fashion, to the behaviour of three-dimensional fluids and yield mean-field critical exponents.

A surfactant molecule combines a water-soluble hydrophilic head with a water-insoluble hydrophobic tail. The molecules adsorb at air–water or oil–water interfaces and form monolayers with the hydrophilic part lying in the water. Surfactants aid in stabilising emulsions of water-insoluble liquids such as oil. If two droplets of oil in an emulsion come into contact, their coalescence would reduce the surface area and is energetically favoured. The emulsion can be stabilised by reducing the interfacial tension, this being achieved by the addition of surfactants.

Surface-active materials can also act as wetting agents and as detergents. A small amount of water poured on a greasy solid surface will take the form of a drop under its surface tension. But if the water contains surfactant molecules, the drop is covered by a layer of such molecules with their hydrophobic tails pointing out and capable to adhere to the solid. Gravitation can then flatten the drop and induce its spreading out on the surface. With regard to detergent action, a particle of dirt sticking to cloth can become progressively covered by a monomolecular layer of surface-active moleules and can thus be detached from the cloth and removed by rinsing water.

Surfactant molecules can also self-assemble in water solutions to hide their hydrophobic part from the water, forming micelles, vesicles, sponge phases, and liquid crystal phases among others.[390] Layers of such amphiphilic molecules are the building blocks of all membranes found in biological systems.[391] The study of mixed aqueous solutions of polymers and surfactants holds much promise both for practical applications and for biology. Following Langevin,[392] the main ideas in understanding the physical chemistry of molecular organisation will be outlined below, with main attention to aqueous surfactant solutions.

Aggregates of surfactant molecules in dilute solutions are in equilibrium with monomers, the monomer concentration staying close to the solubility limit. The issue of the shape of the aggregates can be addressed by means of a parameter (V/AL) where V is the volume of the nonpolar part of the molecule, L its length and A the area occupied by the molecule at the interface. If $(V/AL) < 1/3$ the relative bulkiness of the polar part is dominant and the aggregates show a strong curvature towards water, taking the shape of spherical micelles. If $1/2 > (V/AL) > 1/3$ cylindrical aggregates are favoured, while if $2 > (V/AL) > 1/2$ flat aggregates are favoured. The latter case often corresponds to double chain lipids: there may not be enough surfactant to form a lamellar phase and the portions of lamellae close up into vesicles. For $(V/AL) > 2$ the formation of aggregates into non-polar solvents is favoured, leading to the formation of reversed micelles.

The notion of spontaneous curvature of surfactant layers has been generalized by Helfrich[393] by appeal to curvature elasticity. Flat lamellae are formed when the elastic energy associated with curvature is large compared with the thermal energy k_BT, but if it becomes comparable to k_BT the layers become undulated due to thermal fluctuations and a sponge phase with no long-range order is stabilised. A further elastic energy term regulates the balance between vesicles and the sponge phase. More generally additional terms in the free energy, e.g. from dispersion entropy and from interactions between aggregates, may contribute to determine the actual structure of the aggregate.

Interactions between aggregates play an important role in more concentrated solutions and affect both size and polydispersity of micelles. They can also promote transitions from spherical to cylindrical micelles in ionic surfactant solutions. The cylinders may become entangled when they are sufficiently long and the solution becomes visco-elastic as a semi-dilute polymer solution. In still more concentrated solutions the excluded volume interactions, among others, can promote the appearance of liquid crystalline phases.

Surfactant aggregates are transient because the molecules constantly exchange between them and the solvent. The exchange time turns out to be related to the solubility of the surfactant: it may be of order 10^{-5} seconds for surfactants with chains of 16 carbons and increase up to hours for lipids. The rheological behaviour is strongly affected by the exchanges, an extreme case being the sponge phase for which the viscosity is only slightly greater than that of water although the structure is connected over a macroscopic length scale.

Chapter 12

Turbulence

12.1 Introduction

In 1883 Osborne Reynolds, while studying the flow of liquids through long pipes of uniform circular cross-section, found that the flow would be orderly for velocities up to some critical speed, above which it would abruptly become turbulent at some distance from the inlet. A turbulent state of flow was observed to be the norm above the critical speed, although a metastable state of laminar flow could be maintained by taking care to eliminate disturbances. As already noted in Sec. 1.4, the criterion for the transition from laminar to turbulent flow is formulated in terms of the Reynolds number $\mathrm{Re} = va/\nu$, where a and v are suitable length and velocity scales and ν is the kinematic viscosity (see Sec. 6.5).

Most flows occurring in nature and in engineering practice are turbulent. As defined in the book of Bradshaw,[394] "turbulence is a three-dimensional time-dependent motion in which vortex stretching causes velocity fluctuations to spread to all wavelengths, between a minimum determined by viscous forces and a maximum determined by the boundary conditions of the flow". As remarked in the book by Tennekes and Lumley,[395] on the other hand, "everyone who, at one time or another, has observed the efflux from a smokestack has some idea about the nature of turbulent flow. However, it is very difficult to give a precise definition of turbulence. All one can do is list some of the characteristics of turbulent flows". In the instance of flow past an obstacle, the evolution with increasing Reynolds number is from the laminar fluid motion treated in Sec. 6.4 to the generation of a trail of vortices and to a fully developed turbulent wake involving fluctuations over a wide spectrum of space-time scales.

The development of turbulence is often triggered by one type or another of instability in the state of flow. Some common cases of instability are reviewed in Sec. 12.2. The onset of an instability in a conservative system can be identified by a linear normal-mode analysis, in which one examines small-amplitude distortions of the state of the system and searches for situations in which a frequency of oscillation may become imaginary. In a dissipative system such as a viscous fluid, one may look at small distortions of the form $\xi_n(t) \propto \exp[-i(\omega_n + i\gamma_n)t]$ and search for conditions under which a damping coefficient γ_n changes sign from negative (giving an exponential return to the unperturbed state in a time interval of order γ_n^{-1}) to positive (corresponding to a permanent departure from the assumed stable state).

The path from the first appearance of an instability to the full development of turbulent motions is, however, hard to explore because of the insurgence of nonlinearities. In general terms, an instability brings the system to a new stable state and introduces new characteristic frequencies in its spectrum. Detailed studies of dynamical systems governed by nonlinear equations of motion have shown that their behaviour may become essentially unpredictable through successive period doublings in the phase space trajectory, up to an accumulation point where the trajectory no longer closes on itself. The system is unpredictable because small initial differences may then grow without limit. Ruelle and Takens[396] have shown that the loss of predictability can be the result of the nonlinear interaction of a small number of modes. Intermittency may also arise for certain ranges of nonlinearity parameters, within which the motion is predictable for many periods in succession but is interrupted by bursts of chaos of unpredictable duration. The "route to chaos" in Bénard convection is reviewed in Sec. 12.3.

A characteristic feature of turbulence is indeed its irregularity or randomness, which requires the use of statistical methods. In analogy with other nonlinear systems, a liquid in turbulent flow may show intermittency and the formation of regular structures due to self-organisation of vorticity. Another feature of turbulence is its diffusivity, which causes rapid mixing and high rates of transfer of mass, momentum, and heat. This is the most important property of turbulent flows as far as applications are concerned. Vortex dynamics plays a major role, since turbulence is characterised by high levels of fluctuating vorticity needing a three-dimensional description. This can be based on the nonlinear equations of fluid mechanics, in which the viscous shear stresses perform deformation work which increases the internal energy of the fluid at the expense of the kinetic energy of the turbulence. Turbulence needs a continuous supply of energy to make up for viscous losses.

These features can be illustrated for Newtonian fluids by reference to homogeneous turbulence in incompressible flows, a situation that may be approached experimentally by passing a fluid with uniform speed through a uniform grid of wires. The vortices that are shed by the grid produce a turbulent field downstream which can be regarded as independent of position. This is used in Secs. 12.4 and 12.5 to illustrate the concepts of energy cascade and of diffusivity in turbulence. Turbulent shear flows are introduced in Sec. 12.6, with main attention focused on boundary-free flows. We conclude the chapter with brief sections on the role of compressibility and on turbulence in non-Newtonian fluids. Appendix 12.1 elaborates the parallelism between the Navier–Stokes equation and the Maxwell equations in electromagnetism, which was already appealed to in Sec. 6.5 in introducing vorticity. The mathematical framework for a series solution of the Navier–Stokes equation is presented in Appendix 12.2.

12.2 Instabilities in Fluids

We have indicated in Sec. 12.1 how instabilities in fluids may be identified by a linearised analysis. We first give an example of this in the so-called Rayleigh–Taylor instability. We then turn to convective and vortex-sheet instabilities, of direct relevance to the onset of turbulence.

12.2.1 *The Rayleigh–Taylor instability*

This instability can arise when a vessel containing two liquids, or a liquid and a gas, separated by a planar interface is abruptly turned upside down so that the heavier fluid lies above the lighter one. The two forces at work are the gravitational field and the interfacial tension. We consider a wave-like corrugation of the interface, given by a vertical displacement $\zeta(\mathbf{r},t) = \zeta_{\mathbf{k}}(t)\exp(i\mathbf{k}\cdot\mathbf{r})$ with $\mathbf{r}$ a vector in the interfacial plane. The frequency of such a wave is

$$\omega_{\mathbf{k}}^2 = \frac{-(\rho' - \rho)gk + \gamma k^3}{\rho' + \rho} \tag{12.1}$$

(see Chap. 13), where ρ' and ρ are the mass densities of the two fluids ($\rho' > \rho$), g is the acceleration of gravity and γ the interfacial tension. This result follows from balancing the gain in gravitational energy per unit area, which is $\approx (\rho' - \rho) g \zeta_{\mathbf{k}}$, against the increase in free energy from stretching of the interface, which is $\approx \gamma k^2 \zeta_{\mathbf{k}}$.

We see from Eq. (12.1) that for $\rho' > \rho$ there is a cut-off wave number k_c, given by

$$k_c = \left[\frac{(\rho' - \rho)g}{\gamma}\right]^{1/2}, \tag{12.2}$$

below which a corrugation of the interface can grow in amplitude. Wave-like corrugations with $k < k_c$ have imaginary frequency. The value of k in this range that maximises $-\omega_{\mathbf{k}}^2$, and hence the rate of growth of the instability, is $k_{max} = k_c/\sqrt{3}$ from Eq. (12.1).

The dimensions of the container are in fact relevant in practice. Considering a rectangular container of largest dimension L in the plane of the interface, the smallest wave number that it can admit is π/L so that the inverted configuration will be maintained if π/L is larger than the cut-off wave number in Eq. (12.2).

Other surface tension-controlled instability phenomena are (i) the breaking of a jet into a regular succession of drops, and (ii) viscous fingering. These are discussed in the book by Faber.[4]

12.2.2 *Thermal convection and the Rayleigh–Bénard instability*

In Chap. 7 we introduced heat conduction through Fourier's law setting the heat current as proportional to the temperature gradient and discussed its role in the density fluctuation spectra of a liquid within the framework of linearised hydrodynamics. As the temperature excess at the surface of a warm body in contact with the liquid increases, convective motions become important. They affect the temperature distribution through the liquid and in particular tend to localise the temperature excess within the boundary layer.

A set of equations describing convection in a liquid in steady incompressible flow is given by the so-called Boussinesq approximation. Let the temperature distribution in the liquid be $T + \vartheta(\mathbf{r})$. The gravitational potential as a function of height z is $m\rho(1 - \alpha\vartheta)gz$, where $\alpha = -\rho^{-1}(\partial\rho/\partial T)_p$ is the coefficient of thermal expansion. The Navier–Stokes equation in the form (6.33) is then written as

$$-\boldsymbol{\nabla}\left(\frac{p}{m\rho}+gz\right)+\alpha g\vartheta\boldsymbol{\nabla}z\approx(\mathbf{v}\cdot\boldsymbol{\nabla})\mathbf{v}+\nu\boldsymbol{\nabla}\times\boldsymbol{\omega}\,. \tag{12.3}$$

This equation is combined with the continuity equation for incompressible flow, $\boldsymbol{\nabla}\cdot\mathbf{v}=0$, and with a transport equation for the local excess temperature,

$$(\mathbf{v}\cdot\boldsymbol{\nabla})\vartheta=\kappa\nabla^2\vartheta\,, \tag{12.4}$$

where $\kappa=\lambda/(m\rho c_{\mathrm{p}})$ with λ the thermal conductivity coefficient and c_{p} the specific heat per unit mass. These equations should be solved to find the fields of pressure, temperature, and velocity for given boundary conditions. The ratio $\mathrm{Pr}=\nu/\kappa$ is known as the Prandtl number: it is about 6 for water and rises to $\approx 10^3$ in more viscous non-conducting liquids.

The Rayleigh–Bénard instability can arise when the fluid is confined between two horizontal plates with heat being supplied from below. In this case the hotter fluid near the bottom is less dense and tends to rise, displacing an equal volume of cooler fluid from the top. The result is a cellular pattern of convective currents in the form of an ordered array of convection rolls, as is sketched in Fig. 12.1 (notice the opposite circulation in adjacent rolls). The release of gravitational energy in these motions should at least compensate for viscous dissipation and, since both thermal conductivity and viscosity play a role, the governing parameter should depend on both as well as on the distance h between the two plates. The appropriate dimensionless combination of system parameters is the Rayleigh number $\mathrm{Ra}=\alpha gh^3\Delta T/\nu\kappa$, and the onset of circulation corresponds to a critical value for this number.

The problem was solved analytically by Rayleigh in 1916 for the case of two free-slip boundary conditions (see Drazin and Reid[397] for details). In a

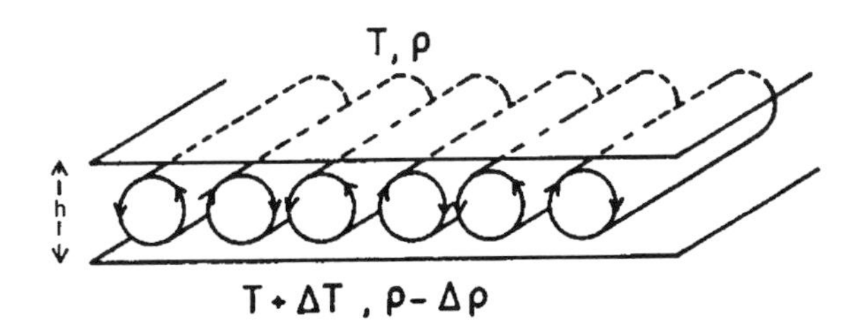

Fig. 12.1. Schematic representation of fluid rolls in Rayleigh–Benard convection.

normal-mode approach one looks for solutions of the Boussinesq equations in the form $v_z(x, z) = f(z)\exp(ik_x x)$ etc., and finds $f(z) = \sin(k_z z)$. Thus, the rolls have extension π/k_x along the x direction and π/k_z in the vertical direction, each having indefinite extension in the third direction as indicated in Fig. 12.1. The smallest value of the Rayleigh number for which the instability sets in corresponds to an aspect ratio $k_x/k_z = 1/\sqrt{2}$ and is equal to $27\pi^4/4 = 657.5$. In the case of two no-slip boundaries the critical Rayleigh number is found by numerical solution to be raised to the value $\mathrm{Ra_c} \approx 1708$ and the aspect ratio of the rolls decreases to about 1. This agrees well with experiment, as discussed for instance in the book of Lesieur.[398]

If the domain is not constrained horizontally, the convective cells will rather be of hexagonal shape with the warm fluid ascending in the centre. The so-called Bénard–Marangoni convection occurs in a thin liquid layer with a free top surface and leads to the formation of hexagonal cells, whose size is of the order of the layer thickness.[399] In this case convection is driven mainly by the release of surface free energy rather than of gravitational energy.

The direction of convection may be reversed in a deep layer of a liquid mixture, when molecular interdiffusion is much slower than thermal conductivity.[398] Let the warm fluid at high concentration c lie above the cold fluid at low c: then a fluid volume displaced upwards will adjust to its new ambient temperature before adjusting its concentration and will feel positive buoyancy and keep rising. This so-called double-diffusive instability leads to formation of hexagonal convective cells with cold fluid of low c ascending in the centre. Such a situation arises in an ocean which is strongly heated on its surface by the sun: evaporation increases the salinity in the surface layers and cold water rises from the depths to the surface.

12.2.3 *The Kelvin–Helmholtz instability*

The third main type of hydrodynamic instability concerns sheets of vortices. The so-called Kelvin–Helmholtz instability occurs in a mixing layer at the interface between two flows of different velocities coming from the trailing edge of a thin plate. This instability is eventually responsible for vortices which pair off and amalgamate in the downstream motion. A detailed representation of the evolution of the flow is given in a set of colour pictures in the book of Lesieur.[398]

The mechanism of vortex formation is described by Eq. (6.34) for vorticity and can be understood as follows.[400] Consider a mixing layer centred in the plane $x_2 = 0$, the field of flow velocity being $v_1(x_2)$ along the x_1 direction and tending to the asymptotic values $\pm U$ far away from the plane (if the asymptotic flow velocities are U_1 and U_2, we set $U_1 - U_2 = 2U$ and adopt a reference frame in which $v_1(x_2 = 0) = 0$). As discussed by Rayleigh for a fluid of uniform density and vanishing viscosity, the velocity profile is represented by $v_1(x_2) = U \text{ tgh}(x_2/\delta_0)$: thus, a strip of width $\approx 2\delta_0$ separates the two regions corresponding to uniform flows at velocity U and $-U$, and a crucial point is that the velocity profile has an inflection (a point of vanishing second derivative) at $x_2 = 0$. In fact, suppose that the strip is perturbed by a sine-wave undulation of wavelength λ_a (Fig. 12.2(a)). Pressure differences between the two layers enhance the amplitude of the disturbance and, since $(d^2v_1(x_2)/dx_2^2) \approx 0$ near the inflection point, the vorticity is convected by the basic flow $v_1(x_2)$ so that the crests of the disturbance (for $x_2 > 0$) and its troughs (for $x_2 < 0$) travel in opposite directions (Fig. 12.2(b)). This steepens the vortex sheet and transforms it into a spiral (Fig. 12.2(c)).

The vortices created in this way have initially a longitudinal wavelength λ_a and they all have the same strength and an indefinite length in the x_3 direction. The wavelength of the perturbation which grows more rapidly is determined by the thickness of the rotational layer and is of order $4\pi\delta_0$. In fact, in real fluids the layer thickness tends to increase with time because viscosity induces diffusion of vorticity as already discussed in Sec. 6.5.

Starting from a regular array of vortices as described above, the subsequent two-dimensional evolution of the layer can be investigated numerically.[398] In essence, the fundamental eddies tend to undergo successive pairings. The

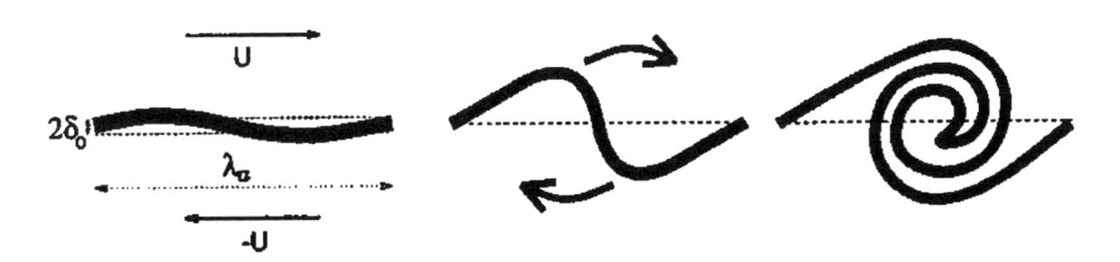

Fig. 12.2. Schematic illustration of the formation of Kelvin–Helmholtz vortices in a mixing layer. (Left to right: (a), (b) and (c)).

first pairing in the row of primary vortices is easily understood in terms of a subharmonic of wavelength $2\lambda_a$. Assuming that the phase of the subharmonic perturbation is such that vortices are alternately lifted into the region $x_2 > 0$ and pushed into the region $x_2 < 0$, the convection of vorticity by the basic flow will draw each vortex in the upper region against its neighbour in the lower region. Each vortex tends at that point to entrain the other in the irrotational motion that it induces outside, so that the vortices roll up about each other. Since the outer parts of each vortex turn more slowly than its inner parts, both vortices develop tails which mix into spirals during their pairing.

Experimentally, the formation of Kelvin–Helmholtz vortices is best studied in the case of a vortex sheet separating two liquids of different density,[401] in which case both gravity and interfacial tension are at work. Instabilities against periodic disturbances may also arise in boundary layers in the absence of an inflection point, their origin being associated with the viscosity of real liquids in flows at very high Reynolds number $\mathrm{Re} = m\rho\delta U/\eta \approx 10^3$ where U is the velocity just outside the boundary layer and δ is the thickness of the layer.

12.3 Evolution of Bénard Convection with Increasing Rayleigh Number

The convection rolls that are formed in a Rayleigh–Bénard cell at the critical value $\mathrm{Ra_c}$ of the Rayleigh number are stable over a considerable range of values of Ra above $\mathrm{Ra_c}$. The insurgence of period doubling is then observed to occur, as in the classical numerical studies of the transition of a nonlinear dynamical system to chaotic behaviour.

Figure 12.3 is taken from experiments by Libchaber *et al.*[402] on Bénard convection in liquid mercury, using a rectangular cell which accommodates four convection rolls at $\mathrm{Ra} > \mathrm{Ra_c} \approx 1700$ (see Sec. 12.2.2). The quantity which is being measured is the temperature of the fluid just above the middle of the lower plate as a function of time: after staying essentially constant up to $\mathrm{Ra} \cong 2\mathrm{Ra_c}$, this starts oscillating in time just as the rolls start oscillating. The period has already doubled once at $\mathrm{Ra} = 3.47\mathrm{Ra_c}$ (top trace in Fig. 12.3) and three further doublings occur at $3.52\mathrm{Ra_c}$, $3.62\mathrm{Ra_c}$ and $3.65\mathrm{Ra_c}$. The sequence of relative separations between successive period doublings approaches an accumulation point at 4.4 ± 0.1 in this experiment, to be compared with the Feigenbaum number ($\cong 4.669$) entering the classical theory of the approach to

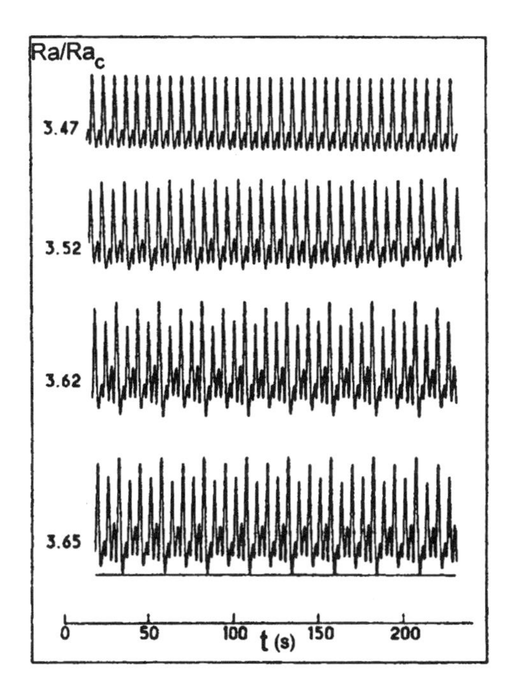

Fig. 12.3. Period doubling cascade in Bénard convection. Direct time recordings of temperature for various stages of the period doubling cascade. (Redrawn from Libchaber *et al.*, Ref. 402.)

chaos.[403] Similar results have been reported by Giglio *et al.*[404] for a transition of water in a Bénard cell to chaotic behaviour *via* a reproducible sequence of period-doubling bifurcations.

Intermittency as a route to turbulence in Bénard convection has been demonstrated experimentally by Bergé *et al.*[405] by measuring the variation in time of the vertical fluid velocity at a point near the middle of a convection cell containing silicone oil. As is shown in Fig. 12.4, the velocity is seen to oscillate in a regular manner at $\mathrm{Ra} = 270\mathrm{Ra}_c$, but at higher Rayleigh numbers erratic bursts of motion disturb these oscillations.

A persistence of coherent structures is observed as the fluid is being driven through successive bifurcations towards chaos and turbulence. These structures are seen for instance in the granulation which is present on the surface of the Sun, at a Rayleigh number of order 10^{20}. The solar granulation consists of hexagonal cells having size of order 10^3 km and lifetime of order 10 minutes.

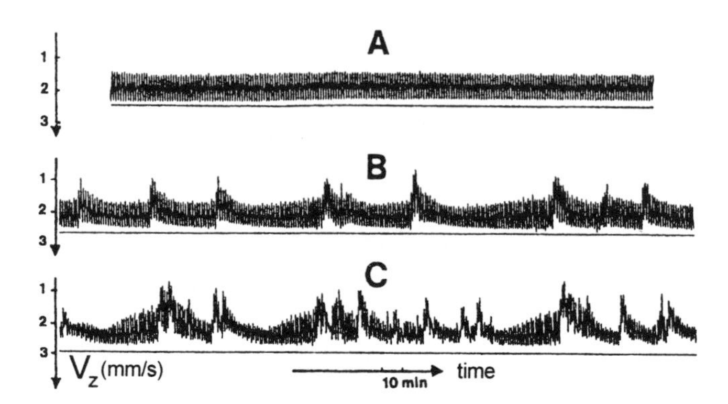

Fig. 12.4. Intermittency in Bénard convection: time dependence of the vertical velocity at the centre of convection cell at Ra/Ra_c = 270 (A), 300 (B) and 335 (C). (Redrawn from Bergé *et al.*, Ref. 405.)

This is an example of a turbulent system where an instability drives the creation of coherent structures which are destroyed from nonlinear interactions and reformed through the instability.[398]

Investigations of Bénard convection in the laboratory have been extended to values of Ra as high as 10^{15}. Figure 12.5 reports the results of such an experiment carried out by Castaing *et al.*[406] on very cold helium gas. The shear viscosity and thermal conductivity of this fluid are very small and can be further reduced by cooling, while its density can be significantly increased under pressure. The data are shown as a plot of normalised heat transfer (giving the so-called Nusselt number Nu) against the Rayleigh number Ra.

In these experiments convection evolves into oscillations with increasing Ra and becomes truly turbulent at $Ra \approx 5 \times 10^5$. The transition from "soft" to "hard" turbulence at $Ra \approx 4 \times 10^7$ is marked by a significant reduction of the observed temperature fluctuations and by a change in the power-law relation $Nu \propto Ra^n$, from $n = 1/3$ to $n \cong 2/7$. The 1/3 law implies that the vertical heat flux is independent of the distance between top and bottom plates, as if it were occurring within thin boundary layers. A quantitative analysis has been given by Faber[4] using the concept of eddy thermal conductivity, paralleling the notion of eddy diffusivity to be introduced in Sec. 12.5.3.

In the regime of hard turbulence, flow visualisation experiments have revealed a complex pattern of intermittent events. A column of fluid rises in a corner of the cell from the boundary layer near the hot plate through fluid

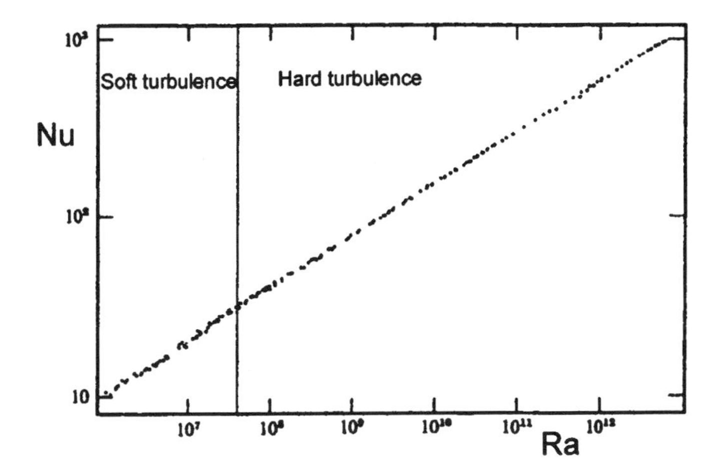

Fig. 12.5. Correlation between reduced heat flow (expressed as the Nusselt number Nu) and the Rayleigh number Ra in Bénard convection. (Redrawn from Castaing *et al.*, Ref. 406.)

which is moving relatively slowly, and on reaching the other boundary layer sets up a wave-like disturbance which triggers the release of a falling column in the opposite corner. This in turn triggers the release of another rising column.

12.4 Energy Cascade in Homogeneous Turbulence

The introduction that we give in the rest of this chapter to the phenomena of turbulence in flows can only be very brief, but is designed to stimulate the reader to further study. We have already referred to useful books on turbulence: here we add references to the more recent book by Pope[407] on turbulent flows and to the works of McComb and of Bohr *et al.*,[408] which focus attention principally on theoretical aspects and modern developments.

The main features of turbulence are not governed by the molecular properties of the fluid in which it occurs. The theoretical framework therefore resides in the nonlinear equations of hydrodynamics: for Newtonian fluids the Navier–Stokes equation combined with the continuity equation for particle density (often reduced to the case of incompressible flow, since compressibility is not essential for turbulent behaviour — but see Sec. 12.7) and the entropy production equations given in Sec. 7.4.2. The linearised form of these equations, valid for very small disturbances, allows us to treat instabilities in flows but cannot deal with the large levels of fluctuation that are met in turbulence. In the opposite limit one may focus on the asymptotic properties of flows at very high Reynolds numbers: this approach is based on a limit process related to vanishing molecular viscosity and is very useful in treating boundary layers.

In intermediate regimes, where turbulence consists of fairly large fluctuations governed by nonlinear equations, simple physical concepts are often of great help in bridging the gap between the equations and the actual flows. Dimensional analysis becomes useful in cases when some aspects of the structure of turbulence depend only on a few independent variables: the form of the relation between dependent and independent variables is then fixed and a solution can be obtained aside from numerical coefficients. More generally, if energy transfer is fast enough that the effects of past events do not dominate the dynamics, the problem may be reduced to treating a state of local dynamical equilibrium which is mainly governed by local parameters such as scale lengths and times. Dimensional methods and similarity arguments can then be very useful.

We illustrate the above notions in the rest of this section and in the following two sections by examining (i) the transfer of energy from large to small eddies in homogeneous turbulence, which is driven by vortex stretching and ends into viscous dissipation of energy near the so-called Kolmogorov microscale; and (ii) the ability of turbulence to transport and mix energy, momentum, heat, and particles of matter.

12.4.1 *Energy cascade and Kolmogorov microscales*

Let us consider a uniform stream of fluid passing at high Reynolds number through a grid of wires having spacing a. The grid feels a drag force and reacts to it by shedding eddies at a uniform rate. Near the grid these have a velocity field distribution which, in Fourier transform, is peaked around a wave number $k_0 \approx a^{-1}$. Further downstream, however, the eddies get twisted around by their neighbours and instabilities develop. The lines of vorticity embedded in the fluid become progressively longer and, as discussed in Sec. 6.5, vorticity is increased by this stretching process. At the same time the lines of vorticity can diffuse because of viscosity and, as they do so, they can come together with other lines of opposite sense and form closed loops which may shrink and collapse. During the evolution the energy initially associated with wave numbers of order a^{-1} is transferred to larger values of k, i.e. to smaller length scales, and is ultimately dissipated as heat. Thus, in this so-called energy cascade the supply of translational kinetic energy for ordered flow has to sustain rotational motions and dissipation by viscosity.

The nonlinear mechanism described above is dissipative because it creates smaller and smaller eddies until the eddy size becomes so small that viscous dissipation of their kinetic energy is very fast. Let us attempt to assess the smallest length and time scales involved in a turbulent flow, where viscosity can be effective in smoothing out the velocity fluctuations and prevent the generation of even smaller scales of motion by dissipating energy into heat. Following Kolmogorov,[409] one assumes that small-scale motions are relatively independent of the mean flow and of the relatively slow large-scale turbulence: they should depend only on the rate of energy supply and on the kinematic viscosity. It is also fair to assume that the rate of energy supply should be equal to the rate of dissipation. With the dissipation rate ε per unit mass (measured in m^2/s^3) and the kinematic viscosity ν (measured in m^2/s) we can form scales of length, time, and velocity as follows:

$$\ell \equiv \left(\frac{\nu^3}{\varepsilon}\right)^{1/4}, \quad \tau \equiv \left(\frac{\nu}{\varepsilon}\right)^{1/2}, \quad v \equiv (\nu\varepsilon)^{1/4}. \tag{12.5}$$

These are referred to as the Kolmogorov microscales. Notice that the Reynolds number formed from ℓ and v is $\ell v/\nu = 1$: thus, small-scale motions are quite viscous. Viscous dissipation can adjust itself to the energy supply by adjusting the length scale.

Let us now try to compare, following Tennekes and Lumley,[395] the large-scale and small-scale aspects of turbulence by an estimate of the rate ε at which the large eddies supply energy to the small eddies. We take the amount of kinetic energy per unit mass in large-scale turbulence as proportional to u^2 and its time scale as d/u, d being the size of the largest eddies as introduced above in discussing flow through a wire grid. The rate of energy supply to the small-scale eddies is therefore of order $u^2/(d/u)$, that is

$$\varepsilon \approx \frac{u^3}{d}. \tag{12.6}$$

By substituting this in Eq. (12.5) we obtain

$$\frac{\ell}{d} \approx \mathrm{Re}^{-3/4}, \quad \frac{\tau u}{d} \approx \mathrm{Re}^{-1/2}, \quad \frac{v}{u} \equiv \mathrm{Re}^{-1/4}, \tag{12.7}$$

where $\mathrm{Re} = ud/\nu$ is the Reynolds number of the flow. We see from Eq. (12.7) that the separation in scales between large and small eddies widens as the Reynolds number increases: that is, a turbulent flow at a relatively low Re has a relatively coarse small-scale structure.

Visual evidence of the small-scale structure of turbulence can be obtained *via* light scattering (see Sec. 7.5.2). Gradients of the index of refraction are steeper if they are associated with smaller eddies, so that any optical system which is sensitive to such fluctuating gradients allows direct observations of the small-scale structure of turbulence. An example is the jittery appearance of the horizon as seen on a very hot day.

We should stress at this point that the Kolmogorov microscales, though much smaller than the typical scales of length and time for large eddies, are still typically larger than molecular scales. The latter can be measured by the molecular mean free path λ_c travelled on average by a molecule between successive collisions in the fluid. Using from kinetic theory the expression $\nu \approx c\lambda_c$ for the kinematic viscosity in terms of the speed of sound c and of the mean free path, we get $\ell/\lambda_c \approx \mathrm{Re}^{1/4}/Ma$ and $\tau c/\lambda_c \approx \mathrm{Re}^{1/2}/Ma^2$ where $Ma = u/c$ is the Mach number. Thus, turbulence may reach down to the scale of molecular motions only if the Mach number is very large — a rare situation. Taking as illustrative orders-of-magnitude for turbulent flows $Ma \approx 1$ and $\mathrm{Re} \approx 10^6$, we get $\ell/\lambda_c \approx 30$ so that a hydrodynamic model is still appropriate. Since the smallest length and time scales in turbulence tend to be appreciably larger than molecular scales, the molecular transport processes are adequately described in terms of hydrodynamic transport coefficients.

12.4.2 *Kinetic energy spectrum*

While we have so far discussed only the length scales of the largest and smallest eddies, dimensional analysis can also give the form of the energy spectrum in the so-called inertial subrange, for values of the wave number k in the range $a^{-1} \ll k \ll \ell^{-1}$. Assuming homogeneous and isotropic turbulence, the energy spectrum is described by a function $E(k)$ defined so that the quantity $E(k)dk$ is the mean kinetic energy per unit mass which is stored in the range of wave number between k and $k + dk$. Since $E(k)$ is measured in $\mathrm{m}^3/\mathrm{s}^2$, the quantity $\varepsilon^2 E^3/k^5$ is a dimensionless combination. This suggests that, if we may assume that E does not depend on other system parameters in any important way, we may write[409]

$$E(k) \cong C_{\mathrm{K}} \varepsilon^{2/3} k^{-5/3} . \tag{12.8}$$

The numerical coefficient C_{K} entering this Kolmogorov $(-5/3)$ power law for the spectrum of kinetic energy is $C_{\mathrm{K}} = 1.7 \pm 0.2$ from a fit to available data.[408]

Experimental evidence on the inertial subrange has been obtained from studies of large-scale turbulence in the oceans or in the atmosphere. Figure 12.6 reports a one-dimensional section of an energy spectrum constructed by Faber[4] from data obtained by Grant *et al.*[410] using a flowmeter towed by a ship in a channel off Vancouver Island. The straight line through the data has a slope equal to $-5/3$ and is seen to hold over almost three orders of magnitude in k. At higher wave numbers viscosity sets in and has the effect of cutting off the spectrum quite rapidly.

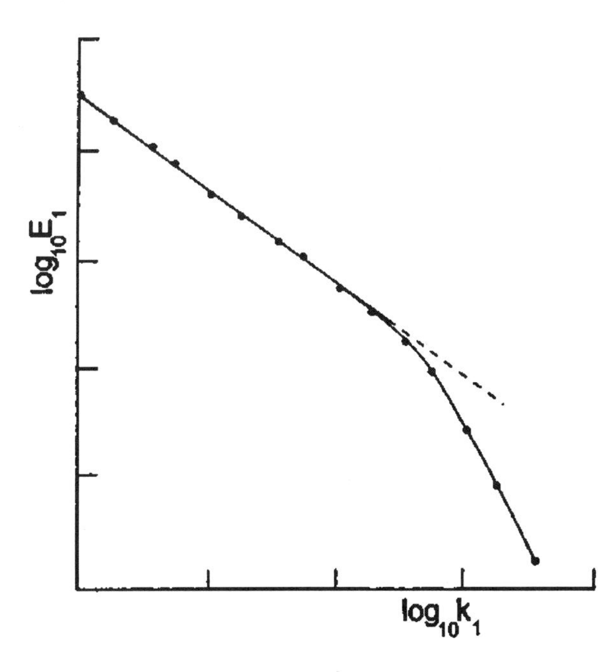

Fig. 12.6. Section of one-dimensional energy spectrum in measured large-scale flow. (Redrawn from Faber, Ref. 4.)

12.4.3 *Energy spectra from renormalisation group approach*

A general result for the energy spectrum of a randomly forced fluid has been obtained by Fournier and Frisch[411] using a renormalisation group (RG) approach. Earlier work by Forster *et al.*[412] had shown that large-distance and long-time properties of randomly stirred fluids which involve strong nonlinearities are amenable to RG techniques. Such machinery has been worked out when the turbulence at small length scales acts like an eddy viscosity.

The case in point concerns three-dimensional incompressible flow under random forcing with a power law spectrum $f(k) = 2Dk^{3-\alpha}$, $f(k)$ being the amount of energy injected per wave number. When α is positive and small, the resulting energy spectrum for the fluid is $E(k) \propto k^{1-2\alpha/3}$. This result had been obtained by Fournier and Frisch[413] by a dimensional argument.

In their later work these authors calculate the coefficient of proportionality in the spectrum and find

$$E(k) = (10\pi^2/3)^{1/3} D^{2/3} \alpha^{1/3} k^{1-2\alpha/3} \,. \tag{12.9}$$

This result was obtained by a quite different approach in the early work of Kraichman.[414]

12.5 Diffusion in Homogeneous Turbulence

12.5.1 *Time and length scales in diffusion*

Turbulence can transport and mix not only kinetic energy as discussed above in Sec. 12.4, but also momentum, heat, and particles. Turbulent diffusivity is by orders of magnitudes faster than molecular diffusion. Following Tennekes and Lumley,[395] we illustrate this fact by simple arguments based on the heat transport equation (12.4).

We consider for our present purposes a space- and time-dependent distribution $\vartheta(\mathbf{r}, t)$ of local temperature and rewrite Eq. (12.4) to account for the time dependence in the form

$$\frac{\partial \vartheta}{\partial t} + (\mathbf{v} \cdot \nabla)\vartheta = \kappa \nabla^2 \vartheta \,. \tag{12.10}$$

We use Eq. (12.10) to discuss a diffusion problem in which a characteristic length a is assigned — this would be the case, for instance, with the heating of a room of given size. If we drop the convective term, then the time scale of the process is the molecular diffusivity time $t_{\rm m}$ which is given by dimensional analysis as

$$t_{\rm m} \approx \frac{a^2}{\kappa} \,. \tag{12.11}$$

The turbulent time scale $t_{\rm t}$ is instead obtained from the two terms on the LHS of Eq. (12.10) as

$$t_{\rm t} \approx \frac{a}{v} \,, \tag{12.12}$$

where v is a characteristic velocity of the turbulent flow. As already remarked in Sec. 12.2.2, the ratio ν/κ is the Prandtl number which is of order unity in a fluid of moderate viscosity. We may therefore replace κ by ν in Eq. (12.11) for an estimate, with the result

$$\frac{t_{\rm t}}{t_{\rm m}} \approx \frac{\nu}{va} = \frac{1}{\rm Re} \,. \tag{12.13}$$

Thus, molecular transport is slower than turbulent transport by a factor of the order of the Reynolds number. As anticipated, this factor may be as large as several orders of magnitude.

The same argument can be used to assess diffusion in the case where a time scale t is assigned, leading to a molecular length scale $a_m \approx (\nu t)^{1/2}$ and a turbulent length scale $a_t \approx vt$. The latter estimate is applied by Tennekes and Lumley[395] to estimate the thickness of the atmospheric boundary layer, where the acceleration of flow by the Coriolis force imposes a time scale of the order of the inverse of the angular velocity of the frame of reference. Typical values at intermediate latitudes are $t \approx 10^4$ s and $v \approx 0.3$ m/s, leading to $a_t \approx 3$ km which is indeed of the same order as the observed layer thickness.

12.5.2 *Stochastic modelling of turbulent diffusion*

A systematic way to assess length and time scales in turbulent diffusion is from the study of appropriate random differential equations. Stochastic modelling plays an important role in many aspects of the physics of the liquid state: for example, the Fokker–Planck stochastic equation entering Kramers' treatment of reaction rates. While the focus here will be on turbulent diffusion, it will be useful to precede this example with some more general material.

At first, it can be asserted mathematically (see, for instance, Vanden Eijnden[415]) that stochastic modelling amounts to the study of a linear partial differential equation for a scalar quantity, say $\rho(\mathbf{r},t)$ evolving in phase space $\{\mathbf{r}\}$ according to

$$\frac{\partial\rho(t)}{\partial t} = \ell\left(\mathbf{r}, \frac{\partial}{\partial\mathbf{r}}; t\right)\rho(t)\,, \tag{12.14}$$

where the operator $\ell(\mathbf{r}, \partial/\partial\mathbf{r}; t)$ is to be taken as random with statistics to be specified. In physical cases, Eq. (12.14) may be the Liouville or the Fokker–Planck equation associated with a set of random nonlinear ordinary differential equations.

A complete solution of Eq. (12.14) would determine the statistics of $\rho(t)$. However, even if only the mean value $\langle\rho(t)\rangle$ is of interest, and in spite of the linear character of the equation, averaging it leads to the highly nontrivial "closure" problem of determining $\langle\ell(t)\rho(t)\rangle$. Vanden Eijden employs the work of Kraichman[416] to approximate the solution of this problem.

In the specific case of turbulent diffusion, Eq. (12.14) is specialised to the form

$$\frac{\partial\rho(t)}{\partial t} = -\mathbf{v}(\mathbf{r},t)\cdot\nabla\rho(t) + D_0\nabla^2\rho(t)\,. \tag{12.15}$$

Here D_0 denotes the molecular diffusion coefficient, while the velocity field $\mathbf{v}(\mathbf{r},t)$ is taken to be a Gaussian random process, statistically isotropic and stationary. The statistics of $\mathbf{v}(\mathbf{r},t)$ are then fully specified by the scalar covariance

$$\langle\mathbf{v}(\mathbf{r}+\mathbf{r}', t+t')\mathbf{v}(\mathbf{r}',t')\rangle \equiv 2\int_0^\infty dk\frac{\sin(kr)}{kr}E(k,t)\,, \tag{12.16}$$

where the quantity $E(k,t)$ is referred to as the energy spectrum. It is normalised as

$$\int_0^\infty dkE(k,0) = \frac{3}{2}v_*^2\,, \tag{12.17}$$

where v_*^2 is the mean square velocity in any direction. The characteristic length and time scales of the velocity field are then defined as

$$\ell_*^2 = \int_0^\infty dk\frac{E(k,0)}{v_*^2k^2}\,, \quad t_* = \int_0^\infty dt\int_0^\infty dk\frac{E(k,t)}{v_*^2}\,. \tag{12.18}$$

Evidently, these scales are determined by the spectrum of the velocity autocorrelations.

Vanden Eijnden discusses the meaning of these length and time scales. If either ℓ_* or t_* is finite, then his approximate solution of the stochastic equation for $\langle\rho(t)\rangle$ corresponds at long times to a diffusion process with an effective diffusion coefficient D_* which, in the limit $D_* \gg D_0$, depends on some combination of v_*, ℓ_* and t_* only. However, if both ℓ_* and t_* are infinite, the asymptotic dynamics is superdiffusive and non-Gaussian. The case $\ell_* = \infty$ means that much of the energy is concentrated in the large scales of the velocity field, while in the case $t_* = \infty$ the velocity field undergoes no effective decorrelation as time goes on.

We refer to the original article[415] the reader who may be interested in further details.

12.5.3 *Eddy diffusivity*

In view of the complex nature of the equations describing turbulent transport, it is tempting to reformulate the problem in terms of a differential equation involving an effective diffusivity. This approach, though treating turbulence as a property of the fluid rather than as a property of the flow, greatly simplifies the mathematical analysis. The problem of particle diffusion in a field of homogeneous isotropic turbulence was addressed in the early work of Batchelor.[417] We summarise his findings in general terms.

A complete description of the statistical history of a group of marked particles would need a large number of parameters. The simplest parameters characterise the position of the group of particles at any time and essentially determine the probability $P(\mathbf{r}, t)$ that any point $\mathbf{r}$ is immersed in marked fluid at time t. The next simplest parameters characterise the shape of the group, and in particular determine the dispersion of marked particles about their centre of mass. Batchelor introduces for this purpose the joint probability distribution $Q(\mathbf{r}, t|\mathbf{r}_0, t_0)$ that the separation between two particles goes from $\mathbf{r}_0$ to $\mathbf{r}$ in the time interval from t_0 to t.

An approximate formulation of the problem of relative diffusion in terms of a differential equation is then sought. Batchelor uses the fact that, after diffusion has been proceeding for some time, the dependence of Q on $\mathbf{r}_0$ is lost. Under this condition the joint distribution is replaced by a function $T(\mathbf{r}, t - t_0 - t_1)$, where t_1 depends on $\mathbf{r}_0$. Under the guidance of the usual diffusion equation as set out in Chap. 5, T is taken to obey the differential equation $DT(\mathbf{r}, t - t_0 - t_1) = 0$, where the differential operator D is given by $D = \partial/\partial t - \nabla \cdot (K\nabla)$ with an effective diffusivity K that may depend on relative displacement and on time.

Batchelor thought it reasonable to relate the effective diffusivity to the mean square relative displacement $\langle \mathbf{r}^2(t) \rangle$. The assumption $K(t) = \alpha(\langle \mathbf{r}^2(t) \rangle)^{2/3}$ then leads to

$$T(\mathbf{r}, \tau) = (2\pi \langle \mathbf{r}^2 \rangle)^{-3/2} \exp\left(-\frac{1}{2} \frac{\mathbf{r}^2}{\langle \mathbf{r}^2 \rangle} \right), \tag{12.19}$$

where $\langle \mathbf{r}^2 \rangle = (2\alpha\tau/3)^3$. Clearly, other forms of the effective diffusivity would allow other physical situations to be treated by this approach.

12.6 Turbulent Shear Flows

We turn at this point to discuss diffusivity of momentum in turbulent shear flows as described by the Navier–Stokes equation. For steady flow of an incompressible fluid with constant viscosity we have

$$(\mathbf{v} \cdot \nabla)\mathbf{v} = -(m\rho)^{-1} \nabla p + \nu \nabla^2 \mathbf{v} \,. \tag{12.20}$$

As a first step it is useful to compare the length scales of momentum diffusivity in laminar against turbulent boundary layers.

12.6.1 *Length scales of momentum transport*

The case of laminar flow over a flat plate with no-slip boundary condition is illustrated in Fig. 12.7, which is taken from the book of Tennekes and Lumley.[395] A characteristic length L and a characteristic velocity U have been attributed to the flow, so that its characteristic time is $t \approx L/U$. The viscous term on the RHS of Eq. (12.20) describes transport of momentum by molecular processes across the main flow: for a fixed time scale, this process defines another length scale $\ell \approx (\nu t)^{1/2} \approx (\nu L/U)^{1/2}$. Evidently, this length is in this case related to the length scale of the main flow by $\ell/L \approx (\nu/LU)^{1/2} = \mathrm{Re}^{-1/2}$. With characteristic velocity fluctuations of order u as indicated in the figure, we can

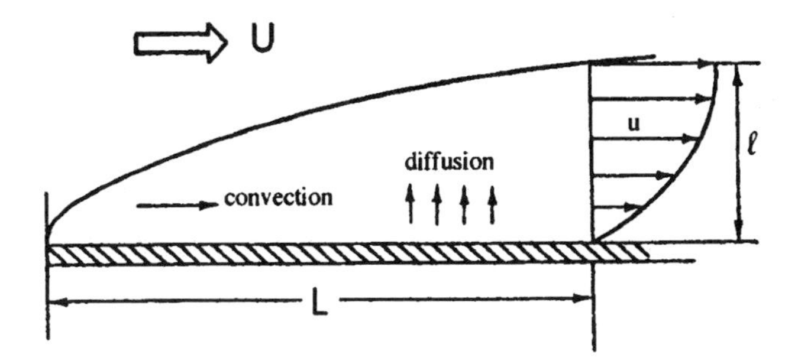

Fig. 12.7. Convection and diffusion in a laminar boundary layer over a flat plate. (Redrawn from Tennekes and Lumley, Ref. 395.)

write $t = \ell/u$ and hence we have the scale relation

$$\frac{\ell}{L} \approx \frac{u}{U} . \tag{12.21}$$

The viscous length ℓ in laminar flow represents the transverse thickness of the boundary layer, inside which the molecular transport of momentum deficit occurs away from the solid surface.

We consider next the case of turbulent flow. The momentum transfer is now dominantly effected by turbulent eddies, and we may estimate the thickness of the boundary layer by again equating the time scale for turbulent diffusivity to the convective time scale, that is $\ell/u \approx L/U$. This merely states that in an imposed flow the turbulence, being part of the flow, must have a time scale commensurate with that of the flow. Thus, Eq. (12.21) still gives the scale relation for the boundary layer. However, the small eddies which are responsible for dissipation (Sec. 12.4.1) have much shorter time scales, tending to make them statistically independent of the main flow.

12.6.2 *Reynolds stresses*

As discussed in detail in Sec. 6.2, the viscous term in Eq. (12.20) arises from the Newtonian stress tensor as determined by the gradient of the velocity field *via* intermolecular collisions. In the regime of fully developed turbulence its role is taken over by turbulent stresses arising from fluctuations in the velocity field, which are known as Reynolds stresses.

Let us consider steady flow at mean velocity $\mathbf{U}$ and let $\mathbf{u} = (u_1, u_2, u_3)$ be a fluctuating velocity field with zero time average. The x_1-component of the momentum flux per unit area across the faces dx_2dx_3 and dx_1dx_3 of a fluid volume element are $m\rho(U_1+u_1)^2$ and $m\rho(U_1+u_1)(U_2+u_2)$, with average values $m\rho(U_1^2+\langle u_1^2\rangle)$ and $m\rho(U_1U_2+\langle u_1u_2\rangle)$ where the brackets denote a time average. The equivalence between momentum flux and stress follows at once from Newton's second law, so that the fluctuating terms in the momentum fluxes above correspond to stresses $\sigma_{11}=-m\rho\langle u_1^2\rangle$ and $\sigma_{12}=-m\rho\langle u_1u_2\rangle$ on the faces of the volume element. Notice that (i) σ_{11} is compressive since $\langle u_1^2\rangle$ is positive, and (ii) the rate at which the x_2-component of momentum passes through the face dx_2dx_3 leads to a shear stress $\sigma_{21}=\sigma_{12}$ on that face, just as with viscous stresses. The conclusion thus is that in turbulent flows, even though the relative root-mean-square fluctuations of the velocity field may be of the order of a percent, the mean motion is not determined only by viscous forces. Explicitly, the x_1-component of Eq. (12.20) yields the mean momentum equation in the form

$$(\mathbf{U}\cdot\nabla)U_1=-(m\rho)^{-1}\frac{\partial\langle p\rangle}{\partial x_1}+\nu\nabla^2U_1 -\left(\left\langle\frac{\partial u_1^2}{\partial x_1}\right\rangle+\left\langle\frac{\partial u_1u_2}{\partial x_2}\right\rangle+\left\langle\frac{\partial u_1u_3}{\partial x_3}\right\rangle\right), \tag{12.22}$$

having used the relation $\langle u_1\nabla\cdot\mathbf{u}\rangle=0$ from the continuity equation. The equations for the other components of the velocity field are obtained by cyclic interchange of indices.

In a mainstream of fully developed turbulence, as already discussed in Sec. 12.5.1, the Reynolds stresses are in fact much larger than the viscous stresses. If one then tries to account for the Reynolds stresses by deriving from the original Navier–Stokes equation additional equations for the velocity autocorrelation functions in Eq. (12.22), one runs into what is known as the closure problem of turbulence: unknown correlation functions such as $\langle u_1u_2^2\rangle$ are generated by the convective term. This problem is typical of all nonlinear stochastic systems.

Many attempts have therefore been made in the literature to find an approximate reduction of the Reynolds stress tensor to a form similar to that of the Newtonian stress involving the velocity gradient, by introducing an "eddy viscosity". A similar attempt in regard to "eddy diffusivity" has been illustrated in Sec. 12.5.3. There are some special cases in which the turbulent diffusivities depend simply on the velocity and length scales of the flow, as in the Couette flow between a fixed wall and a moving wall (see the discussion given in the book of Bradshaw[394]). In general, however, there are two profound differences between turbulent stresses and viscous stresses: (i) turbulent stresses are continuous whereas molecules collide only at intervals, and (ii) the dimensions of turbulent eddies are not small relative to those of the flow. The reader may find critical assessments of the so-called mixing length for eddy viscosity in the books of Faber[4] and of Tennekes and Lumley.[395]

More recently, great progress has been made in understanding turbulence through the solution of the nonlinear Navier–Stokes equation by direct numerical methods.[418] We have met an example of such results in the study of the evolution of a layer of Kelvin–Helmholz vortices in Sec. 12.2.3. The so-called lattice Boltzmann method, that we introduce immediately below, has known rapid expansion in the late eighties and has been progressively refined and extended to the point where it is a competitive technique to treat a variety of nontrivial flows.

12.6.3 *Lattice Boltzmann computing*

The structure of the Navier–Stokes equation is quite independent of the details of the underlying microscopic dynamics, which only determine the numerical values of the transport coefficients. This university motivates the use of microdynamical models which, while giving up as much irrelevant detail as possible, still retain the basic aspects of the physics of fluids. Lattice gas models are within such a class of models. Their aim consists in the definition of a simplified microworld which allows one to recover in the macroscopic limit the equations of fluid dynamics. A proper choice of the symmetry of the lattice is crucial for this purpose.[419] Once the correct symmetries of the lattice are chosen, there are two possible ways of defining the evolution rules for the system under study. These are known as the lattice gas automata and the lattice Boltzmann method.

In lattice gas automata the variables are the boolean populations and the evolution is defined by a set of collision rules. The method is also suitable to analyse fundamental issues such as the long-time tail problem in diffusion phenomena (see Sec. 5.5.1). In particular, one can consider the Boltzmann approximation to the dynamics, by making the same assumption which leads to the Boltzmann equation in continuous kinetic theory: particles entering a collision are uncorrelated.

The development of lattice Boltzmann (LB) methods[420] was originally motivated by the need to overcome the statistical noise problem plaguing the lattice gas automata method. The main LB equation has a Chapman–Enskog form for the evolution of a particle population distribution $f_i(\mathbf{r}, t)$ occupying at time t the node $\mathbf{r}$ of the lattice along a direction specified by a velocity variable $\mathbf{c}_i$ (with $i = 1, \ldots, b$):

$$f_i(\mathbf{r} + \mathbf{c}_i, t+1) - f_i(\mathbf{r}, t) = \sum_{j=1}^{b} A_{ij}(f_j - f_j^{\mathrm{eq}})\,. \tag{12.23}$$

The matrix A_{ij} is symmetric and its form is chosen so as to satisfy the sum rules coming from mass and momentum conservation. The function f_i^{eq} is determined by the local flow field.

This approach ensures that the Navier–Stokes equation is recovered in the hydrodynamic limit, defined as the limit in which the particle mean free path is much smaller than typical scales of macroscopic variation in the properties of the system. Focal points for further development at the time of writing concern the ability to deal with complex geometries and to incorporate existing models for turbulence. An example of advanced results is the demonstration of space-time intermittency in channel flow turbulence[421]: homogeneous isotropic turbulence is observed near the centre of the channel, but intermittency is found to grow as one moves towards the channel boundaries, in parallel with an intensified presence of ordered vortical structures.

12.7 Turbulence in Compressible Fluids

A common simplification in the study of turbulence is that its general behaviour seems to be unaffected by compressibility, as long as the pressure fluctuations within the turbulent flow do not become comparable with the average pressure. In most instances the velocity fluctuations are small compared with the speed of sound. Similarly, density fluctuations due to temperature differences in the flow are often small enough to have no direct effect.

A systematic discussion of the role of compressibility in turbulence has been given by Moyal.[422] Turbulent motion in a compressible fluid presents two different physical aspects: (i) the breakdown of laminar flow and the creation of fluctuating eddy motion (what Moyal calls "eddy turbulence"), and (ii) the existence of fluctuating compressional waves, corresponding to random noise.

The state of a real fluid is fully specified by its velocity, density, pressure and temperature fields, and all of these quantities fluctuate when the fluid is in turbulent flow. Since these fluctuations are connected by dynamical and state equations, the phenomena (i) and (ii) above must be closely related. Indeed Moyal shows that the velocity spectrum can be analysed into two components, which are to be interpreted as the spectra of eddy turbulence and of noise respectively. He also indicates how these are connected through the equations of motion.

Random noise and proper eddy turbulence interact most strongly at high levels of turbulence and large Reynolds numbers. Under conditions such as these, it can be anticipated that eddy turbulence will act as a source of noise, and *vice versa.* In contrast, at low turbulence levels and low Reynolds numbers the two phenomena will tend to proceed independently.

The component of the velocity field associated with eddy turbulence has much the same character as that found for the total velocity field when the fluid is taken to be incompressible. The fact that pressure changes are transmitted with a finite velocity is of little consequence in the regime of low Mach numbers. Moyal argues that probably the main effect of compressibility on eddy turbulence is to introduce an additional source of energy dissipation through production of disordered acoustic waves which are transmitted or absorbed at the boundaries.

12.8 Turbulent Behaviour of Non-Newtonian Fluids

There is much practical interest in turbulence in non-Newtonian fluids and viscoelastic materials. As the separation in length scales widens between the large energy-containing eddies and the small eddies affected by molecular properties, the turbulent behaviour will become similar to that of Newtonian fluids as it becomes independent of the precise mechanism for energy dissipation. However, non-Newtonian effects will appear near solid boundaries even if they are absent from the main part of the turbulent motion, since the length scale of the largest eddies in the boundary layer is of the order of its thickness (see Sec. 12.6).

Most viscoelastic substances, such as long-chain polymer solutions, have a rather low yield stress beyond which the strain-dependent part of the stress ceases to rise. In fact, solutions of organic chain molecules, even at very high dilution, show significantly smaller drag in turbulent pipe flow than the pure solvent. An explanation offered by Bradshaw[394] is that the increase in turbulent intensity and shear stress with distance from the walls is inhibited by distortions of the molecules, leading to a larger velocity gradient for a given shear stress.

The anisotropy of liquid crystals makes them especially suited to flow visualisation by birefringence techniques (see Sec. 9.5). It also has significant consequences on their behaviour in thermal convection.[423] Consider a nematic layer confined between horizontal plates which favour a planar orientation of the director along the x direction, say. The orientation of the convection rolls cannot be arbitrary in the (x, y) plane, but is related to the orientation of the unperturbed director $\mathbf{n}_0$ by a coupling between orientation and flow. This coupling vanishes by symmetry when the roll axis is parallel to $\mathbf{n}_0$ and is at its maximum when the rolls are perpendicular to $\mathbf{n}_0$.

The effect is treated by Manneville[423] in a one-dimensional model where orientation fluctuations are confined to the (x, z) plane, using an anisotropic heat-conduction coefficient. The dynamical equilibrium of the director is determined by a balance between the elastic torque of bending and a viscous torque due to convective flow. The latter results from molecular rotations induced by the vertical shear and from damping due to viscous friction. The disorientation of the molecules induces a supplementary horizontal heat flux which tends to reinforce temperature fluctuations (see Fig. 12.8) and hence to lower the stability threshold.

A similar account can be given for the Carr–Helfrich electrodynamic instability in nematics inserted in a parallel-plate condenser,[423] using the similarity in structure between Ohm's law and the thermal conduction law. The control parameter in this case is the potential drop applied to the capacitor and the new variable is the electric charge density.

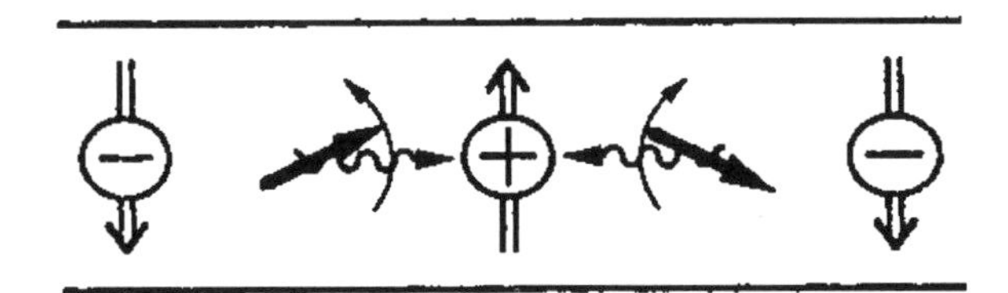

Fig. 12.8. Schematic illustration of heat focalisation in Rayleigh–Bénard convection by anisotropy in a nematic liquid crystal. (Redrawn from Manneville, Ref. 423.)

Appendix 12.1 Navier–Stokes Equation: Analogy with Maxwell's Equations

In dealing with Stokes' law in Sec. 6.4.1 we met an instance of vorticity-free, incompressible flow and we saw that in this case the velocity field can be written as the gradient of a potential function obeying the Laplace equation. A mapping into magnetostatics can thus be set up, with the velocity field being mapped into the magnetic induction **B** and fluid sources and sinks being mapped into the effective magnetic poles which appear at the ends of long thin current-carrying wires. More generally, the Navier–Stokes equation shows analogies with Maxwell's equations of electrodynamics. Here we summarise the findings of Marmanis.[535]

The equations of motion obtained from the Navier–Stokes equations for an incompressible Newtonian flow can be written in the form

$$\frac{\partial \mathbf{v}(\mathbf{r},t)}{\partial t} = -\nabla\Phi(\mathbf{r},t) - \mathbf{l}(\mathbf{r},t) + \nu\nabla^2\mathbf{v}(\mathbf{r},t)\,, \tag{A12.1.1}$$

where $\Phi = \mathbf{v}^2/2 + (p/m\rho)$ is the Bernoulli energy function, $\mathbf{l} = \mathbf{v}\times\omega$ is termed the Lamb vector with $\omega = \nabla\times\mathbf{v}$ being the vorticity, and ν is the kinematic viscosity. The equation of continuity reduces in this case to the condition that the velocity field is divergenceless,

$$\nabla\cdot\mathbf{v}(\mathbf{r},t) = 0\,. \tag{A12.1.2}$$

Equations (A12.1.1) and (A12.1.2) constitute a system of coupled partial differential equations, to be, of course, supplemented by appropriate boundary and initial conditions.

As Marmanis notes, the above system is considered to be an adequate representation of flows at high Reynolds number. However, he also stresses that turbulent flows are characterised by many spatio-temporal scales produced and sustained by a continuous transfer of energy from the larger scales to the smaller ones. Under such circumstances, one must be content with a description of average quantities. The basic aim is to construct a theoretical framework which will allow one to find the average values of the velocity and pressure fields at high Reynolds numbers and will be useful for both homogeneous and inhomogeneous cases. His proposal consists of a set of equations for the average values of the vorticity field and the Lamb vector.

After some calculation, the above program can be summarised in four equations:

(i) Description of evolution of vorticity,

$$\frac{\partial \omega}{\partial t} = -\nabla \times \mathbf{l} + \nu \nabla^2 \omega\,; \tag{A12.1.3}$$

(ii) Divergenceless vorticity,

$$\nabla \cdot \omega = 0\,; \tag{A12.1.4}$$

(iii) Equation for "turbulent charge density" $n(\mathbf{r}, t)$,

$$\nabla \cdot \mathbf{l} = -\nabla^2 \Phi \equiv n(\mathbf{r}, t)\,; \tag{A12.1.5}$$

(iv) Evolution of Lamb vector,

$$\frac{\partial \mathbf{l}}{\partial t} = \mathbf{v}^2 \nabla \times \omega - \mathbf{j} + \nu \nabla n - \nu \nabla^2 \mathbf{l} \tag{A12.1.6}$$

where $\mathbf{j}$ is termed the turbulent current vector and is given by

$$\mathbf{j} = n\mathbf{v} + \nabla \times (\mathbf{v} \cdot \omega)\mathbf{v} + \omega \times \nabla(\Phi + \mathbf{v}^2) + 2(\mathbf{l} \cdot \nabla)\mathbf{v}\,. \tag{A12.1.7}$$

In reaching Eq. (A12.1.6), a "second-order correction" $\nu^2 \nabla^4 \mathbf{v}$ in the viscosity has been dropped, relative to the term $\nu \nabla^2 \mathbf{l}$ retained there.

Thus, Marmanis has constructed a system of four equations which are valid locally in any part of the fluid, to a certain useful approximation as far as viscous "corrections" are concerned. He stresses the correspondence between Maxwell's equations and the set of Eqs. (A12.1.3)–(A12.1.6) for the case when the kinematic viscosity is put to zero:

$$\begin{aligned}
\text{Vector potential} &\leftrightarrow \mathbf{v}(\mathbf{r}, t)\\
\text{Scalar potential} &\leftrightarrow \Phi(\mathbf{r}, t)\\
\text{Magnetic field} &\leftrightarrow \omega(\mathbf{r}, t)\\
\text{Electric field} &\leftrightarrow \mathbf{l}(\mathbf{r}, t)\\
\text{Charge density} &\leftrightarrow n(\mathbf{r}, t)\\
\text{Current vector} &\leftrightarrow \mathbf{j}(\mathbf{r}, t)
\end{aligned}$$

Marmanis then invokes a spatial filtering method proposed by Russakoff[536] in the context of Maxwell's equations in order to derive, after some reasonable approximations on averages, equations for both the homogeneous and the inhomogeneous case.

Appendix 12.2 Series Solution of Navier–Stokes Equation

Wyld[537] appears the first to make systematic use of perturbation theory to treat isotropic turbulence in incompressible fluids. Lee[538] subsequently generalized Wyld's formulation to treat hydromagnetic turbulence. In this Appendix, however, we choose to summarise the method of Phythian,[539] who constructs a series solution of the Navier–Stokes equation. The self-consistent procedure of Phythian yields the direct-interaction approximation of Kraichnan[540] as the simplest nontrivial approximation.

For simplicity of presentation, we follow Phythian in using the Burgers model equation to derive the desired expansion. As he notes, the whole procedure can be carried through in the same fashion for the Navier–Stokes equation when the pressure term has been eliminated. The Burgers equation was introduced[541] as a one-dimensional model of turbulence and is relevant to a number of problems, including the formation of large-scale structures in the Universe.[542]

The velocity field $v(x,t)$ obeys the equation

$$\frac{\partial v}{\partial t} = \nu \frac{\partial^2 v}{\partial x^2} - v\frac{\partial v}{\partial x} + f\,. \tag{A12.2.1}$$

Here $f(x,t)$ is an applied random "force" field which has the properties that (i) it is statistically homogeneous and stationary, and (ii) has a Gaussian distribution with zero mean. The resulting statistical distribution of the velocity field is obtained by Fourier analysis, with the fluid enclosed in a space-time box of volume VT and satisfying periodic boundary conditions. The Fourier components $v(k,\omega)$ and $f(k,\omega)$ are readily shown to obey the equation

$$\begin{aligned}(i\omega + \nu k^2)v(k,\omega) &= f(k,\omega)\\ &\quad + \frac{\lambda}{VT}\sum M(k,\omega;k_1,\omega_1;k_2,\omega_2)v(k_1,\omega_1)v(k_2,\omega_2)\,,\end{aligned} \tag{A12.2.2}$$

where the function M has the form $M = -(ik/2)\delta_{k_1+k_2,k}\delta_{\omega_1+\omega_2,\omega}$ and for later convenience an expansion parameter λ has been introduced.

Phythian now bases a perturbation expansion on the form

$$i\omega v(k,\omega) = \alpha(k,\omega)v(k,\omega) + g(k,\omega)\,, \tag{A12.2.3}$$

where $\alpha(k,\omega) = -\nu k^2$ and $g(k,\omega) = f(k,\omega)$ when λ is set to zero. The corrections to this zeroth-order results are to be chosen so that certain statistical properties of the velocity field agree from Eq. (A12.2.3) and from the full Eq. (A12.2.2). These corrections are written explicitly up to second order on the assumption that they can be expanded in powers of λ. Phythian stresses that the convergence properties of his series expansion are unknown.

The spectral function $U(k,\omega)$ of the velocity field is introduced so that

$$\langle v(x,t)|v(x',t')\rangle = \frac{1}{(2\pi)^2}\int dk \int d\omega U(k,\omega)\exp[ik(x-x') + i\omega(t-t')]\,. \tag{A12.2.4}$$

Assuming that the random function $g(k,\omega)$ has a spectral function $\beta(k,\omega)$, according to

$$\langle g(k,\omega)g(k',\omega')\rangle = VT\delta_{k+k',0}\delta_{\omega+\omega',0}\beta(k,\omega)\,, \tag{A12.2.5}$$

Phythian finds

$$U(k,\omega) = S(k,\omega)S^*(k,\omega)\beta(k,\omega)\,, \tag{A12.2.6}$$

where $S(k,\omega)$ is the response function $S(k,\omega) = [i\omega - \alpha(k,\omega)]^{-1}$.

These results are identical to those of the direct-interaction approximation of Kraichan.

Chapter 13

Liquid–Vapour Interface

13.1 Background and Empirical Correlations

In this Chapter, we move away from the theme of bulk liquid properties and focus on the liquid–vapour interface. A number of basic concepts concerning surface properties of liquids and related applications[424] have already been introduced in Chap. 1: here the main emphasis will be on the atomistic aspects of surface tension. For the most part we shall concentrate on the simplest case of a planar interface, which we take to be perpendicular to the z axis. Then one of the central aims of the theory must be to predict the density profile as a function of z through the interface. In a monatomic liquid such as argon, this density, $\rho(z)$ say, must evidently reach the homogeneous bulk liquid density ρ_l deep into the dense liquid phase and tend far into the vapour phase to the vapour density ρ_v (with $\rho_\mathrm{v} \ll \rho_\mathrm{l}$ except in the vicinity of the liquid–vapour critical point). In between, we must expect a decrease from ρ_l to ρ_v over a characteristic distance which will be a measure of the surface thickness. A schematic illustration is given in Fig. 13.1.

The recognition that the liquid–vapour interface will have associated with it a characteristic length has derived from the example above of a simple monatomic system. This example will be made quantitative below, but the idea of such a characteristic length has much wider validity.

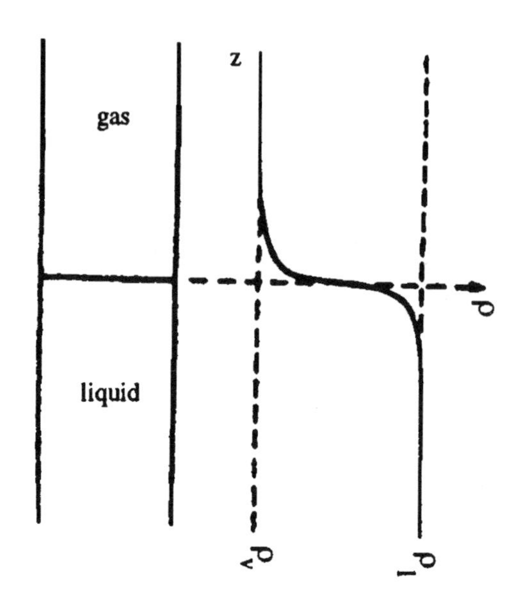

Fig. 13.1. Schematic illustration of the particle density profile across the liquid–vapour interface near the triple point.

13.1.1 *Relation between surface tension and bulk properties: Organic liquids near 298 K*

One consequence of knowing the density profile $\rho(z)$ through the liquid–vapour interface, which was already familiar to van der Waals, is that an (approximate) calculation can then be made of the surface tension γ. Being a thermodynamic property, γ is available from experiment for a wide variety of liquids. In this section, we shall focus on data for organic liquids.[425]

In Table 13.1, values of the surface tension are recorded for organic liquids at 298 K. For comparison purposes, four other liquids are recorded there, i.e. CS_2, H_2O, argon at 87 K and xenon at 165 K. From the second column of this Table 13.1, γ is seen to range, in the units specified there, from about 11 for $C(CH_3)_4$ and Ar to about 70 for H_2O or about 40 for $C_6H_5CH_2$. Thus, for this rather wide variety of liquids, surface tension varies by a factor of up to 7.

There has been a long-standing interest in relating such a thermodynamic surface property to bulk liquid properties.[426,427] In this spirit the isothermal compressibility K_T of the bulk liquid is recorded in the third column of Table 13.1. From purely dimensional considerations, since γ is a force per

Table 13.1. Relation between surface tension γ and isothermal compressibility K_T of (mainly) organic liquids at 298 K.

Liquid	Surface tension γ (10^{-3} kg/s^2)	Compressibility K_T (10^{-10} m s^2/kg)	Product γK_T (10^{-12} m)
$C(CH_3)_4$	11.5	—	—
n-C_5H_{12}	15.6	18	28
n-C_6H_{14}	18.4	16.7	31
n-C_7H_{16}	19.6	14.4	28
n-C_8H_{18}	21.2	12.8	27
n-$C_{10}H_{22}$	23	11.0	26
n-$C_{12}H_{26}$	25	9.9	25
C_6H_6	28	9.7	27
$C_6H_5CH_3$	28	9.3	26
$C_6H_5NH_2$	42	4.7	20
$C_2H_5OC_2H_5$	17	19.7	33
CH_3COCH_3	23	13	30
CH_3OH	22	12.6	28
C_2H_5OH	22	11.4	25
CH_3COOH	27	9.4	25
tr-$C_2H_2Cl_2$	23	11.2	25
C_2H_5Br	24	13.6	33
C_6H_5Br	36	6.7	24
C_2H_5I	29	9	26
CS_2	32	9.5	30
H_2O	72	4.6	33
Ar (87 K)	11	21.8	24
Xe (165 K)	18	17.7	32

unit length while K_T has the dimensions of inverse pressure, it follows that the product γK_T has the dimensions of a length. We shall formalise this by writing

$$\gamma K_T = L\,. \tag{13.1}$$

This length L is given in the last column of Table 13.1 and, as Freeman and March[425] stress, has the average value 0.27 ± 0.03 Å for the entire set of liquids considered. The constancy of the product γK_T is remarkable, bearing in mind the spread of values in the separate tabulations of γ and of K_T. In fact, the approximate constancy of the product γK_T extends to metals and molten salts, in addition to insulating molecular liquids.[426]

Later in this chapter, but now for monatomic liquids such as Ar and Xe, we shall discuss the meaning to be attached to the length L. Suffice it to say here that knowledge that the product γK_T for classes of liquids in standard thermodynamic conditions has an approximately constant value is evidently sufficient, in conjunction with the bulk liquid compressibility, to make quite impressive "predictions" of the liquid–vapour interfacial tension.

13.2 Definition of a Surface and its Thermodynamic Properties

Let us now give a formal definition of a surface and its thermodynamic properties. A surface is the region between two bulk phases in which the properties of the matter comprising the two phases are characteristic of neither phase, but approach those of each of the two phases in regions remote from the other. The surface which is the focus of the present chapter exists between a dilute vapour phase and a dense liquid phase, separated by a planar region due to the effects of gravity and of molecular cohesion. The distance over which this region extends into the bulk phases is not precisely defined, although as we shall see further below, useful characteristic lengths with the order of magnitude of the surface thickness are not difficult to obtain.

13.2.1 *Gibbs surface*

In the studies of Gibbs[428] it was shown that the properties of a system with a surface could be described without reference to the entire heterogeneous region. The construction of Gibbs was to choose a dividing plane which passes through all points which are similarly situated with respect to the conditions of adjacent matter. While the choice of such a dividing surface is at one's disposal, it turns out that a specific choice commends itself in single-component systems.

The so-called Gibbs surface separates the system into three parts, i.e. two three-dimensional phases and a two-dimensional interface region. A thermodynamic quantity such as the free energy F can be expressed as the sum of contributions from the three parts:

$$F = F_L + F_v + F_S \,. \tag{13.2}$$

Here F is the observable value of the free energy for the entire system, while F_{L} and F_{v} are the values of the free energy corresponding to homogeneous bulk phases occupying the volumes bounded by the dividing surface. The value of F_{S} depends on the choice of the dividing surface, as discussed further below. Let us turn in this light to the definition of surface tension.

13.2.2 *Surface tension*

The surface tension of a liquid (or in fact a solid too) is defined as the work required to reversibly create unit area of new surface in an isothermal process without changes in structure. The reversible work done on a single-phase system equals the change in the Helmholtz function. In a two-phase system with a surface, however, one must take account of the position of the division (cf. Brown and March[427]). For the general case of a multi-component fluid, the surface tension γ is related to the surface Helmholtz function F_{S} by

$$\gamma A = F_{\mathrm{S}} - \sum_i \mu_i \Gamma_i \,, \tag{13.3}$$

A being the interfacial area. While γ does not depend on the choice of the Gibbs surface, F_{S} in Eq. (13.3) does. In this equation μ_i is the chemical potential of the ith component of the two-phase system and Γ_i is its adsorption, which must then depend on the choice of the dividing surface. If the z axis is taken normal to the planar surface, with bulk liquid conditions obtaining in the limit $z \to -\infty$ and bulk vapour for $z \to \infty$, and the Gibbs surface is taken to lie at $z = z_{\mathrm{S}}$, the definition of Γ_i is

$$\Gamma_i = \int_{-\infty}^{\infty} dz [\rho_i(z) - \rho_i^{\mathrm{liquid}} \vartheta(z_{\mathrm{S}} - z) - \rho_i^{\mathrm{vapour}} \vartheta(z - z_{\mathrm{S}})] \,. \tag{13.4}$$

Here ρ_i denotes the number density of the ith component. The Heaviside step function $\vartheta(x)$ entering Eq. (13.4) is equal to unity for positive argument and to zero otherwise.

We now emphasise that, in the special case of a single-component system, a choice of z_{S} can be made such that the integral arising in Eq. (13.4) is zero. This so-called surface of zero adsorption is the most natural choice of dividing surface in such systems, as it ensures the equality of the surface tension γ and the surface Helmholtz function F_{S} through Eq. (13.3).

13.2.3 *Surface entropy*

For a single-component system with the choice of the dividing surface taken at the Gibbs equimolecular surface, from $F_S = \gamma A$ we get the differential of the surface free energy as $dF_S = \gamma dA + A d\gamma$. On the other hand, we can also write the change in surface free energy associated with changes dT in temperature and dA in surface area as $dF_S = -S_S dT + \gamma dA$. Thus, the thermodynamic definitions of surface tension and surface entropy for a pure liquid are

$$\begin{cases} \gamma = \left(\dfrac{\partial F_S}{\partial A}\right)_T \\ S_S = -A\left(\dfrac{\partial \gamma}{\partial T}\right)_A \end{cases} . \tag{13.5}$$

It is easily seen that these formulae agree with the definition of the surface internal energy given earlier in Sec. 1.6.2.

The surface excess entropy per unit area may be viewed as reflecting the state of order of the interface relative to the bulk. Typically, the surface tension in a system like argon decreases monotonically with increasing temperature and vanishes at the liquid–gas critical point, with S_S/A showing a similar behaviour. This is consistent with the idea that the interface progressively spreads out with temperature, to disappear at the critical point. There are some exceptions to this rule, however: one example being that of the liquid metals Zn, Cd and Cu, where the surface tension shows a maximum in its temperature dependence. This behaviour is taken to reflect a quasi-crystalline state of order of the first few surface layers in these liquids near to the freezing point, which also appears through a visible faceting of solidified specimens formed from sessile liquid drops.[23]

Studies of the temperature dependence of the surface tension of water have shown that its surface entropy is sizably smaller than that of typical nonpolar liquids, the entropy deficit being $\approx 1.7\ k_B$ per molecule in the surface. This fact has been attributed to a preferred orientation of the molecules in the surface layers, implying the presence of surface polarisation. The orientation of the molecular dipoles is ultimately due to the role of the molecular quadrupole moment in exerting a torque on a molecule in the interfacial region.[429] The quadrupolar moment modifies the electric field lines and the molecules minimise the free energy by orienting themselves so that their electric fields lie as much as possible in the region of high dielectric constant occupied by the dense liquid.[430] Indeed, no surface polarisation could result if the molecular dipole moment were symmetrically located relative to the molecular centre.

13.3 Phenomenology

Having discussed the thermodynamics of liquid surfaces, and after giving a very practical route to relate γ to the compressibility K_T of the bulk liquid, it will be useful to proceed to the microscopic theory of the surface tension of liquids *via* a phenomenological approach.

As a preliminary let us define the free-energy density $\psi(\rho, T)$ for the homogeneous fluid at number density ρ and temperature T and let us for a moment suppose that the surface tension could be expressed through this function as

$$\gamma = \int_{-\infty}^{\infty} dz[\psi(\rho(z)) - \psi(\rho_l)\vartheta(-z) - \psi(\rho_v)\vartheta(z)], \tag{13.6}$$

having taken the Gibbs dividing surface at $z = 0$. The function $\psi(\rho, T)$ below the liquid–vapour critical point has the typical shape shown in Fig. 13.2, presenting two minima which allow one to locate the two coexisting phases by a common-tangent construction. Equation (13.6) expresses the surface tension

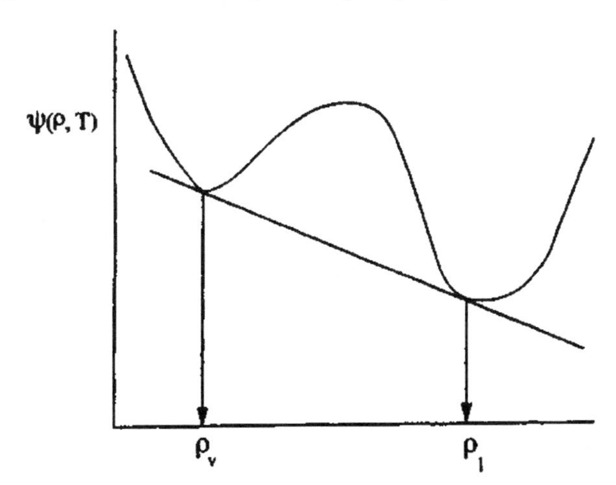

Fig. 13.2. Schematic drawing of the excess free energy density as a function of the particle density for a fluid well below the critical point.

as an explicit function of the density profile $\rho(z)$: however, it is evident that the profile which minimises this functional is a discontinuous jump from ρ_l to ρ_v at $z = 0$, leading as a result to $\gamma = 0$. This result implies that the expression (13.6) is wrong: the relevant free-energy density cannot be a purely local function of the particle density.

The simplest approximation which allows us to account for non-locality of the free-energy density in the interfacial region is to supplement the integrand in Eq. (13.6) by terms which depend on the density gradient. This idea appears to go back as far as van der Waals, but the way in which it was first phenomenologically implemented is usually associated with the names of Cahn and Hilliard.[431] The presentation of this approach outlined below follows closely that of Widom.[432]

13.3.1 *Free energy from inhomogeneity*

From fluctuation theory it has been established that, if in a substantial subvolume v a small density fluctuation $\Delta\rho$ occurs in the number of molecules per unit volume ρ, then there is a free energy increase F_1 which is given by

$$F_1 = \frac{\alpha v (\Delta\rho)^2}{\rho^2 K_\mathrm{T}} . \tag{13.7}$$

Here, as above, K_T is the isothermal compressibility of the bulk liquid while α is a constant which, for genuine fluctuations in a liquid, is precisely $1/2$.

The essence of the first step in the Cahn–Hilliard approach is then to extrapolate Eq. (13.7) to describe the free energy associated with any inhomogeneity occurring in a substantial volume of a system in which we pass to the thermodynamic limit. The point to be emphasised is that Eq. (13.7) is still to be used when the inhomogeneity is that which occurs at the interface between liquid and vapour phases. Of course, this is using the form (13.7) outside its strict range of validity. One further point needs to be noted: what should one take for the product $\rho^2 K_\mathrm{T}$ in Eq. (13.7). At the liquid–vapour interface, this will be assumed to refer to the bulk liquid.

If F_1 were the entire free energy associated with the inhomogeneity of the density across the interface, then the total interfacial free energy per unit area, to be identified with the surface tension, would be $\gamma_1 = \alpha L (\Delta\rho)^2/\rho^2 K_\mathrm{T}$ where L is used as simply a measure of the thickness of the interface. But due to the presence of a density gradient across the interface, there is an additional contribution to the interfacial free energy.

13.3.2 *Density gradient contribution to free energy*

This further contribution to the interfacial free energy is written as

$$F_2 = \alpha' v \left(\frac{\Delta\rho}{L}\right)^2 , \tag{13.8}$$

where α' is yet another constant. The surface tension is now the sum of the free energies F_1 and F_2 per unit area, $\gamma = \alpha L(\Delta\rho)^2/\rho^2 K_T + \alpha' L(\Delta\rho/L)^2$, and the surface thickness L can be determined by minimising γ with respect to L. This yields $L = \sqrt{\alpha'\rho^2 K_T/\alpha}$ and $F_1 = F_2$, i.e. equal contributions to the interfacial free energy from the "density fluctuation" and from the density gradient. Hence it follows that

$$\gamma = \frac{2\alpha L(\Delta\rho)^2}{\rho^2 K_T} . \tag{13.9}$$

It is reasonable at this point to take as an estimate for $\Delta\rho$ the difference between the bulk liquid density ρ and the vapour density ρ_v. But near the triple point we have $\rho \gg \rho_v$ and hence it follows that $\gamma = 2\alpha L/K_T$, or $\gamma K_T \approx L$. The product γK_T is therefore a measure of the thickness of the surface region (compare Sec. 13.1.1).

13.3.3 *Extension to binary alloys and surface segregation*

The Cahn–Hilliard approach has been extended to binary alloys by Bhatia and March.[433] In their approach the surface tension can be written in the form

$$\gamma \approx \frac{L}{K_T}\left[1 + \frac{\delta^2 S_{cc}(0)}{\rho k_B T K_T}\right] \tag{13.10}$$

involving, in addition to the concentration-dependent compressibility, the size factor δ and the long-wavelength value $S_{cc}(0)$ of the concentration–concentration structure factor (see its definition given in the context of molten salts in Sec. 8.3). These two quantities are given for an alloy at concentration

c by

$$\delta = \frac{v_1 - v_2}{cv_1 + (1-c)v_2} , \tag{13.11}$$

where v_1 and v_2 are the partial molar volumes of the two species, and by

$$S_{\rm cc}(0) = \rho k_{\rm B} T \left(\frac{1}{V} \frac{\partial^2 G}{\partial c^2} \right)^{-1} , \tag{13.12}$$

where G is the Gibbs free energy for volume V.

Let us briefly report from the work of Bhatia and March some examples of how the size factor in Eq. (13.10) influences the surface tension of some liquid–metal alloys. A remarkable case is that of amalgams formed by dissolving alkali metals into liquid mercury, where the surface tension has a very strong dependence on the alkali-metal content at high dilution. From the size factor one estimates that $-(d\ln\gamma/dc)|_{c=0}$ is of order 100 for dilute solutions of K in Hg, while for Cs a value three or four times larger results. The experimental observations on amalgams are thus explained in a natural physical manner.

The size-factor term may also lead to a change in sign from negative to positive in the concentration dependence of γ. This behaviour is indeed observed, for example, in the Mg–Sn and Mg–Pb systems.

Finally, one expects in general that the surface layers will be enriched in solute if in the pure state the solute element has a lower surface tension. Equation (13.10) is an adequate basis to discuss surface segregation in dilute alloys, an issue of importance in materials science and in some aspects of catalysis. It indicates that solute segregation in the surface layers will be favoured if the compressibility increases with solute addition, the correlations of this proposed trend with the available data being exposed in the work of Bhatia and March.

13.4 Microscopic Theories: Direct Correlation Function

A formally exact approach to the evaluation of the particle density profile $\rho(z)$ and of the surface tension γ associated with a planar liquid–vapour interface in a monatomic liquid will be set out in this section. It involves the direct correlation function $c(\mathbf{r}, \mathbf{r}')$ in the presence of the inhomogeneity in the particle density across the interface (see Appendix 4.1 for the generalised Ornstein–Zernike relation between $c(\mathbf{r}, \mathbf{r}')$ and the inhomogeneous two-body correlation function).

The results will then be compared with those coming from a low-order density gradient expansion of the free-energy density of the inhomogeneous fluid.

13.4.1 *Density profile and surface tension*

The approach that we sketch in this section is based on fluctuation theory and its main result was known to Yvon, although the formal development is recorded in the literature through the work of Triezenberg and Zwanzig.[434] Actually, in the absence of gravitation a planar liquid–vapour interface can undergo long-wavelength fluctuations with an amplitude that diverges in the thermodynamic limit (see Sec. 13.6 below). It is therefore assumed that the fluid lies in a vertical gravitational field, which is sufficiently weak that its only effect is to stabilise the interface.

A density fluctuation is now assumed to occur so that the density change is given by

$$\delta\rho(\mathbf{R}, z) = \rho(\mathbf{R}, z) - \rho(z) \,. \tag{13.13}$$

Here, the vertical direction is the z axis of a coordinate system and $\mathbf{R}$ is a vector in the plane of the interface. As the density fluctuates, the location of the Gibbs surface also fluctuates and this determines a change in surface area. The corresponding free-energy change is related to the direct correlation function $c(R; z, z')$ by the method presented in Appendix A4.1. If one assumes that $c(R; z, z')$ can be Fourier-transformed with respect to the variable $\mathbf{R}$ and expanded in even powers of the corresponding Fourier variable, the calculation leads to an integro-differential equation for the equilibrium density profile,

$$\frac{d\rho(z)}{dz} = \rho(z) \int_{-\infty}^{\infty} dz' c_0(z, z') \frac{d\rho(z')}{dz'} \tag{13.14}$$

and to the expression

$$\gamma = \frac{k_B T}{4} \int_{-\infty}^{\infty} dz \int_{-\infty}^{\infty} dz' \frac{d\rho(z)}{dz} c_2(z, z') \frac{d\rho(z')}{dz'} \,, \tag{13.15}$$

for the surface tension. Here we have defined

$$\begin{cases} c_0(z, z') = \displaystyle\int d^2\mathbf{R} c(R; z, z') \\ c_2(z, z') = \displaystyle\int d^2\mathbf{R} R^2 c(R; z, z') \,. \end{cases} \tag{13.16}$$

Though there are alternative statistical mechanical approaches which are formally exact (see Sec. 13.5 below), these involve the force field describing the interparticle interactions in the fluid together with the inhomogeneous two-body density across the interface. In the theory embodied in Eqs. (13.14)–(13.16) the interactions are subsumed in the direct correlation function.

13.4.2 *Density gradient expansion: Pressure through interface*

At this point we appeal to an approximate approach, which is based on gradient expansions in the density $\rho(z)$. This yields the low-order density gradient (ldg) result

$$\gamma_{\text{ldg}} = \int_{-\infty}^{\infty} dz A(\rho(z)) \left(\frac{d\rho(z)}{dz} \right)^2 , \tag{13.17}$$

a formula which appears to go back to van der Waals, but which is given quite explicitly by Yang *et al.*[435] As a consequence of the density gradient expansion, in Eq. (13.17) the function $A(\rho)$ is determined by the direct correlation function $c(r;\rho)$ of the homogeneous fluid, but is needed over the whole range of density sampled by the inhomogeneous density profile $\rho(z)$:

$$A(\rho) = \frac{1}{6} k_{\text{B}} T \int d\mathbf{r} r^2 c(r;\rho) . \tag{13.18}$$

One can now compare the formally exact result (13.15) with the approximate model (13.17). To bring the two forms into contact, the simplest assumption is to approximate $c_2(z,z')$ by a local form $c_2(z,z') \approx A(\rho(\bar{z}))\delta(z-z')$ with $\bar{z} = (z+z')/2$. A possible improvement on Eq. (13.17) may thus be obtained[436] by broadening the above delta function into a Gaussian function with a width determined by the surface thickness L.

The expression for the chemical potential obtained in the density gradient expansion leading to Eq. (13.17) is

$$\mu_{\text{ldg}} = \mu(\rho(z)) - \frac{1}{2\rho'(z)} \frac{d}{dz} [A(\rho(z))\rho'^2(z)] , \tag{13.19}$$

with $\rho'(z) \equiv (d\rho(z)/dz)$, and integrating it gives for the pressure p the result

$$p_{\text{ldg}} = \mu_{\text{ldg}} \rho(z) - \psi(\rho(z)) + A(\rho(z))\rho'^2(z) . \tag{13.20}$$

In these equations $\mu(\rho)$ and $\psi(\rho)$ are the chemical potential and the free energy density of a homogeneous fluid of constant density ρ, with the local density $\rho(z)$ then inserted for ρ. These approximate expressions serve to evaluate the particle density profile and again can be obtained[436] by a suitable local approximation to $c_0(z, z')$ in Eq. (13.14). Broadening of the off-diagonal dependence on $(z - z')$ may again be effected through a Gaussian approximation.

Using these results Bhatia and March[437] have derived an alternative expression for the surface tension in the low-order density gradient approximation, that is

$$\gamma_{\rm ldg} = \int_{-\infty}^{\infty} dz[p - \mu\rho(z) + \psi(\rho(z))]\,. \tag{13.21}$$

By an appropriate expansion of this expression around the bulk liquid density they were then able to show that $\gamma_{\rm ldg}$ is connected to the isothermal compressibility of the bulk liquid by a relation of the form (13.1), with L having the meaning of a surface thickness.

13.4.3 *Critical behaviour of surface tension*

The work of Fisk and Widom[438] was directed at the same problem, with the main focus on surface tension in the critical region. They predicted that the interfacial thickness L, that they wrote as

$$L = \frac{1}{c\beta^2}\frac{\gamma\rho^2 K_{\rm T}}{(\rho_{\rm l} - \rho_{\rm v})^2}\,, \tag{13.22}$$

where β is the usual critical exponent which describes the shape of the coexistence curve and c is a parameter of order unity, may be identified with the correlation range ξ for temperatures near to $T_{\rm c}$. An immediate consequence of the assumption $L = \xi$ is the Widom equality[439]

$$\mu + \nu = 2\beta + \gamma'\,, \tag{13.23}$$

where μ is the critical exponent characterising the temperature variation of the surface tension, while ν and γ' are the exponents for the correlation range and for $\rho^2 K_{\rm T}$ (see Sec. 4.8).

In their light scattering experiments on Xenon in the critical region, already introduced in Sec. 7.5.2, Zollweg *et al.*[203] measured the temperature

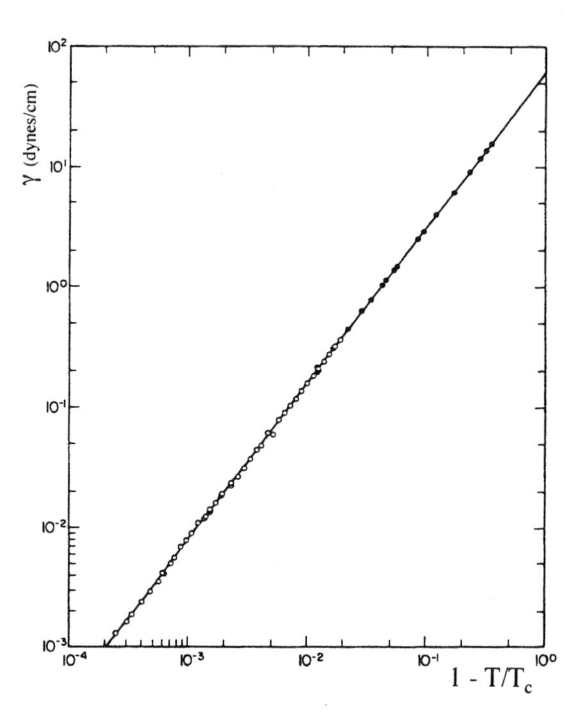

Fig. 13.3. Surface tension of xenon as a function of the reduced temperature approaching the critical point. (Redrawn from Zollweg *et al.*, Ref. 203.)

dependence of the surface tension as

$$\gamma = (63 \pm 2)\left(1 - \frac{T}{T_c}\right)^{1.302\pm0.006} \quad \text{dyn/cm} \tag{13.24}$$

(see Fig. 13.3). Using the value $\mu = 1.302 \pm 0.006$ and taking from experiment the values $\nu = 0.57 \pm 0.05$, $\gamma' = 1.21 \pm 0.03$ and $\beta = 0.345 \pm 0.01$, they found $2\beta + \gamma' - \mu - \nu = 0.03 \pm 0.06$. The temperature dependence of L is therefore consistent with that of the correlation length within experimental error. Zollweg *et al.* also estimated $c = 0.83 \pm 0.15$.

13.4.4 *Application to nucleation theory*

An important application of the density functional approach in liquid state theory, of which the low-order density gradient expansion presented in

Sec. 13.4.2 above is a specific practical realisation, has been to the evaluation of the free energy barrier for the formation of a small heterogeneous nucleus inside a bulk phase, leading to estimations of the nucleation rate. In particular, vapour–liquid nucleation plays an essential role in phenomena such as cloud formation and climatic changes, and in various processes of technological relevance.

The classical theory of nucleation relies on the so-called capillarity approximation (see Sec. 10.2): that is, the assumption that the free energy of a small germ of a new phase can be described by a bulk free energy difference, plus a surface term in which the effects of curvature are ignored. Oxtoby and coworkers[440] have used density functional theory to develop a more refined approach, which allows the order parameters to vary through the cluster so as to give the lowest possible barrier for nucleation. Evaluations of processes such as liquid condensation from the vapour and crystallisation from the melt show significant deviations from classical nucleation theory and yield nucleation rates that are orders of magnitude closer to experiment.

The key idea behind the density functional approach is that a droplet must be characterised not only by its radius in conjunction with macroscopic bulk-phase parameters, but by its full particle density profile. The quantity that has to be calculated is the grand potential for an inhomogeneous fluid of given density profile: rigorous statistical mechanics[441] ensures that the exact functional contains complete information on the equilibrium states of the fluid, which are minima of the functional. In the liquid–vapour nucleation problem one is instead concerned with clusters that lie at the saddle point in function space between the uniform gas and the uniform liquid. There must be some critical cluster density that lies higher in free energy than the vapour but lower than any other saddle point, and that provides the lowest barrier to nucleation. At the saddle point the first functional derivative of the grand potential is zero, but, unlike the case of a true equilibrium case, the second derivative is not uniformly positive. In particular, the matrix of second derivatives has one negative eigenvalue, corresponding to the fact that the free energy decreases with gain or loss of particles by the cluster.

The essence of the theory thus is the realisation of approximate forms of the density functional which preserve the main physics of the nucleation problem and yet are amenable to explicit calculations. In Oxtoby's approach the attractive part of the molecular interactions appears as a perturbation relative to a repulsive hard sphere term. Significant progress has been made by Gránásy *et al.*[442] by using a gradient theory proposed by Iwamatsu and Horii[443] for Yukawa attractions, which introduces the hard sphere chemical potential as the order parameter. This has allowed an analytic approximation to the results of the density functional approach, which reproduces the density profile and the free energy of critical cluster fluctuations with high accuracy.

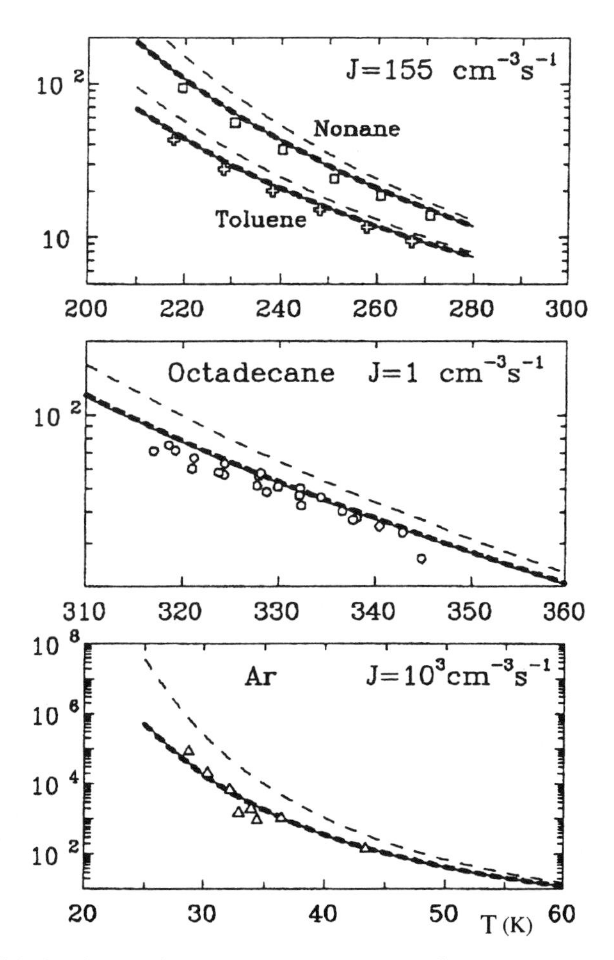

Fig. 13.4. Calculated critical supersaturations corresponding to a given nucleation rate J from a density functional approach (full and thick-dashed lines), compared with experiment (various symbols) and with classical nucleation theory (thin-dashed lines). (Redrawn from Gránásy *et al.*, Ref. 442.)

Figure 13.4 compares the results of Gránásy *et al.* for the predicted critical supersaturations corresponding to given nucleation rates J in insulating fluids with experiment and with the results of classical nucleation theory.

13.5 Microscopic Theories: Two-Particle Distribution Function

An alternative formally exact approach to the evaluation of surface tension, which applies to the case of an assumed pair-potential interaction, was based by Kirkwood and Buff[444] on the two-particle distribution function in the inhomogeneous fluid. Taking as usual the dividing surface at the interface as lying in the $z = 0$ plane, their result involves a calculation of the tangential pressure $p_t(z)$ normal to an element of area parallel to the z axis. Away from the interface $p_t(z)$ becomes equal to the equilibrium bulk pressure p and indeed the Kirkwood–Buff expression gives back the virial pressure reported earlier in Eq. (4.17). We can now argue that the pressure deficit $[p - p_t(z)]$ in the surface layer manifests itself macroscopically as a tension exerted by the fluids on the walls of the container. The integrated magnitude of this tension per unit length is evidently the surface tension,

$$\gamma = \int_{-\infty}^{\infty} dz [p - p_t(z)] \,. \tag{13.25}$$

This agrees, of course, with the concept of surface tension as the stress needed to stretch the interface.

13.5.1 *Tangential pressure deficit and surface tension*

For a liquid surface in the x–y plane, we calculate the pressure $p_t(z)$ along the normal to the y–z plane at height z by starting from the x component of the force exerted on the molecules in a volume element $d\mathbf{r}_1$ by the molecules in a volume element $d\mathbf{r}_2$. Integration over all pairs of particles gives the force transmitted across the y–z plane as

$$F_x(z_1) = \int_{-\infty}^{0} dx_1 \int_{(x_2>0)} d\mathbf{r} \frac{x_2 - x_1}{r} \phi'(r) \rho^{(2)}(z_1, \mathbf{r}) \,, \tag{13.26}$$

where $\mathbf{r} = \mathbf{r}_2 - \mathbf{r}_1$, $\phi'(r) = d\phi(r)/dr$ with $\phi(r)$ the pair potential, and $\rho^{(2)}(z_1, \mathbf{r})$ is the two-particle distribution function giving the probability of finding two particles at relative position $\mathbf{r}_2 - \mathbf{r}_1$ in the inhomogeneous fluid. The integration over x_1 is easily carried out and, using the fact that the integrand becomes an even function of $x_2 - x_1$, we can write

$$F_x(z) = \frac{1}{2} \int d\mathbf{r} \frac{(x_2 - x_1)^2}{r} \phi'(r) \rho^{(2)}(z, \mathbf{r}) . \tag{13.27}$$

To this we must add the kinetic part of the stress, given by the momentum transport $\rho(z) k_B T$ across the unit area. The result is

$$p_t(z) = k_B T \rho(z) - \frac{1}{2} \int d\mathbf{r} \frac{(x_2 - x_1)^2}{r} \phi'(r) \rho^{(2)}(z, \mathbf{r}) . \tag{13.28}$$

As already remarked, this expression reduces to the virial equation of state (4.17) far away from the interface, where $\rho(z)$ becomes the constant density ρ and $\rho^{(2)}(z, \mathbf{r})$ reduces to $\rho^2 g(r)$.

Insertion of Eq. (13.28) in Eq. (13.25) yields after some simple algebra

$$\gamma = \frac{1}{2} \int_{-\infty}^{\infty} dz \int d\mathbf{r} \frac{(x_2 - x_1)^2}{r} \phi'(r) [\rho^{(2)}(z, \mathbf{r}) - \vartheta(-z) \rho_l^{(2)}(r) - \vartheta(z) \rho_v^{(2)}(r)] . \tag{13.29}$$

13.5.2 *The Fowler approximation: Relation of surface tension to shear viscosity*

Equation (13.29) is formally exact for a one-component fluid with pair interactions. However, it is obviously very difficult to evaluate $\rho^{(2)}(z, \mathbf{r})$ in the presence of the interface. The calculations based on this approach have therefore had to approximately relate this function to the density profile and to the bulk pair function.

The most simple (though drastic) assumptions are (i) that the gas has negligible density, and (ii) the liquid is homogeneous up to the Gibbs surface. This yields

$$\gamma = \frac{\pi}{8} \rho^2 \int_0^{\infty} dr\, g(r) r^4 \phi'(r) , \tag{13.30}$$

where $g(r)$ is the pair distribution function of the bulk liquid.

Various authors, and especially Egry,[445] have noticed that the structural integral entering the approximate expression (13.30) for surface tension also enters an approximate formula given long ago by Born and Green[446] for the shear viscosity η of a dense monatomic liquid in terms of the bulk pair function $g(r)$ and a pair potential $\phi(r)$.

Dividing γ by η therefore eliminates the structural integral and yields

$$\frac{\gamma}{\eta} \approx \bar{v}\,, \tag{13.31}$$

where $\bar{v}$ is the thermal velocity $(k_{\mathrm{B}}T/m)^{1/2}$.

Tests for liquid metals at freezing[447] show that the order of magnitude predicted by Eq. (13.31) is correct, within a spread of roughly a factor of 3.

13.5.3 *Computer studies: Role of interatomic forces in condensed rare-gas elements*

From a number of fully qualitative computer studies of surface properties in simple fluids, we briefly refer at this point to the calculations carried out by Barker[448] on argon, krypton and xenon near to their triple point. He used the Monte Carlo technique to investigate the role of the interatomic forces in determining surface tension, using models both with and without the Axilrod–Teller–Muto three-body forces. The three-body forces were treated perturbatively.

The main conclusions reached by Barker were that pair potentials alone give values of surface tension which are higher than the experimental values by some 20% for argon and 35% for xenon. However, when the three-body forces are included, the calculated values of the surface tension are within about 2% of the measured values for all three condensed rare gases.

13.6 Interfacial Dynamics

From the above presentation of surface thermodynamics and structure we turn to discuss some aspects of interfacial dynamics and transport in this and in the following section.

13.6.1 *Surface waves*

We consider an ideal non-viscous liquid of mass density ρ occupying the half-space $z < 0$ and in equilibrium with a gas at $z > 0$. We shall be concerned in this section with the waves that in a linear regime can propagate in such an interface, with wave vector $\mathbf{k}$ along the x direction say, under the combined action of a vertical gravitational field and of surface tension.

Assuming for the present that the density of the gas is negligible, the velocity field $\mathbf{v}$ is confined to the $z < 0$ region and vanishes there as one goes away from the interface deep into the liquid, this being the boundary condition appropriate to surface motions. In the absence of viscosity and vorticity, $\mathbf{v}$ is described at all times by a potential function as $\mathbf{v} = \nabla\phi$, with $\phi(x,z)$ obeying the Laplace equation $\nabla^2\phi = 0$ (see Chap. 6). Writing $\phi(x,z) = \phi_{\mathbf{k}}(z)\exp(ikx)$, the Laplace equation with the boundary condition specified above immediately yield $\phi_{\mathbf{k}}(z) \propto \exp(kz)$ for $z < 0$.

The interface at $z = 0$ moves up and down as the surface wave propagates. We describe this motion by means of a vertical displacement $\zeta(x,t)$ such that $(D\zeta/Dt) = v_z(x,\zeta;t)$ or, after linearisation, $\dot{\zeta}(x,t) \cong v_z(x,0;t)$. The change in free energy density associated with a local displacement ζ is $\rho g\zeta + \gamma\frac{\partial^2\zeta}{\partial x^2}$, where the first term is due to the gravitational potential and the second is the free energy density of the strain arising from the stretching of the surface area. We thus find the local equation of motion of the interface by equating the acceleration $\dot{v}_z$ to the vertical force per unit mass:

$$\ddot{\zeta} = -\left(\frac{k}{\rho}\right)\left[\rho g\zeta + \gamma\frac{\partial^2\zeta}{\partial x^2}\right] . \tag{13.32}$$

Setting $\zeta(x,t) = \zeta_{\mathbf{k}}\exp[i(kx - \omega t)]$ we finally get the dispersion relation of surface waves,

$$\omega_{\mathbf{k}}^2 = gk + \left(\frac{\gamma}{\rho}\right)k^3 . \tag{13.33}$$

If the density of the gas is not negligible compared with that of the liquid, it is easy to show that Eq. (13.33) becomes

$$\omega_{\mathbf{k}}^2 = \frac{(\rho_{\mathrm{l}} - \rho_{\mathrm{v}})gk + \gamma k^3}{\rho_{\mathrm{l}} + \rho_{\mathrm{v}}} . \tag{13.34}$$

This equation was already used in Sec. 12.2.1 in discussing the Rayleigh–Taylor instability, arising when the locations of the two fluid phases are interchanged.

A nonlinear relation between frequency and wave number, as in Eq. (13.33), implies that beat patterns are generated whenever two such waves, having equal amplitude but different wave numbers, are made to propagate in the same direction starting from a common origin. The two waves combine into a single travelling wave of intermediate wave number and modulated amplitude, with the crests and troughs of the modulation travelling at the group velocity $d\omega/dk$.

Equation (13.33) involves a critical wavelength $\lambda^* = 2\pi/k^*$, with $k^* = (\rho g/\gamma)^{1/2}$, which separates two different propagation regimes. The dispersion of the surface waves changes from $\omega_{\mathbf{k}} = (gk)^{1/2}$ for $\lambda \gg \lambda^*$ (the so-called "gravity waves") to $\omega_{\mathbf{k}} = (\gamma/\rho)^{1/2}k^{3/2}$ for $\lambda \ll \lambda^*$ (the so-called "ripples"). Further discussion of gravity waves in situations different from those considered here (e.g. waves in shallow ponds or so-called "Stokes waves" of amplitude becoming comparable to the wavelength) are discussed in the book of Faber.[4]

13.6.2 *Capillary waves and surface fluctuations*

The mean square amplitude u^2 of the fluctuations in the position of the interface due to capillary waves described by Eq. (13.34) is given by[424]

$$u^2 \propto \frac{k_B T}{\gamma} \int_{k_{\min}}^{k_{\max}} k\, dk \frac{1}{\Lambda^{-2} + k^2} \,. \tag{13.35}$$

We have introduced a lower and an upper cut-off in the integral over the surface modes and we have defined

$$\Lambda = \left[\frac{g(\rho_l - \rho_v)}{\gamma} \right]^{-1/2} \,. \tag{13.36}$$

The lower cut-off in the integral in Eq. (13.35) is inversely proportional to the system size and therefore vanishes in the thermodynamic limit.

The quantity Λ plays the role of a transverse correlation length for fluctuations of the location of the Gibbs surface and, if the gravity field is allowed to vanish so that this length diverges, the integral in Eq. (13.35) diverges logarithmically at the lower cut-off in the thermodynamic limit. This was anticipated in Sec. 13.4.1.

In fact, a common method for measuring the surface tension consists in measuring the capillary length in a capillary rise experiment, as already sketched in Fig. 1.7. Assuming zero contact angle, the relation $\Lambda = (hr/2)^{1/2}$ holds, where h is the height of the capillary rise and r the radius of the capillary tube. Typically, the order of magnitude of Λ is $\Lambda \approx 1.5$ mm for the liquid–vapour interface at 0°C.

13.6.3 *Interface reflectivity and diffuse interface in a critical fluid mixture*

When a wave of any kind is incident on an interface between two systems having different indices of refraction, a fraction of the wave will be reflected, the reflection being specular if the interface is planar. Experimentally, the structure of the interface can thus be studied in measurements of the reflectivity of photons or neutrons. Since the refractive index is simply related to density and composition, specular reflection can probe the system in a direction normal to the interface, over a length scale which is determined by the component of the wave vector perpendicular to the interface.[449]

The reflectivity is given by

$$R = (n_1 + n_2)^{-2} \left| \int_{-\infty}^{\infty} dz\, n'(z) e^{ikz} \right|^2 , \qquad (13.37)$$

where n_1 and n_2 are the indices of refraction of the bulk phases and $n'(z)$ is the gradient of the refractive index in the interface. The wave number k in this expression is $k = (4\pi/\lambda)\sin\theta$, where λ is the wavelength of the probe and θ the semi-angle of scattering. The Fresnel reflectivity $R_F = (n_1 - n_2)^2/(n_1 + n_2)^2$ is recovered from Eq. (13.37) in the case of a step-like interfacial density profile.

In the opposite limit the reflectance vanishes if the interface becomes very diffuse, as would be the case if the root-mean-square amplitude u of interfacial fluctuations diverges. As an example we quote the work of Gilmer *et al.*,[450] who have measured the reflectivity of the interface between the two phases in equilibrium for a cyclohexane-methanol fluid mixture just below the critical consolute temperature of 45.22°C. The observed reflectivities fall far below the Fresnel values expected of corresponding sharp interfaces, with greater attenuation at shorter wavelengths, precisely as would be expected of a diffuse interface.

Figure 13.5 reports from the work of Gilmer *et al.* the measured temperature dependence of the effective interfacial thickness L on the approach to the critical temperature. These data yield $L \propto (T_c - T)^{-\mu}$ with $\mu = 0.76 \pm 0.1$, in clear disagreement with the value $\mu = 1/2$ derived for an interface in a classical fluid by Cahn and Hilliard. Gilmer *et al.* also obtain the temperature

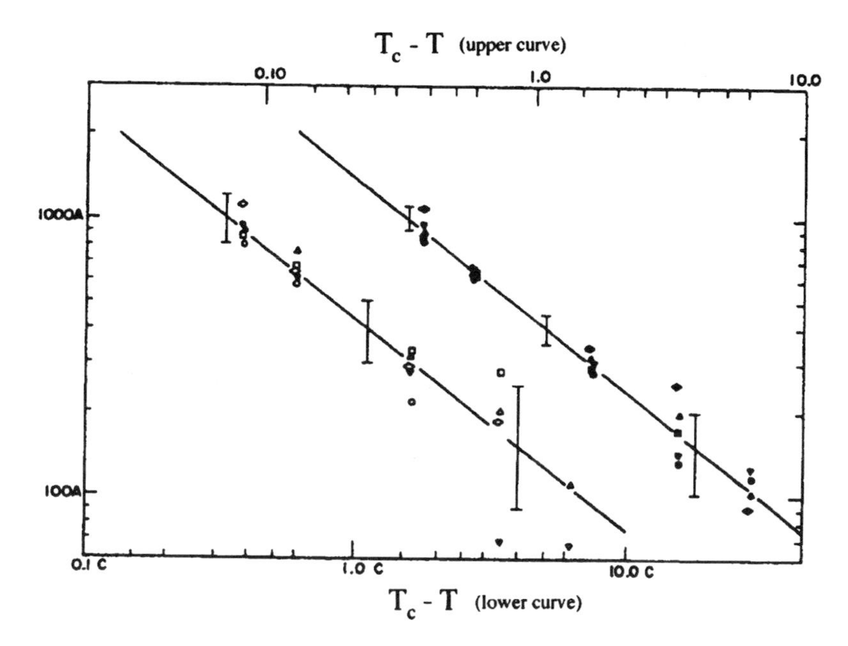

Fig. 13.5. Temperature dependence of the effective interface thickness for a cyclohexane-methanol mixture near the critical point for demixing, as measured in reflectivity experiments using light at various wavelengths. (Redrawn from Gilmer *et al.*, Ref. 450.)

dependence of the miscibility gap for the two liquids from the measured difference of the indices of refraction along the coexistence curve, finding the value $\beta = 0.347 \pm 0.008$ for the appropriate critical exponent.

For an extensive presentation of experimental techniques for probing fluid surfaces and interfaces, with special emphasis on polymeric materials, the reader may refer to the book of Jones and Richards.[451]

13.7 Interfacial Transport and Rheology

Transport phenomena in the interfacial region between two fluids form an area of considerable current interest at the time of writing. Interfacial rheology studies the response of mobile interfaces to strain.

Fluid interfacial motions induced by surface tension play a fundamental role in many natural and industrial contexts. Examples arise not only in studies of capillarity and of low-gravity flows, but also in hydrodynamic and Bénard instabilities, in cavitation, and in the dynamics of droplets in clouds and in fuel sprays inside combustion engines. One can also cite applications to interfacial turbulence, thin-liquid-film hydrodynamics and stability, and the rheology and stability of foams and emulsions. The reader wishing to go more deeply into this area could consult the edited book of Edwards *et al.*[452]

To begin a discussion of interfacial transport, one is concerned with theory, measurement and applications of interfacial hydrodynamics. One can start by adopting the classical view of fluid interfaces as idealised two-dimensional singular surfaces. The adsorption of molecular or macromolecular surfactants imparts intrinsic rheological properties to the interface such as interfacial shear and dilatational viscosities and Gibbs elasticity, which indicates the change in interfacial tension with area. Gradients in surfactant concentration and temperature cause interfacial tension gradients. With regard to experiment, light scattering techniques have acquired considerable importance (see e.g. Dorshow and Turkevich[453]), as has the deep channel viscometer for measuring interfacial shear viscosity. This is defined as the ratio between the interfacial shear and the strain rate.

Numerical simulation of multiphase flows is also of considerable significance.[454] A most challenging part of the simulation of such flows is to allow the modelling of interfaces between different phases and the associated problem of surface tensions. An interface is regarded as a transition region where physical properties can vary significantly, but smoothly. When macroscopic equations (to be specific, Euler and Navier–Stokes) are used to study multiphase flows, one must confront the need for their generalisation, due to the fact that flow in an interface region evolves differently from behaviour in homogeneous regimes owing, in essence, to the phenomena of surface tension.

As background, it should be noted that several approaches have been proposed for incorporating surface tension into the macroscopic continuum equations for multiphase flows (see e.g. Chang *et al.*[455] and earlier references there). Our main concern here is to illustrate the approach proposed in the work of Nadiga and Zaleski[456] and in the later study of Zou and He.[457] These workers insert in the Navier–Stokes equation, in parallel with the usual viscous stress tensor (see Sec. 6.2), an additional contribution $\Pi^{(S)}$ representing surface tension. This is expressed in the form

$$\Pi^{(S)} = K\left[\left(\frac{1}{2}|\nabla\rho|^2 + \rho\nabla^2\rho\right)\mathbf{I} - \nabla\rho\nabla\rho\right] . \tag{13.38}$$

In Eq. (13.38) the parameter K measures the strength of the effect of surface tension.

Though the setting up of a generalised Navier–Stokes equation by insertion of a surface-related stress tensor is essentially phenomenological, Zou and He demonstrate that the continuum equation (13.38), together with the usual equation of continuity for the fluid density, can be directly obtained from kinetic theory starting out from the Boltzmann equation for the single-particle distribution, including a collisional term as set out by Chapman and Cowling.[132] They also use the molecular theory of capillarity[6] to relate the parameter K in Eq. (13.38) to the interfacial tension γ *via* the density profile $\rho(z)$. The approximate result then emerging is

$$\gamma = K \int_{-\infty}^{\infty} dz \left[\frac{\partial \rho(z)}{\partial z} \right]^2 . \tag{13.39}$$

This is reminiscent of the expression that can be obtained from Eq. (13.15) by the Fisk–Widom approximation[438] $c_2(z, z') \approx (3\ell^2/2)\delta(z - z')$, that is

$$\gamma = \frac{1}{6} k_{\mathrm{B}} T \ell^2 \int_{-\infty}^{\infty} dz \left[\frac{\partial \rho(z)}{\partial z} \right]^2 . \tag{13.40}$$

Chapter 14

Quantum Fluids

We turn in this last chapter to fluids of particles in the quantum regime. All important in this connection is the ratio between the thermal de Broglie wavelength λ_{dB} and the mean interparticle spacing a, $\lambda_{\mathrm{dB}} = h/p_{\mathrm{dB}}$ being defined from Planck's constant h and from the momentum of a particle of mass m with kinetic energy equal to the thermal energy $k_{\mathrm{B}}T$. If $\lambda_{\mathrm{dB}} \ll a$ (or equivalently $\rho\lambda_{\mathrm{dB}}^3 \ll 1$, with ρ the particle number density), then a classical description is valid and Planck's constant merely determines the size of a cell in phase space. As the value of the de Broglie wavelength approaches the interparticle distance, however, quantum interference associated with the wave-like nature of the particles emerges. The parameter $\rho\lambda_{\mathrm{dB}}^3$ already enters to determine the leading quantum deviation from the equation of state $pV/Nk_{\mathrm{B}}T = 1$ of the classical ideal gas,[52] with a sign which reflects the statistics of the gas (positive for fermions and negative for bosons).

It is evident, therefore, that quantum effects become important for light particles in dense fluids at very low temperatures. The main focus of the present discussion will be on the Helium liquids, on the Bose–Einstein condensates of alkali atoms, and on fluids of electrons.

14.1 Ideal Fermi and Bose Gases

An ideal gas is a thermodynamic system of particles whose mutual interactions may be neglected.[52] The energy levels of the gas can be written in terms of the single-particle energies ε_k and of the number n_k of particles in a single-particle state at energy ε_k,

$$E = \sum_k n_k \varepsilon_k \,, \tag{14.1}$$

k being an index which labels the single-particle states (for a gas inside a macroscopic box, this index subsumes a momentum $\hbar\mathbf{k}$ and a spin index S).

Below we consider two classes of particles appearing in nature[a]: (i) bosons, having zero or integer spin, for which n_k can take all non-negative integer values; and (ii) fermions, with semi-integer spin, for which n_k can take only the values 0 or 1 according to the Pauli exclusion principle. Maximisation of the entropy of the gas yields the two equilibrium statistics:

$$n(\varepsilon, T) = \frac{1}{e^{(\varepsilon-\mu)/k_B T} \pm 1} \,, \tag{14.2}$$

where $n(\varepsilon, T)$ is the mean occupation number of a single-particle state of energy ε in the gas at temperature T and chemical potential μ, and the $+(-)$ sign refers to fermions (bosons).

14.1.1 *The Fermi surface*

A cell of volume h^3 in phase space can contain a maximum of $(2S+1)$ fermions, so that at zero temperature the fermions must have a spread of momenta in a range up to a maximum momentum p_F. For N fermions in a volume V we can write the density $\rho = N/V$ as

$$\rho = \frac{\pi}{3}(2S+1)\left(\frac{p_F}{h}\right)^3 . \tag{14.3}$$

The quantity $k_F \equiv p_F/\hbar$ is the radius of a spherical surface in wave number space (the "Fermi surface"). In the common case $S = 1/2$ we have $k_F = (3\pi^2\rho)^{1/3}$, increasing with the density ρ. The energy of a fermion having momentum on the Fermi surface is the Fermi energy $\varepsilon_F = \hbar^2 k_F^2/2m$ and coincides with the chemical potential of the gas at $T = 0$.

The notion of a Fermi sphere allows a vivid picture of ground state and excitations in a normal Fermi fluid. At $T = 0$ the states inside the Fermi

[a]Particles with fractional statistics ("anyons") are met in some two-dimensional condensed matter systems, most notably those exhibiting a fractional quantum Hall effect (see, e.g. Lerda[458]).

sphere are fully occupied by the fermions and the states outside it are empty. Thermal excitations at finite temperature bring fermions from states inside to states outside the Fermi sphere, leaving "holes" behind and lowering the chemical potential.

14.1.2 *Bose–Einstein condensation*

There is no restriction on the state occupation numbers for bosons and at zero temperature they can all condense in the state of zero momentum and zero energy. That is, in the thermodynamic limit ($N \to \infty$)

$$N = \lim_{T\to 0} n(0,T) = \lim_{T\to 0} \frac{1}{e^{-\mu/k_B T} - 1} . \qquad (14.4)$$

In this limit μ must have approached zero from below, since state occupancy cannot be negative.

As was shown by Einstein in 1925, a macroscopic occupation of the zero-energy state persists up to a critical temperature T_0 given (in the common case $S = 0$) by the condition

$$\rho \lambda_{\rm dB}^3 |_{\rm crit} = 2.612 . \qquad (14.5)$$

Since $\lambda_{\rm dB} \propto T^{-1/2}$, at temperatures $T < T_0$ the fraction N_0/N of particles in the condensate is

$$\frac{N_0}{N} = 1 - \left(\frac{T}{T_0}\right)^{3/2} . \qquad (14.6)$$

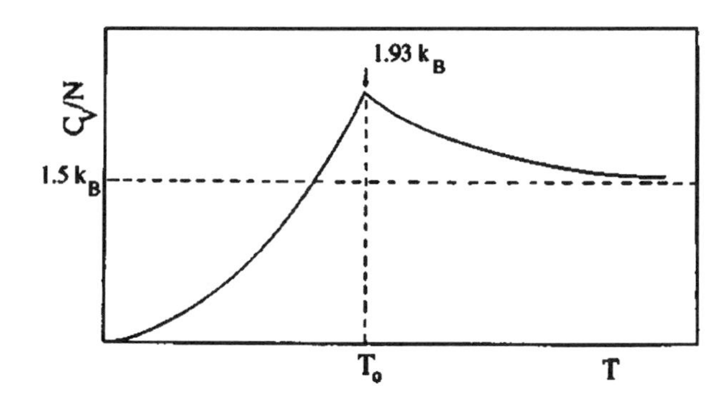

Fig. 14.1. Specific heat C_v of the ideal Bose gas at low temperature.

Evidently, all bosons are in the condensate for $T \to 0$ and are excited out of it for $T \geq T_0$.

Excitations with momentum $\hbar\mathbf{k}$ have energy $\varepsilon_{\mathbf{k}} = \hbar^2 k^2/2m$. Figure 14.1 shows the behaviour of the heat capacity of the ideal Bose gas as a function of temperature. The appearance of a Bose–Einstein condensate is marked by a singularity and below T_0 the specific heat is determined by the thermal excitations out of the condensate, giving $C_V/N \propto T^{3/2}$.

14.2 Boson Fluids

14.2.1 *The weakly interacting Bose gas (WIBG)*

A gas of bosons interacting *via* a weak Fourier-transformable repulsion is the simplest example of a bosonic superfluid. The theory of this model was given by Bogoliubov.[459]

The basic assumption of Bogoliubov was that a macroscopic number N_0 of particles have collected into a single quantum state, forming a Bose–Einstein condensate. Even at $T = 0$, N_0 does not necessarily coincide with the total number N of particles in the system (although Bogoliubov's theory implies that N_0/N is not far from unity). At variance from classical systems, in a quantum system the interactions drive a gain in the total energy through a gain in potential energy and a simultaneous loss in kinetic energy: thus, the interactions modify the ground state of the ideal Bose gas by also causing a depletion of the condensate.

Starting from this assumption, controlled approximations allow the Hamiltonian of the WIBG to be diagonalised by means of a canonical transformation into the form

$$H = E_0 + \sum_{\mathbf{k} \neq 0} \hbar\omega_{\mathbf{k}} b_{\mathbf{k}}^{\dagger} b_{\mathbf{k}} \,, \tag{14.7}$$

where $b_{\mathbf{k}}^{\dagger}(b_{\mathbf{k}})$ is an operator which creates (annihilates) an excitation with momentum $\hbar\mathbf{k}$, while the operator $b_{\mathbf{k}}^{\dagger} b_{\mathbf{k}}$ counts the number of excitations that are present in the gas and has expectation value zero in the ground state. In Eq. (14.7) E_0 is a constant and $\hbar\omega_{\mathbf{k}}$ is the energy of an excitation, which is given within Bogoliubov's theory by the relation

$$\hbar\omega_{\mathbf{k}} = \sqrt{(2\rho_0 v_{\mathbf{k}} + \varepsilon_{\mathbf{k}})\varepsilon_{\mathbf{k}}} \,. \tag{14.8}$$

Here, $\rho_0 = N_0/V$ is the condensate density, $v_{\mathbf{k}}$ is the Fourier transform of the interaction potential and $\varepsilon_{\mathbf{k}} = \hbar^2 k^2/2m$ are again the single-particle kinetic energies.

The crucial point of Eq. (14.8) is that excitations out of the condensate in the WIBG have, instead of an energy spectrum given by the kinetic energies $\varepsilon_{\mathbf{k}}$ as in the ideal Bose gas, an energy-momentum dispersion relation which becomes linear at low momenta:

$$\lim_{k\to 0} \omega_{\mathbf{k}} = ck\,, \quad c = \sqrt{\frac{\rho_0 v_0}{m}}\,. \tag{14.9}$$

That is, the long-wavelength excitations are phonons, i.e. quantised density waves propagating through the gas. An important feature of the WIBG is the equivalence between single-particle excitations and collective density excitations, the excitation of a density wave in the condensate being necessarily associated with emission or injection of a Bose particle and *vice versa.*

Related to the above basic result the WIBG was shown to have the following properties:

(i) the structure factor $S(k)$ has a linear behaviour at small k,

$$\lim_{k\to 0} S(k) = \frac{\hbar k}{2mc}\,; \tag{14.10}$$

(ii) the momentum distribution $n_{\mathbf{k}}$ has a divergence at small k,

$$\lim_{k\to 0} n_{\mathbf{k}} = \frac{mc}{2\hbar k}\,, \tag{14.11}$$

a relation between $n_{\mathbf{k}}$ (a one-body property) and $S(k)$ (a two-body property) being in fact a manifestation of the equivalence between single-particle and collective excitations; and

(iii) the dynamic structure factor $S(k,\omega)$ has a single peak at each value of k, at a frequency

$$\omega_{\mathbf{k}} = \frac{\hbar k^2}{2mS(k)}\,. \tag{14.12}$$

This expression has become familiar as the result of Feynman's theory[460] for the single-mode excitations of superfluid liquid ^{4}He (see Sec. 14.2.2 immediately below). However, rotons are not contained in the dispersion relation (14.8) of the WIBG.

The effects of temperature are easily included in Bogoliubov's theory of the WIBG by appropriate insertions of the Bose thermal distribution function, allowing an evaluation of the condensate density ρ_0 and of the superfluid density ρ_s as functions of temperature. Given the assumption that there exists a single-particle state $\psi_0(\mathbf{r}, t)$ which is macroscopically occupied, the conceptual basis for superfluidity is simple.[461] We write $\psi_0(\mathbf{r}, t) = \sqrt{\rho_0(\mathbf{r}, t)} \exp[i\varphi(\mathbf{r}, t)]$ and define the superfluid velocity $\mathbf{v}_s(\mathbf{r}, t)$ by the prescription

$$\mathbf{v}_s(\mathbf{r}, t) = \left(\frac{\hbar}{m}\right) \nabla\varphi(\mathbf{r}, t) . \tag{14.13}$$

This embodies the property $\nabla \times \mathbf{v}_s(\mathbf{r}, t) = 0$, i.e. the superfluid flow is irrotational. Also, since no entropy is associated with a single quantum state, the entropy is entirely carried by the particles occupying states other than ψ_0. Furthermore, from the fact that the phase of ψ_0 must be single-valued modulo 2π, one obtains the Onsager–Feynman quantisation condition on superfluid circulation along a closed circuit. This quantisation condition was already given in Eq. (7.48) and discussed there in connection with quantum vortices.

More appropriately, the "condensate wave function" $\psi_0(\mathbf{r}, t)$ should be thought of as the order parameter of the Bose–Einstein-condensed fluid. It should be stressed that the superfluid density ρ_s is generally different from the condensate density $\rho_0 = |\psi_0|^2$. For instance in bulk superfluid liquid ^{4}He, to which we turn below, at the lowest temperatures the whole liquid is superfluid whereas the condensate fraction is of the order of 10% of the total particle number (see the discussion given in Sec. 7.7).

14.2.2 *Superfluid liquid ^{4}He*

We have already given in Sec. 7.7 an introduction to a number of characteristic phenomena which are observed in superfluid He-II and to its transport properties and dynamical spectrum. Here we focus on some aspects of the relevant theory, with some attention to the connections that exist between properties of the dense liquid and those of the dilute, weakly interacting gas.

The ground state of the dense, strongly interacting boson fluid is described to useful accuracy by a wave function of the Jastrow–Bijl form,

$$\Psi(\mathbf{r}_1, \mathbf{r}_2, \ldots, \mathbf{r}_N) = \prod_{i<j} \exp\left[\frac{1}{2}u(r_{ij})\right] , \tag{14.14}$$

where $u(r_{ij})$ arises from correlations between pairs of particles. Forming $|\Psi|^2$ and hence obtaining the various distribution functions, it can be seen that there is analogy with classical statistical mechanics, with $-\phi(r)/k_B T$ in the classical theory replaced by $u(r)$. In particular the result in Eq. (14.10), which was shown by Gavoret and Noziéres[462] to be exact for sufficiently small k, leads to the asymptotic expression $u(r) \approx -(mc/\hbar\pi^2\rho r^2)$ for $u(r)$ at large r. Evidently, long-wavelength phonons induce very long range correlations between the particles.

Wu and Feenberg[463] used experimental data on the structure factor $S(k)$ within the hypernetted-chain approximate theory (see Sec. 4.7) to extract the function $u(r)$, finding that it is strongly negative in the region of the atomic core. The mean kinetic energy $\langle K \rangle$ could then be calculated from $u(r)$ and from the pair distribution function $g(r)$ by the expression

$$\langle K \rangle = N \frac{\hbar^2 \rho}{8m} \int dr g'(r) u'(r) \,. \tag{14.15}$$

This replaces the classical kinetic energy $3Nk_B T/2$ in Eq. (4.15) for the internal energy U. For more details the work of Reatto and Chester[464] and of Feenberg[465] may be referred to.

Let us now turn to dynamics and see how the Feynman expression in Eq. (14.12) follows by general considerations from the assumption of a single-mode dynamical spectrum. That is, we write the van Hove dynamic structure factor in the form

$$S(k,\omega) = S(k)\delta(\omega - \omega_{\mathbf{k}}) \,, \tag{14.16}$$

which already satisfies the zeroth-moment sum rule, $S(k) = \int d\omega S(k,\omega)$. The first moment of the spectrum in the quantum fluid is given by

$$\int d\omega\omega S(k,\omega) = \frac{\hbar k^2}{2m} \,, \tag{14.17}$$

and hence Eq. (14.12) immediately follows by inserting Eq. (14.16) into Eq. (14.17). Using $S(k)$ from experiments, the dispersion curve (14.12) has the qualitative features of the Landau phonon and roton excitations, and in particular shows the roton minimum near the wave number of the main peak in $S(k)$. However, agreement with the measured dispersion relation of single-mode excitations from neutron inelastic scattering, that we discussed in

Sec. 7.7.2, is quantitatively poor and, of course, the multi-mode part of the spectrum is missing.

It was early realised by Feynman and Cohen[466] that the Feynman picture of independent elementary excitations had to be extended to account for their interactions. The current associated with a Feynman excitation (or, more precisely, with a localised excitation built as a wave packet of Feynman excitations) involves a contribution from the backflow of the fluid around it. Feynman and Cohen showed that, in the simple case of a particle tearing through the liquid at a given velocity, the pattern of induced longitudinal currents far away from the particle has the form of a dipolar backflow (see also the book of Pines and Noziéres[467]). By accounting for backflow Feynman and Cohen obtained a marked improvement in the one-phonon dispersion curve. Very similar results were obtained by a more conventional treatment of the interactions between Feynman excitations by Jackson and Feenberg[468] and these are shown in Fig. 14.2 in comparison with the measured dispersion curve from the neutron inelastic scattering experiments of Woods and Cowley.[224]

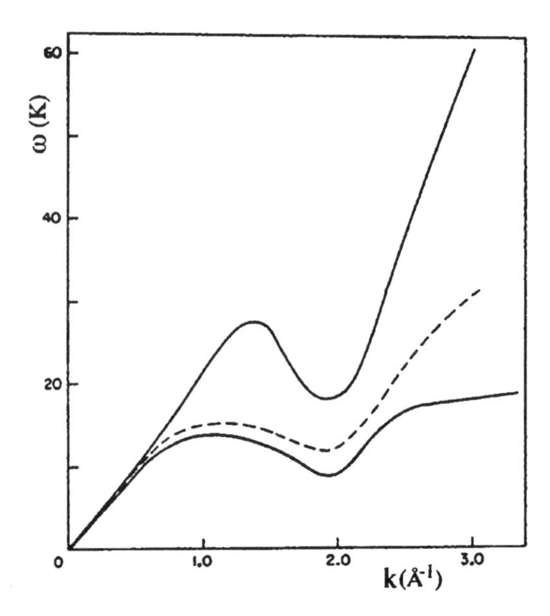

Fig. 14.2. Comparison between the measured dispersion relation for elementary excitations in liquid ^{4}He (bottom curve) and the theoretical results of Feynman (top continuous curve) and of Jackson and Feenberg (– – –).

Attention has been drawn by a number of authors[469] to the remarkable similarity that exists between the excitation spectra exhibited by $S(k,\omega)$ in superfluid ^{4}He at low temperatures and in the quantum (strongly anharmonic) solid ^{4}He. As stressed by Griffin,[470] it is important to bear in mind the subtle differences that exist between elementary excitations and density fluctuations, the latter being measured in inelastic neutron scattering. The precise nature of the atomic motions in the roton region is still a question of considerable interest at the time of writing.[471]

To conclude the discussion of He-II we shall make some brief remarks on vortices. It has been amply demonstrated experimentally[472] that quantised vortices and rings can be excited in superfluid ^{4}He, with the circulation associated with superfluid flow being $\pm h/m$ since vortices with higher circulations are usually unstable against decay into these. Apart from the vortex core, the flow field associated with the vortex is similar to the case of a classical vortex. Feynman[459] was the first author to show how a wave function could be constructed to lead to the velocity field as a classical vortex line, in the presence of strong interatomic correlations. Subsequent work by Chester *et al.*[473] led to the construction of a model wave function for a quantised vortex line and ring. A topical review of quantised vortices in confined Bose–Einstein condensates can be found in an article by Fetter and Svidzinsky.[474]

14.2.3 *Bose–Einstein condensates*

At the time of writing Bose–Einstein condensation has been realised in various monatomic alkali gases (^{87}Rb, ^{23}Na and ^{7}Li) and also in gases of ^{1}H and ^{4}He (for reviews see Cornell *et al.*,[475] Ketterle *et al.*,[476] Kleppner *et al.*,[477] and Ketterle and Cornell[478]; see also Bradley *et al.*[479] and Pereira Dos Santos *et al.*[480]). Counting both nucleons and electrons, these atoms possess an even number of fermions and obey Bose–Einstein statistics. The condition (14.5) for condensation is extremely severe: it implies having to work with metastable gases kept under magnetic confinement at temperatures T below 0.1 μK and densities $\rho \approx 10^{12}$–10^{14} atoms/cc.

The critical de Broglie wavelength and the mean interatomic spacing are of order 0.1 μm, i.e. much larger than the range of the interatomic forces: the gas is therefore so dilute that only binary collisions matter and these can be described by a contact (delta-function) interaction pseudopotential, $v(\mathbf{r}) = (4\pi\hbar^2 a_s/m)\delta(\mathbf{r})$ where a_s is the s-wave scattering length. The interaction is repulsive in all aforementioned systems except in the ^{7}Li gas, where the condensate is stable against collapse driven by the interatomic attractions only if it contains a limited number of atoms ($N \approx 10^3$). A common method of observation is by measuring the optical density of the atomic cloud after turning off the magnetic trap and allowing the cloud to ballistically expand for a controlled length of time. Interest in these systems comes from two different perspectives: (i) they are new quantum fluids on which to study many-body physics within the simple model of contact interactions containing a single (and tunable) parameter, i.e. the scattering length; and (ii) they are coherent atomic assemblies on which to study the optics of matter waves, leading up to the realisation of matter-wave lasers.

The magnetic confinement is usually well represented as an external harmonic potential $V(\mathbf{r})$ (real magnetic traps are anisotropic, either pancake-shaped or cigar-shaped, but for the sake of simplicity we shall below have in mind a spherical confinement, $V(\mathbf{r}) = m\omega^2 r^2/2$). Condensation then occurs in both momentum and coordinate space: in particular, the atoms would condense into the ground state of the three-dimensional harmonic oscillator if their mutual interactions were negligible. In this case the critical temperature T_0 can be estimated for a given number N of atoms in the trap by setting $\frac{1}{2}k_B T_0 \approx \frac{1}{2}m\omega^2\ell^2$, where ℓ is the linear size of the cloud, and by using Einstein's criterion (14.5) in the form $(N/\ell^3)\lambda_{\mathrm{dB}}^3|_{\mathrm{crit}} \approx 2.6$: this yields $k_B T_0 \approx \hbar\omega N^{1/3}$ and hence the condensate fraction at temperature $T \leq T_0$ is

$$\frac{N_0}{N} = 1 - \left(\frac{T}{T_0}\right)^3 . \tag{14.18}$$

Comparison with Eq. (14.6) shows that the confinement modifies in an essential fashion the thermodynamic behaviour of the condensed phase.[481]

The condensate is very dilute as already remarked, but nevertheless is in a regime of strong coupling.[481] The size of the cloud can be estimated by equating the harmonic potential energy $m\omega^2\ell^2/2$ to the mean interaction energy, which is of order $(4\pi\hbar^2 a_s/m)(N/\ell^3)$ since the contact interactions attribute to each atom a potential energy proportional to the local atomic density. This yields

$$\frac{\ell}{a_{\mathrm{ho}}} \approx \left(\frac{8\pi N a_s}{a_{\mathrm{ho}}}\right)^{1/5} , \tag{14.19}$$

where $a_{\text{ho}} = (\hbar/m\omega)^{1/2}$ is called the harmonic-oscillator length. The coupling strength of the gas is measured by the ratio between the average potential energy and the average kinetic energy per atom, that is

$$\frac{\langle E_{\text{pot}} \rangle}{\langle E_{\text{kin}} \rangle} = \frac{m\omega^2 \ell^2/2}{\hbar^2/(2m\ell^2)} \propto N^{4/5} \tag{14.20}$$

and this ratio increases rapidly with the number of atoms. Condensates with $N \approx 10^7$ atoms have been routinely available in many laboratories. In practice, the kinetic energy density becomes appreciable only in the outer regions of the condensed cloud.

A large number of experimental and theoretical studies have been carried out on these systems since the first realisation of Bose–Einstein condensation in gases of ^{87}Rb in 1995, and rapid progress is still taking place at the time of writing. Further richness in behaviour has been achieved by placing the condensate inside an "optical lattice", i.e. by superposing onto the magnetic trap a detuned standing wave of laser light which acts on the atoms as a periodic external potential. Here we can only cite a few of these results. Among the early quantitative tests that condensation was in fact being achieved in the confined clouds of alkali atoms were the comparisons between experiment and theory in regard to the frequencies of the shape-deformation modes of the condensed cloud[483] and to the internal energy of the cloud as a function of temperature.[484] Macroscopic coherence has been directly demonstrated from the interference patterns that are formed in the overlap region of two expanding condensates released from a divided magnetic trap.[485] Emission of coherent matter waves from a condensate, both in pulsed and in quasi-continuous form, has been demonstrated.[486] A transition from superfluid to dissipative behaviour has been seen to occur across a local velocity threshold when a condensate is driven through an optical lattice by a harmonic force.[487] Finally, the formation of vortices in a stirred condensate and their arrangement into triangular configurations have been observed.[488]

14.3 Normal Fermion Fluids

14.3.1 *Liquid ^{3}He in the normal state*

In contrast to ^{4}He, the ^{3}He atoms carry a nuclear spin of 1/2 and obey Fermi statistics. The Fermi degeneracy temperature of liquid ^{3}He, calculated as

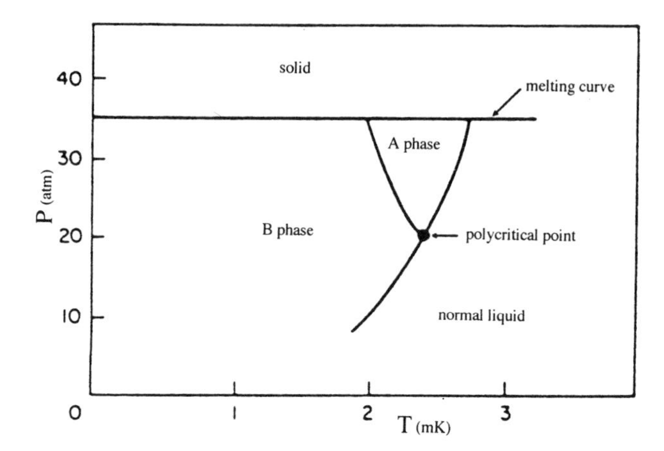

Fig. 14.3. Schematic phase diagram of liquid ^{3}He at low temperatures and zero magnetic field.

$T_F = \varepsilon_F / k_B$ from the fluid density and the atomic mass, is about 1 K. Below T_F, quantum effects become dominant at $T \approx 100$ mK and a phase transition to a superfluid state, to be discussed in Sec. 14.4, occurs at $T \approx 3$ mK. Again, solidification is achieved only under applied pressure. Figure 14.3 reports a schematic phase diagram for ^{3}He at low temperatures. Here we are concerned with the liquid in the normal state below 100 mK.

The measured specific heat C_V of liquid ^{3}He in the normal state shows a linear dependence on temperature T. This linear behaviour is characteristic of the ideal Fermi gas, where it can be explained as follows. According to the Fermi distribution in Eq. (14.2), at temperature T only those fermions lying within an energy strip of order $k_B T$ around the Fermi surface have access to empty states into which they can be thermally excited. The fermions occupying lower energy states in the Fermi distribution cannot be excited because of the Pauli exclusion principle. Thus, only a fraction of order T/T_F of the total number N of fermions can absorb thermal energy and, if we attribute to each of these its full classical thermal energy $3k_B T/2$ we get the thermal energy of the gas as $E_{th} \approx (3k_B T/2)(NT/T_F)$. Hence $C_V = (\partial E_{th}/\partial T)_V \propto T$. A full quantum calculation yields for the ideal Fermi gas

$$C_V = Nm \left(\frac{\pi}{3\rho} \right)^{2/3} \frac{k_B^2 T}{\hbar^2}, \tag{14.21}$$

the proportionality factor being determined by the atomic mass m and by the number density ρ.

It is somewhat surprising that a linear relationship between specific heat and temperature should be observed in liquid ^{3}He, in view of the strong interactions that are expected to lead to substantial departures from ideality. The explanation of this and other unexpected properties has been given by Landau[489] (for a detailed account of Landau's theory of normal Fermi liquids, the reader may refer to the book by Pines and Nozières[467]). Landau assumed that in a Fermi liquid the net effect of the interactions would be to "dress" each atom excited out of the ground state with a cloud of surrounding atoms, building up what is termed a quasi-particle. Once one deals with quasi-particles rather than with "bare" atoms, the excitations of the liquid can be described as an almost ideal gas of quasi-particles. In Landau's theory, a crucial point is that the quasi-particles obey Fermi statistics, so that at low temperatures we can describe the low-lying excited states in terms of the excitation of quasi-particles across the Fermi surface. A linear law for the specific heat of liquid ^{3}He follows.

There is, however, one major difference to be emphasised between quasi-particles and bare particles, and this is that a quasi-particle has an effective mass m^* which is different from the mass m of a ^{3}He atom. Given an atom which is tearing through the dense liquid, its motion is accompanied by a backflow of the other atoms and this is responsible for a shift of the bare atom mass into an effective mass. The value of m^* can be measured from the ratio C_V/T (see Eq. (14.21), where m should be replaced by m^*) and turns out to be about three times that of a ^{3}He atom at atmospheric pressure, increasing to almost a factor six near the solidification pressure.

The quasi-particle picture is only valid if momentum and energy are not very different from p_F and ε_F, i.e. it is useful to describe phenomena occurring near the Fermi surface. With this restriction it explains not only the observed thermodynamic properties of liquid ^{3}He in its normal state, but also some collective dynamical behaviours which can be associated to small distortions of the Fermi surface away from its equilibrium size and shape. Considering first a wave of ordinary hydrodynamic sound which propagates through the liquid, this is associated with periodic expansions and compressions of the liquid density and hence can be viewed as a local "breathing" of the Fermi sphere. From the quasi-particle picture Landau inferred that other types of periodic deformation (anisotropic ones, in fact) could be sustained by the Fermi sphere in the absence of dissipative collisions. The waves propagating through the liquid in association with these anisotropic oscillations of the Fermi sphere were called by Landau "zero sound". In the quantum Fermi liquid this type of high-frequency collective motion is the analogue of the "fast sound" that we have discussed in Sec. 6.7.3 for water and glass-forming classical liquids.

Figure 14.4 reports from the work of Abel *et al.*[490] data on sound velocity and sound attenuation in liquid ^{3}He as functions of temperature. A transition

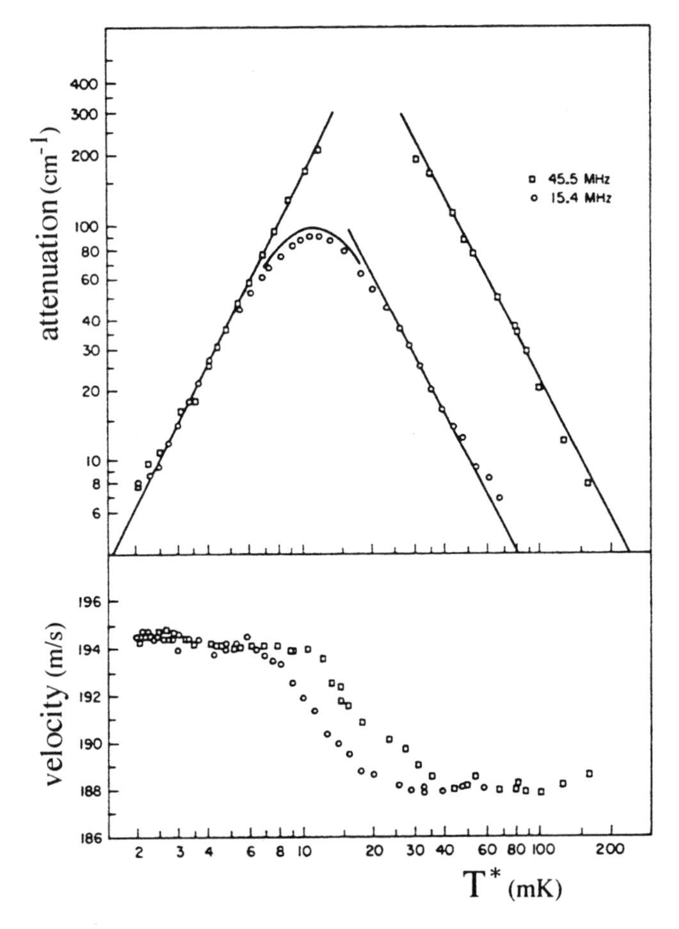

Fig. 14.4. Measured amplitude attenuation coefficient and speed of sound waves in liquid ^{3}He, for two values of the driving frequency. The straight lines through the attenuation points are from the Landau theory. (Redrawn from Abel *et al.*, Ref. 490.)

in sound wave behaviour, from hydrodynamic to zero sound, is seen to occur as the liquid is cooled through a temperature of $\approx$ 16 mK. In the limit of vanishing interactions the speed of hydrodynamic sound is $v_F/\sqrt{3}$ and that of zero sound tends to v_F, where $v_F \equiv \hbar k_F/m$ is the velocity of a fermion on the Fermi surface. As is seen from Fig. 14.4, a consequence of the strong interactions obtaining in liquid ^{3}He is that the two speeds of sound are much closer to each other.

14.3.2 *Electron fluids*

Conduction electrons in normal metals form a highly degenerate Fermi liquid, the ratio T/T_F being typically of order 10^{-2} in laboratory conditions. Here we focus on the so-called jellium model, which smears out the underlying ionic assembly into a uniform background of positive charge.[491] The model is also useful to describe fluids of electronic carriers in doped semiconductors, after rescaling the units of length and energy by introduction of an effective carrier mass and a background dielectric constant. We shall in later sections turn to liquid metals as ion-electron systems and to ion-electron plasmas such as are met in the giant planets.

As a starting point we return to the ideal Fermi gas and enquire about the electron–electron distribution function $g(r)$ in this model. Whereas $g(r) = 1$ everywhere in the classical ideal gas, the probability of finding two fermions with the same spin at a distance r vanishes as $r \to 0$ because of the Pauli exclusion principle. That is, each electron of given spin induces a local depletion of the density of electrons with the same spin. This so-called Pauli hole is given by

$$g(r) = 1 - \frac{9}{2}\left[\frac{j_1(k_F r)}{k_F r}\right]^2 , \qquad (14.22)$$

where $j_1(x) = (\sin x - x\cos x)/x^2$ is the first order spherical Bessel function (see Fig. 14.5).

No correlations between electrons with antiparallel spins are present in the ideal Fermi gas. Further depletion of the local electron density around each electron, and especially for electrons having opposite spin, arises when the Coulomb repulsive interactions are switched on. An exact property of the so-called Pauli–Coulomb hole is that the total local depletion of electron density corresponds to taking away one electron from the neighbourhood of

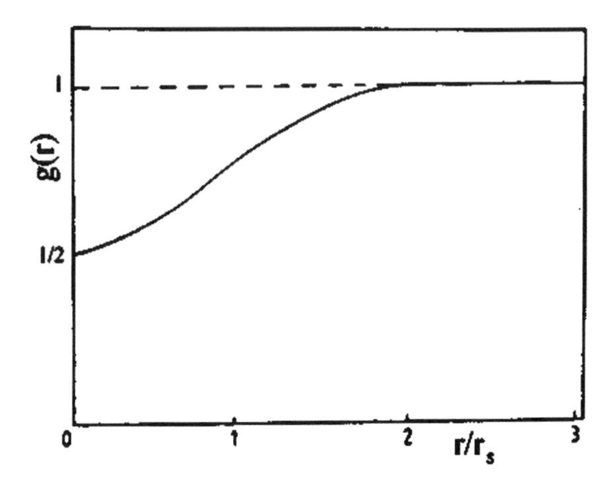

Fig. 14.5. Exchange hole around an electron in the electron gas.

each electron. This is, in fact, a microscopic manifestation of Faraday's law according to which a constant electric field cannot penetrate in the depth of a conductor. The bare potential generated by each electron is on average completely screened by a local rearrangement of the surrounding electron density, so that the effective potential that an electron generates decays over a microscopic distance.

We introduce screening in a semiclassical way by considering a static impurity with charge Ze placed at the origin inside a fully degenerate electron plasma.[b] The impurity and the induced redistribution of electronic charge create a potential $V(r)$ obeying the Poisson equation

$$\nabla^2 V(r) = 4\pi e^2[Z\delta(\mathbf{r}) - \rho(r) + \rho]\,, \tag{14.23}$$

where $\rho(r)$ is the electron density at distance r from the impurity and ρ is the average electron density (equal to the density of the uniform positive background). An additional relation between $V(r)$ and $\rho(r)$ follows from the equilibrium condition expressed by the constancy of the electrochemical potential, $\mu(\rho(r)) + V(r) = \mu(\rho)$. An analytic solution can then be obtained for $|Z| \ll 1$,

[b]The classical equivalent of this problem is the Debye–Hückel theory of electrolyte solutions (see Sec. 8.1). The Debye screening length can be calculated by the argument given here if one uses the relationship between density and chemical potential which holds for a classical ideal gas.

when we have $\rho(r) - \rho \approx -(\partial\rho/\partial\mu)V(r)$ by Taylor expansion of $\mu(\rho(r))$ around the homogeneous state described by $\mu(\rho)$. Solving Eq. (14.23) by Fourier transforms we get

$$V(r) = -\left(\frac{Ze^2}{r}\right) e^{-\kappa_{\mathrm{TF}} r}\,, \tag{14.24}$$

where $\kappa_{\mathrm{TF}}^{-1}$ is known as the Thomas–Fermi screening length and is given by

$$\kappa_{\mathrm{TF}}^{-1} = \left[4\pi e^2 \left(\frac{\partial\rho}{\partial\mu}\right)\right]^{-1/2} = \left(\frac{6\pi\rho e^2}{\varepsilon_{\mathrm{F}}}\right)^{-1/2}. \tag{14.25}$$

This length is typically of order 1 Å in the sea of conduction electrons in a metal.

The above semiclassical argument does not take account of the quantum wave nature of the electrons. While the Poisson equation remains valid, the equilibrium condition relating $V(r)$ to $\rho(r)$ needs microscopic modification. We consider for an illustrative purpose in Appendix 14.1 a one-dimensional box with free particles moving between $x = 0$ and $x = L$, in the limit $L \to \infty$. The electron density associated with filling the lowest N levels singly is seen to be

$$\frac{\rho(x)}{\rho} = 1 - \frac{\sin(2k_{\mathrm{F}}x)}{2k_{\mathrm{F}}x}\,. \tag{14.26}$$

That is, long-range oscillations of wavelength π/k_{F} are induced in the one-dimensional Fermi gas by the perturbation due to a wall located at $x = 0$, the other wall having gone to infinity.

The induction of microscopic density oscillations of wavelength π/k_{F} is a general property of perturbations on a Fermi gas, descending from the discontinuity in the momentum distribution across the Fermi surface. For a test charge in an electron gas, and transcending the weak-perturbation restriction, the asymptotic behaviour of the displaced electron density is

$$\rho(r) - \rho \propto \frac{\cos(2k_{\mathrm{F}}r + \varphi)}{r^3} \tag{14.27}$$

at large distance from the charged impurity (see again Appendix 14.1). These oscillations in the displaced charge density are known as Friedel oscillations.[492,493] Therefore, the effective potential created by an ion inserted in the electron plasma (and indeed the effective potential with which an electron in the plasma acts on other electrons) is an oscillating function of distance — showing both repulsive and attractive regions. This result has crucial consequences in the electron theory of metals (see Sec. 14.5).

The notion of screening provides the microscopic basis for the construction of the Landau quasi-particles accounting for the thermodynamic properties of the sea of conduction electrons in metals. In regard to its dynamical properties, we may instead recall the argument given by Langmuir[494] for the longitudinal plasma resonance — as already reported in Sec. 8.1. A bodily shift of the plasma by an amount ξ relative to the background sets up a surface charge density and hence a restoring electric field given by Gauss's law. The plasma is driven into harmonic motion at frequency ω_p given by

$$\omega_p = \left(\frac{4\pi\rho e^2}{m}\right)^{1/2} . \tag{14.28}$$

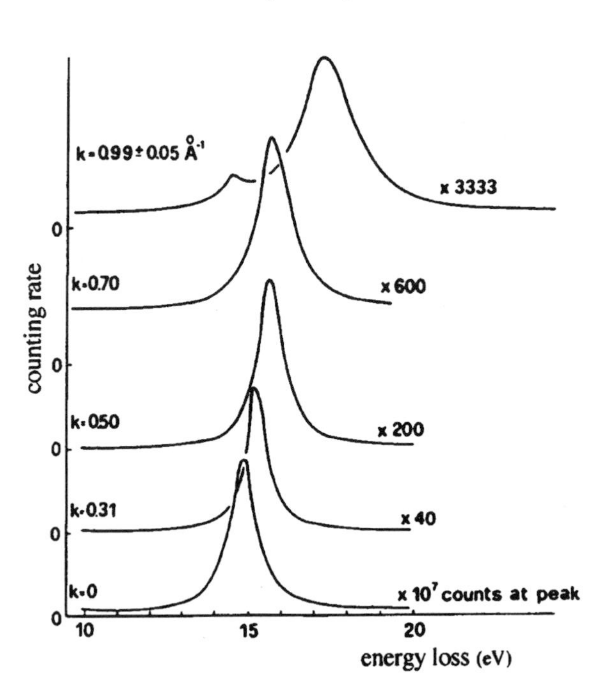

Fig. 14.6. Electron energy loss spectra of Al in the plasmon energy range at various values of the wave number k. (Redrawn from Gibbons *et al.*, Ref. 495.)

Plasmon excitations in metals are observed in inelastic scattering experiments, using either a beam of fast electrons (EEL, or electron energy loss experiments) or a beam of X-rays. An example of such spectra taken on Al metal[495] is reported in Fig. 14.6. The various spectra in this figure refer to different scattering momenta and clearly show the dispersion and broadening of the collective plasma excitations as their wave number k increases.

14.3.3 *Wigner crystallisation*

The coupling strength of the jellium model is measured by the ratio

$$\frac{\langle E_{\rm pot}\rangle}{\langle E_{\rm kin}\rangle} \approx \frac{e^2/a}{\hbar^2/(2ma^2)} \propto r_{\rm s}\,, \tag{14.29}$$

where we have introduced the average interparticle spacing a through the relation $4\pi a^3/3 = \rho^{-1}$ and used the notation $a = a_{\rm B} r_{\rm s}$ with $a_{\rm B} = \hbar^2/me^2$ being the Bohr radius. As the dimensionless length $r_{\rm s}$ increases (i.e. as the electron density *decreases*) the kinetic energy is becoming less and less important relative to the potential energy associated with the Coulomb interactions.

In the thirties Wigner[496] had already noticed that an optimal value is obtained for the potential energy of jellium if the electrons are placed on the sites of a crystalline lattice having a body-centred-cubic structure. The problem of "Wigner crystallisation" has become a classic in many-body physics. Localisation of the electrons raises their kinetic energy and the crystalline state may become favoured only at very low density, where the potential energy becomes dominant. Of course, in the crystal the electrons are not strictly localised on the lattice sites, but execute vibrational motions around them. Strong anharmonicity and relatively high concentrations of lattice defects are expected in this quantum solid near melting.

Quantum computer simulations[497] have indicated that with increasing $r_{\rm s}$ the ground state of three-dimensional (3D) jellium undergoes a continuous transition from the paramagnetic (spin-disordered) fluid state to a ferromagnetic (spin-aligned) fluid state and then a first-order transition to a ferromagnetic crystal at $r_{\rm s} \approx 65$. Similar studies of 2D jellium[498] indicate a first-order transition from a paramagnetic to a ferromagnetic fluid and crystallisation into a triangular lattice at $r_{\rm s} \approx 35$. The search for Wigner crystallisation in the laboratory has been addressed to quasi-2D assemblies of electronic carriers, mostly in man-made semiconductor structures.

In fact, the first unambiguous observation of Wigner crystallisation concerned a fluid of electrons in a quasi-classical regime, floating on top of a liquid ^{4}He substrate.[499] Metal-insulator transitions due to interparticle correlations have been subsequently reported in quasi-2D systems of carriers in semiconductor structures subject to very strong magnetic fields[500] or in the presence of disorder,[501] and also in high-purity samples having exceptionally high carrier mobility that seem to approach the expected behaviour of the ideal jellium model.[502]

14.4 BCS Superconductivity and Superfluidity in Fermion Fluids

14.4.1 *The superconducting state*

The electrical resistance of a metal arises because the travelling waves representing conduction electrons are scattered by the vibrating ions and by impurities (see Sec. 14.5). In 1911 Kamerlingh Onnes observed that Hg at about 4 K lost all its electrical resistance. He later found that resistance is restored in a sufficiently strong, temperature-dependent magnetic field. Similar behaviours have been observed in many, though not all, metal and alloys (for an easily readable account emphasising analogies with superfluids, the reader may refer to the book of Tilley and Tilley[503]). Critical temperatures T_c in excess of 100 K have been achieved after 1986 in doped cuprate compounds (the so-called high-T_c or ceramic superconductors).

The most basic property of the bulk superconducting state is the opening of a gap in the density of states around the Fermi surface: that is, no state are available for electrons in an energy strip from $\varepsilon_F - \Delta$ to $\varepsilon_F + \Delta$, where $2\Delta \approx k_B T_c$. Experimental manifestations of the opening of an energy gap are the appearance of transparency in the far infrared and of an exponential fall-off of the specific heat with decreasing temperature below T_c,

$$\frac{C_V}{\gamma T_c} \approx a e^{-bT_c/T}, \tag{14.30}$$

where γT is the linear electronic specific heat of the normal metal and the quantities a and b, though weakly temperature dependent, are about 9 and 1.5 for all superconductors.

The BCS theory of the homogeneous superconducting state[504] accounts for the energy gap through the formation of bound electron pairs in a zero-spin state. Cooper[505] considered a pair of electrons with opposite momenta and spins at the Fermi energy ε_F and studied the consequences of switching on an effective retarded attraction between them. A mechanism for such time-dependent interaction could be the coupling of conduction electrons with lattice vibrations: in essence, an electron moving through the positive-ion lattice could locally initiate induction of a lattice polarization, which would be fully formed in a characteristic lattice-vibration period $T_D \approx 10^{-13}$ seconds and could then be enjoyed by a second electron passing through the same region of space with a time delay of order T_D relative to the first electron. Including the direct Coulomb repulsion, this model yields a net attractive interaction between the two electrons over an energy range $\hbar\omega_D$, where $\omega_D = 2\pi/T_D$ is the cut-off "Debye frequency" of the vibrational spectrum.

Cooper showed that formation of a bound electron pair leads to an energy gain given by

$$E - 2\varepsilon_F \approx -2\hbar\omega_D e^{-1/\nu_F V}, \tag{14.31}$$

where ν_F is the density of electronic states on the Fermi surface in the normal metal and V is the strength of the effective electron–electron attraction. He also showed that the mean distance ξ between the two partners in the bound pair is of order of a few hundred Ångstroms — indeed, the electron–ion mechanism can overcome the direct Coulomb repulsion only if the two electrons travel in a correlated manner far away from each other.

In the limit $\xi \to 0$ a Cooper pair would be a point-like boson and the transition to the superconducting state could be viewed as an instance of Bose–Einstein condensation in a dense fluid of such pairs. Since, however, $\xi \approx 300$ Å as noted above, there is massive interpenetration between different pairs when a macroscopic number of them is formed. It remained for the BCS theory to account for the superconducting state as a dense fluid of paired electrons closely resembling Cooper pairs. Pairing lowers the ground-state energy by $-0.76\nu_F\Delta^2$ and the elementary excitations out of the BCS ground state have the dispersion relation

$$\hbar\omega_{\mathbf{k}} = \sqrt{\Delta^2 + \xi_{\mathbf{k}}^2}, \tag{14.32}$$

where the gap parameter Δ is given by $\Delta \approx 2\hbar\omega_D \exp(-1/\nu_F V)$ (see Eq. (14.31)) and $\xi_{\mathbf{k}}$ are the electronic excitation energies in the normal state referred to the Fermi energy, i.e. $\xi_{\mathbf{k}} = |\varepsilon_{\mathbf{k}} - \varepsilon_F| \approx \hbar|\mathbf{v}_F \cdot (\mathbf{k} - \mathbf{k}_F)|$. The measured gap corresponds to the breaking of a pair and is equal to 2Δ, decreasing with increasing temperature in a positive-feedback fashion because of interference between the excited electrons and the remaining bound pairs.

The reason for attributing antiparallel spins to the two partners of an electronic Cooper pair is that in this way the Meissner effect is immediately accounted for. Since the pairs carry zero spin, the effect of an applied magnetic field (below the threshold set by the critical field) is purely to induce diamagnetic supercurrents flowing in the "skin" of the sample and completely screening the applied field. In the Meissner effect it is indeed observed that only the tangential component of the field can penetrate into the surface layers of the superconducting sample, over a finite penetration length $\lambda \approx (4\pi\rho e^2/mc^2)^{-1/2}$ with c the speed of light.

In summary, a Fermi liquid is unstable against the formation of Cooper pairs, provided that some attraction, no matter how weak, is present between fermions on the Fermi surface. Appeal to the idea of Cooper pair formation has also been made to explain the transition of liquid ^{3}He to superfluid phases at very low temperatures (see Sec. 14.4.3), as well as in connection with proton and neutron matter inside atomic nuclei and with neutron stars.[506]

14.4.2 *Flux quantisation and Josephson effects*

An inhomogeneous superconductor may be described by means of a position-dependent gap parameter $\Delta(\mathbf{r})$ that under suitable conditions can be viewed as the order parameter $\psi(\mathbf{r}) = |\psi(\mathbf{r})| \exp[i\varphi(\mathbf{r})]$ of the superconducting state.[507] We are back to the notion of a condensate wave function for the assembly of Cooper pairs, having density $\rho_s = |\psi(\mathbf{r})|^2$ and carrying a supercurrent density $\mathbf{j}_s(\mathbf{r})$ given by

$$\mathbf{j}_s = \left(-\frac{i\hbar e_s}{2m_s}\right)(\psi^*\nabla\psi - \psi\nabla\psi^*) = \left(\frac{\hbar e_s}{m_s}\right)\rho_s\nabla\varphi\,. \tag{14.33}$$

We have attributed a charge e_s and a mass m_s to the carriers of supercurrent and have made use of the momentum operator $\hat{\mathbf{p}} = -i\hbar\nabla$ to express the current density in terms of the gradient of the phase. Equation (14.33) is immediately extended to a superconductor in a magnetic field described by a vector potential $\mathbf{A}(\mathbf{r})$ through the transformation $\hat{\mathbf{p}} \to \hat{\mathbf{p}} - (e_s/c)\mathbf{A}$, yielding

$$\mathbf{j}_s = \left(\frac{\hbar e_s}{m_s}\right)\rho_s\nabla\varphi - \left(\frac{e_s^2}{m_s c}\right)\rho_s\mathbf{A}\,. \tag{14.34}$$

In a simply connected sample the vector potential is defined to within an additive term given by the gradient of a scalar and the phase term in Eq. (14.34) can therefore be eliminated by a gauge transformation ($\mathbf{A} \to \mathbf{A} + (\hbar c/e_s)\nabla\varphi$). The phenomenological constitutive equation proposed by London (1938) to account for the Meissner effect is recovered from Eq. (14.34),

$$\mathbf{j}_s = -\left(\frac{c}{4\pi\lambda^2}\right)\mathbf{A}\,, \tag{14.35}$$

with λ being the penetration length. Considering instead a thick ring, we evaluate the integral

$$\oint \mathbf{j}_s(\mathbf{r}) \cdot d\mathbf{l} = \left(\frac{e_s\rho_s}{m_s}\right) \oint \left[\hbar\nabla\varphi - \left(\frac{e_s}{c}\right)\mathbf{A}\right] \cdot d\mathbf{l} \tag{14.36}$$

over a circular circuit deep inside the ring. The integral is zero because the current vanishes in the depth of the ring, while its two components are given by $\oint \mathbf{A} \cdot d\mathbf{l} = \int \mathbf{B} \cdot d\mathbf{S} = \Phi$ and by $\oint \nabla\varphi \cdot d\mathbf{l} = 2n\pi$: in the first component we have used a theorem in vector analysis to transform the line integral into a surface integral giving the magnetic field flux threading the area embraced by the integration circuit, while in the second we have used the fact that the total change in phase along the circuit must be an integer multiple of 2π. The result is that the magnetic field flux embraced by the ring is quantised in integer multiples of the flux quantum hc/e_s,

$$\Phi = n\left(\frac{hc}{e_s}\right)\,. \tag{14.37}$$

Measurement of the flux quantum confirms the Cooper-pair picture by giving $e_s = 2e$.

Let us now turn with Josephson[508] to a junction between two superconductors separated by a layer of insulating material, which is sufficiently thin to allow tunnelling of Cooper pairs across the junction. The equation for the order parameter in superconductor 1 is written as

$$i\hbar\frac{\partial\psi_1}{\partial t} = \mu_1\psi_1 + K\psi_2\,, \tag{14.38}$$

where μ_1 is the chemical potential and the constant K represents a coupling between the two superconductors allowing for the transfer of carriers across the insulating barrier. Writing a similar equation for ψ_2, a simple calculation yields the following results:

(i) $(\partial \rho_{s1}/\partial t) = -(\partial \rho_{s2}/\partial t) = (4\pi K/\hbar)(\rho_{s1}\rho_{s2})^{1/2} \sin \varphi$, i.e. the current $I \propto -(\partial \rho_{s1}/\partial t)$ through the junction is a sinusoidal function of the phase difference $\varphi \equiv \varphi_2 - \varphi_1$;

(ii) the time dependence of the phase difference is determined by the difference in chemical potentials, $(\partial \varphi/\partial t) = (\mu_1 - \mu_2)/\hbar$.

Thus, a continuous current from an external source can be driven across the junction with no potential drop appearing across it ($\mu_1 - \mu_2 = 0$), provided that the intensity of the current is below some maximum value. This is known as the d.c. Josephson effect and has a number of important applications, including the realisation of extremely sensitive magnetometers. On the other hand, if a given potential drop $\mu_1 - \mu_2$ is applied across the junction, the a.c. Josephson effects are observed: (i) carriers may tunnel along the potential drop by emitting photons at frequencies given by $n\omega = (\mu_1 - \mu_2)/\hbar$; or (ii) carriers may be made to tunnel against the potential drop by a process of absorption of n photons provided by shining on the junction an electromagnetic wave of frequency $\omega = (\mu_1 - \mu_2)/\hbar n$.

14.4.3 *Superfluidity in liquid ^{3}He*

We have seen in Fig. 14.3 a schematic phase diagram for ^{3}He in the (p, T) plane, showing that as the liquid is cooled along the melting line a transition occurs from the normal Fermi liquid phase to a new phase termed A. On cooling to still lower temperatures a second transition is observed to a phase termed B. Both the A and the B phase are superfluid.[509]

The theory[510] that has been invoked to explain ^{3}He superfluidity parallels the BCS theory for superconductivity in metals, summarised in Sec. 14.4.1. In ^{3}He a natural source of attractive interactions is provided by the van der Waals interatomic forces. However, an effective attractive interaction can lead in this fluid to the formation of bound atom pairs only if the resulting pair wave function vanishes at short distances inside the atomic core diameter, where the interactions become strongly repulsive. A way of achieving this is by building pair wave functions with nonzero total angular momentum L.

While the total angular momentum and the total spin of a Cooper pair in the BCS theory for the electron fluid are zero, the evidence indicates that in the case of ^{3}He one gets in the A and B phases bound-atom pairs of p character ($L = 1$). Since the radial part of the pair wave function is therefore antisymmetric under exchange of the two partners, the spin part must be symmetric to comply with the Pauli principle. That is, the two partners have parallel spins.

There are three symmetric ($S = 1$) spin states that can be built from two particles, each with spin 1/2, and these correspond to the values $S_z = 1$, 0 and -1 for the component of the total spin along a chosen quantisation axis. The occurrence of two superfluid phases can be accounted for by invoking two different admixtures of these three components. In both phases two privileged directions can then be identified: one in spin space corresponding say to the direction $\mathbf{d}$ along which the component of the total spin is zero, and the other in coordinate space corresponding say to the direction $\mathbf{l}$ along which the orbital angular momentum has component 1.

While in the absence of other interactions these two directions would be uncorrelated, a coupling between $\mathbf{l}$ and $\mathbf{d}$ is provided by the interaction between the nuclear magnetic moments. This is very small within each Cooper pair, but its effects are enhanced by the coherence of the macroscopic condensate of pairs. These interactions between orbital momentum and spin, which are responsible for the existence of two superfluid phases, have been investigated in Nuclear Magnetic Resonance experiments: the applied field tends to vary the orientation of the spins and exerts on them an additional force which may change the position and shape of the resonance line. A variety of collective excitations have been predicted, and in part observed, in superfluid ^{3}He, corresponding to different ways of breaking the complex order that exists in it.

14.5 Electron Theory of Liquid Metals

Here we consider some properties of simple liquid metals having conduction electrons in s and p states, that specifically reflect their nature as two-component liquids of ions and electrons. These properties are (i) the effective interaction between pairs of ions as determined by screening of their bare Coulomb repulsions by the conduction electrons; (ii) the structural correlation functions involving the conduction electrons and supplementing the nuclear structure factor $S(k)$ in a full description of the liquid-metal structure; and (iii) the theory of electrical resistivity and viscosity of liquid metals. Full references are given in an article by March.[511] For a general account of liquid metals the book of March[512] may be consulted.

14.5.1 *Interatomic forces from liquid structure factor $S(k)$*

Classical theories of liquid structure aim to calculate $S(k)$, or its $\mathbf{r}$-space equivalent $g(r)$, from a given force law (see Chap. 4). Johnson and March proposed to reverse this approach and invoked experimental diffraction data to extract a pair potential $\phi(r)$ for ions in liquid Na near its freezing point. They used for this purpose the Born–Green theory of liquid structure presented in Appendix 4.3, based on an approximate decoupling of three-body correlations.

Currently, computer simulation is being used to bypass the need for an approximate theory of liquid structure.[513] This has led to the extraction of the so-called "diffraction potential", the example of liquid Na being shown in Fig. 14.7 in comparison with the results obtained by Perrot and March from electron theory. It is remarkable that all major features of the diffraction potential are being reproduced by the theory, though quantitative discrepancies remain.

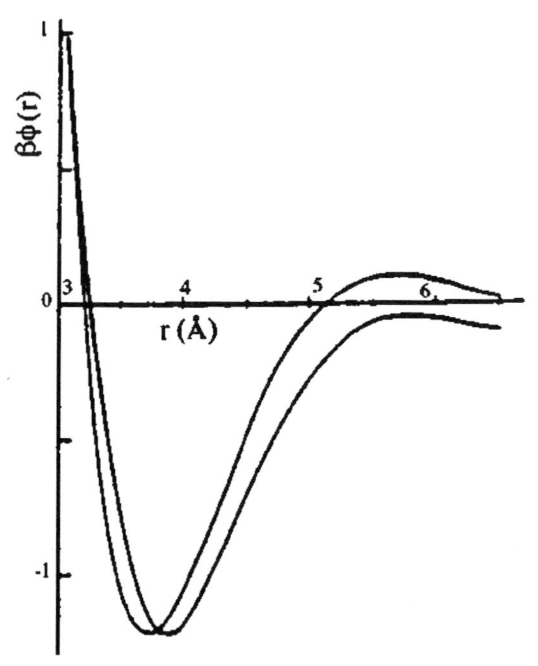

Fig. 14.7. Diffraction pair potential (in units of k_BT) in liquid Na near freezing. For comparison the electron-theory pair potential of Perrot and March (upper curve at large r) is also shown.

It should be emphasised that the pair potential $\phi(r)$ in a liquid metal, being determined by electron screening, depends on electron density in an important way. The extensive work of Hensel and coworkers[514] on expanded liquid alkali metals, especially on liquid Cs taken up the liquid–vapour coexistence curve towards the critical point, has shown that as the liquid density is reduced the coordination number decreases rapidly while the near-neighbour distance remains remarkably constant. The coordination number obtained from the data by extrapolation to the critical density approaches the value 2, suggesting formation of chain-like structures as the fluid is driven towards a metal-insulator transition. The shape of the diffraction potential from data on expanded Cs still resembles that shown in Fig. 14.7, i.e. a steep repulsive core followed by an attractive well in the region of first neighbours.

14.5.2 *Diffractive scattering from two-component plasmas*

From the argument given in Sec. 4.3, the intensity of X-rays scattered from a liquid metal is

$$I_X(k) = F(\langle\rho(\mathbf{r})\rho(\mathbf{r}')\rangle) \tag{14.39}$$

where $\rho(\mathbf{r})$ is the electron density operator and F denotes the Fourier transform with respect to $\mathbf{r}-\mathbf{r}'$. Egelstaff *et al.*[515] assumed that the total electron density can be decomposed into the sum of contributions from core electrons and valence (conduction) electrons, i.e. $\rho(\mathbf{r}) = \rho_c(\mathbf{r}) + \rho_v(\mathbf{r})$ with the core electrons being rigidly attached to their own nuclei. Introducing the core scattering factor $f_c(k)$, Eq. (14.38) can be rewritten as

$$I_X(k) = f_c^2(k)S(k) + 2f_c(k)S_{ie}(k) + F(\langle\rho_v(\mathbf{r})\rho_v(\mathbf{r}')\rangle) \tag{14.40}$$

where $S(k)$ is the nucleus–nucleus structure factor, as directly accessible *via* neutron diffraction experiments, and $S_{ie}(k)$ results from interference between waves scattered by the ionic cores and by the valence electrons. The last term on the RHS of Eq. (14.40) is proportional to the valence electron structure factor, denoted by $S_{ee}(k)$.

The idea behind the approach of Egelstaff *et al.* was that one could in principle extract the electronic correlation functions in a real liquid metal from three diffraction experiments using X-ray, neutron, and electron beams. Progress in implementing this proposal has been slow, but significant advance has come from the study of de Wijs *et al.*[516] who used computer simulation to obtain $S_{\mathrm{ie}}(k)$ for liquid Mg and liquid Bi. In the case of liquid Mg a simple model can be constructed to give a good account of the simulation data, leading to

$$\frac{S_{\mathrm{ie}}(k)}{S(k)} = Z^{1/2}\left[1+\left(\frac{k}{\kappa_{\mathrm{TF}}}\right)^2\right]^{-1}\cos(kR_c) \qquad (14.41)$$

where Z is the ionic valence ($Z = 2$ for Mg), κ_{TF} is the Thomas–Fermi inverse screening length (see Eq. (14.25)) and R_{c} is the radius of the ionic core. The results of the model are reproduced in Fig. 14.8. Notice that in the limit $k \to 0$ the exact property $S_{\mathrm{ie}}(0) = Z^{1/2}S(0)$ follows from perfect screening of the ions by the valence electrons, the corresponding result for the electron–electron structure factor being $S_{\mathrm{ee}}(0) = ZS(0)$. Of course, $S(0)$ is related to the isothermal compressibility of the liquid metal by Eq. (4.6), $S(0) = \rho k_{\mathrm{B}}TK_{\mathrm{T}}$.

The liquid Bi data of de Wijs *et al.*, on the other hand, are an example of strong electron–ion interactions (for instance, Bi on freezing takes on semi-metallic character) and therefore the simple model used for Mg is not

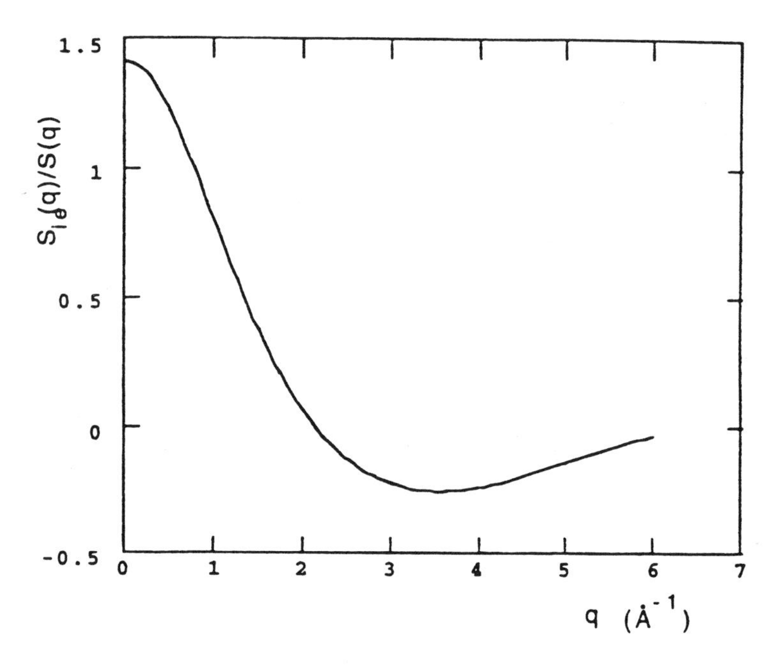

Fig. 14.8. Result of model in Eq. (14.41) for the ratio $S_{\mathrm{ie}}(k)/S(k)$ in liquid Mg.

appropriate. In the case of Bi $S_{\mathrm{ie}}(k)$ shows a positive peak in phase with the main peak in $S(k)$, whereas in Mg it has a deep negative region in (anti)phase with $S(k)$ (see Fig. 14.8). We are meeting again the structural features that were emphasised for two-component ionic fluids in Chap. 8, that is Coulomb ordering in liquid Mg and intermediate-range ordering in liquid Bi.

14.5.3 *Transport coefficients*

The electrical resistivity ρ_{e} of simple liquid metals in an ideally pure state is determined by the quasi-elastic scattering of electrons on the Fermi surface against (screened) fluctuations in the ionic density. Baym[517] expressed ρ_{e} in terms of the nucleus–nucleus dynamic structure factor $S(k,\omega)$ by exploiting the fact that, in the regime of validity of the (weak scattering) Born approximation, the scattering cross-sections for electrons and neutrons are both proportional to $S(k,\omega)$. His expression for ρ_{e} involves an integration over all energy transfers $\hbar\omega$ to the ionic system and, on account of the vast difference in the energy scales for ionic and electronic motions, this integration can be carried out to yield the result

$$\rho_{\mathrm{e}} = \frac{12\pi m^2}{\hbar^3 Z\rho e^2 k_{\mathrm{F}}^2}\int_0^1 d\left(\frac{k}{2k_{\mathrm{F}}}\right)|V_{\mathrm{ie}}(\mathbf{k})|^2 S(k)\left(\frac{k}{2k_{\mathrm{F}}}\right)^3, \tag{14.42}$$

where $V_{\mathrm{ie}}(\mathbf{k})$ is the screened electron–ion potential. This formula goes back to Krishnan and Bhatia (1945) and was brought to full fruition by Ziman[518] (1961) using pseudopotentials to treat the bare electron–ion interaction. An example of the calculated electrical resistivity as a function of temperature, in comparison with experimental data, is shown in Fig. 14.9 for the liquid alkali metals.

The electronic contribution is also dominant in the thermal conductivity κ of liquid metals. Indeed, the Wiedemann–Franz law asserts the constancy of the quantity $\rho_{\mathrm{e}}\kappa/T$. Theory relates this constant to the so-called Lorenz number $L \equiv \pi^2 k_{\mathrm{B}}^2/3e^2$ through

$$\frac{\rho_{\mathrm{e}}\kappa}{T} = L. \tag{14.43}$$

Empirically, this relation is well satisfied by a number of liquid metals near their freezing point, within a scatter of about $\pm 10\%$.

The electronic contribution is instead minor in determining the longitudinal viscosity $\eta_l = \frac{4}{3}\eta + \zeta$, with η and ζ the shear and bulk viscosities

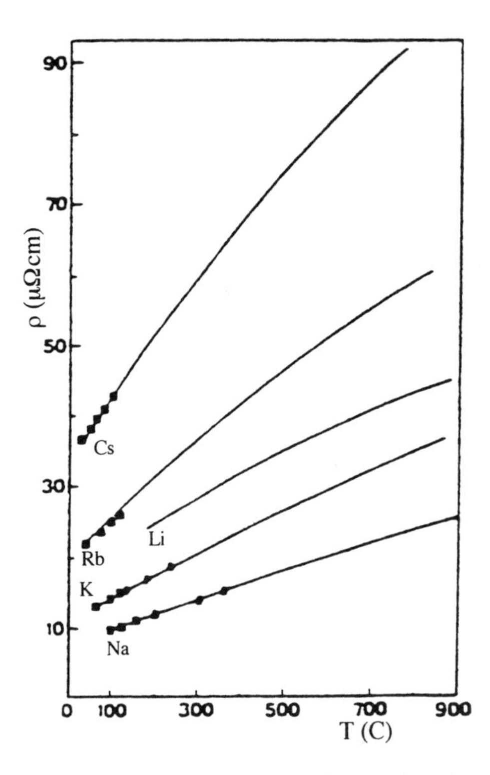

Fig. 14.9. Electrical resistivity of liquid alkali metals as a function of temperature from electron-theory, compared with experimental data (■ and •).

respectively, and hence the attenuation of sound waves propagating in the liquid metal. Treatment of the dynamic structure factor $S(k,\omega)$ within an electron–ion plasma model in the Green–Kubo limit leads to the result

$$\eta_l = \rho\left(Mf_i + \frac{4}{15}Zmv_{\mathrm{F}}\ell_{\mathrm{e}}\right), \tag{14.44}$$

where M is the ionic mass, f_i is an ionic friction coefficient, v_{F} is the Fermi velocity and ℓ_{e} the electronic mean free path. Numerical evaluation of Eq. (14.43) shows that the first term on its RHS is dominant, the electronic term being only a small correction to the ionic friction term.

From Eq. (14.44) with $\eta_l = \frac{4}{3}\eta + \zeta$ and $\eta \gg \zeta$, these results lend support to a formula proposed by Andrade[519] for the shear viscosity of liquid metals at freezing (see Sec. 5.6.3).

14.6 Liquid Hydrogen Plasmas and the Giant Planets

14.6.1 *Exploring the phase diagram of hydrogen*

Hydrogen is the most abundant element in the Universe and, notwithstanding the simplest constitution of the atom, has a very rich phase diagram of which nature provides various examples over a wide range of thermodynamic conditions.[520] Cooling the gas of diatomic molecules under standard laboratory conditions brings hydrogen into a molecular liquid where the diatomic molecules are weakly coupled by interactions including a long-range quadrupolar contribution, and then into a rotational crystalline solid in which the molecular angular momentum remains a remarkably good quantum number in spite of the dense crystalline environment. Compression at low temperature eventually brings the rotational crystal into a new phase where the molecular rotations are hindered, and ultimately leads into a further crystalline phase whose strong infrared activity indicates a mainly dipolar distortion of the electronic charge distribution. As hydrogen is brought into the region of high temperature and density which is of main interest to astrophysics, molecular dissociation can occur by thermal excitations or through compression, as the electrons are forced into high energy states by the increasing chemical potential. Dynamic compression experiments[521] report densities corresponding to $r_s \approx 1.5$ and temperatures of up to 4400 K in dense hydrogen, and to $r_s \approx 1.73$ and even higher temperatures for deuterium. In these experiments a continuous transition from a semiconducting to a metallic state has been reported to occur at 3000 K and 140 GPa.

A theoretical approach to the prediction of the phases of fluid hydrogen has been to include several species, such as diatomic molecules, atoms, protons, and electrons, and to examine their chemical equilibria. In the model of Saumon and Chabrier,[522] a first-order transition from a molecular phase to a partially ionised atomic gas has been predicted. A path-integral Monte Carlo study covering the density range $1.75 \leq r_s \leq 2.2$ has provided general support for these findings, although the quantitative details differ.[523] This study shows, first of all, that a molecular gas forms spontaneously from a neutral system of protons and electrons as the temperature is lowered from 10^5 K to 5000 K, with a molecular bond length which contracts with increasing density apparently as a consequence of stiff intermolecular repulsions. Molecular dissociation is then seen to occur as the temperature is raised or as a result of isothermal compression. At high density ($r_s < 2.0$) thermally activated dissociation is

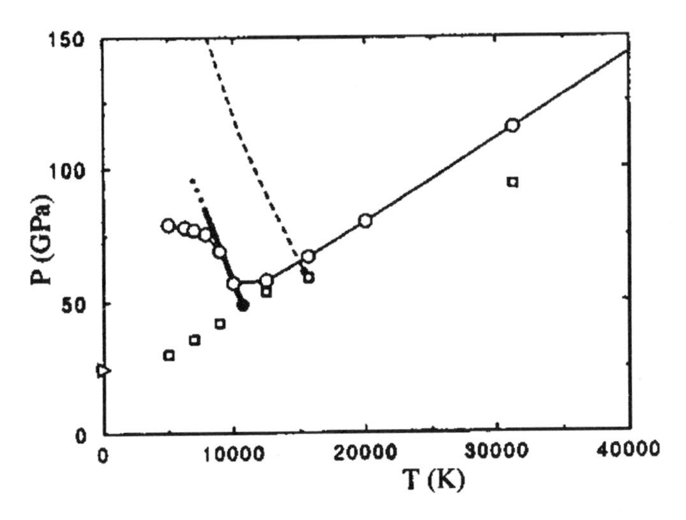

Fig. 14.10. Computed pressure from simulation (○) and from chemical picture (□) versus temperature for hydrogen at $r_s = 2.0$. — and – – – show the corresponding results for phase coexistence. (Redrawn from Magro *et al.*, Ref. 523.)

accompanied by decreasing pressure, signalling the presence of a first-order transition and a critical point (see Fig. 14.10). A proposed explanation for this transition lies in the increase of electronic kinetic energy associated with bond formation: in essence, this increase derives from angular localisation as the electrons leave spherical atomic-like orbitals to go into molecular bonding orbitals.

14.6.2 *Hydrogen–helium mixtures and the constitution of giant planets*

The dissociation transformation revealed by these studies, in which molecular hydrogen transforms not directly into a fully ionised plasma but first into a partially ionised atomic fluid, is certainly relevant to modelling of the interior of the giant planets such as Jupiter or Saturn. However, the situation there is more complex because of the coexistence of hydrogen and helium as main component elements.[524]

It is estimated that these two elements are subjected to pressures up to 4500 GPa and temperatures up to 24000 K in Jupiter, and about 1000 GPa and 10000 K in Saturn. These planets are thought to consist of three main layers: an outer layer of molecular hydrogen and atomic helium, a middle layer of metallised atomic hydrogen and helium, and a rocky core compressed to as high as 10^4 GPa at the centre of Jupiter. In Saturn an internal energy source has been proposed, associated with demixing and gravitational separation of the hydrogen–helium fluid at pressures below 1000 GPa. The existence of this phase transition is very sensitive to the equations of state for the two elements.

Appendix 14.1 Density Profiles in the Perturbed Electron Gas

We consider a one-dimensional box with free electrons moving between $x = 0$ and $x = L$ and having wave functions $\psi_n(x) = (2/L)^{1/2}\sin(n\pi x/L)$ with $n = 1, 2, 3, \ldots$. The electron density $\rho(x)$ associated with filling the lowest N levels singly is given by

$$\rho(x) = \sum_{n=1}^{N} |\psi_n(x)|^2 = \frac{N + 1/2}{L} - \frac{\sin[2\pi(N + 1/2)x/L]}{2L\sin(\pi x/L)}\,. \qquad \text{(A14.1.1)}$$

The summation can be completed exactly as indicated, but as L gets very large we can replace the summation by an integration. The one-dimensional wave vector space is occupied from $-k_F$ to k_F and the area of occupied phase space is $2k_FL$. Since each cell in phase space has area h and can accommodate one electron, the number of cells needed for N electrons in singly occupied levels is $2k_FL/h = N$. After integrating from $-k_F$ to k_F, we obtain the result

$$\rho(x) = \frac{N}{L}\left[1 - \frac{\sin(2k_Fx)}{2k_Fx}\right] \qquad \text{(A14.1.2)}$$

showing that oscillations of wavelength π/k_F are induced in a free electron gas facing an impenetrable wall at $x = 0$.

After this simple example we turn to the problem of the displaced electron density around a charged impurity in the three-dimensional electron gas (Sec. 14.3.2). The electron density to be used in the Poisson equation (14.23) is $\rho(r) = 2\sum_{\mathbf{k}} |\psi_{\mathbf{k}}(\mathbf{r})|^2$, where the factor 2 comes from the spins and $\psi_{\mathbf{k}}(\mathbf{r})$ are one-electron orbitals obeying the Schrödinger equation

$$\nabla^2\psi_{\mathbf{k}}(\mathbf{r}) + \left[k^2 - \left(\frac{2m}{\hbar^2}\right)V(r)\right]\psi_{\mathbf{k}}(\mathbf{r}) = 0\,. \qquad \text{(A14.1.3)}$$

$V(r)$ is the effective potential generated by the impurity and we have imposed that the orbitals should reduce to plane waves with energy $\hbar^2k^2/2m$ far away from the impurity. A perturbative solution of Eq. (A14.1.3) is obtained by replacing $\psi_{\mathbf{k}}(\mathbf{r})$ in its second term by a plane wave $\exp(i\mathbf{k}\cdot\mathbf{r})$ (March and Murray[493]), with the result

$$\rho(r) = \rho - \frac{mk_{\mathrm{F}}^2}{2\pi^3\hbar^2}\int d\mathbf{r}'\frac{j_1(2k_{\mathrm{F}}|\mathbf{r}-\mathbf{r}'|)}{|\mathbf{r}-\mathbf{r}'|^2}V(\mathbf{r}'), \tag{A14.1.4}$$

where $j_1(x) \equiv (\sin x - x\cos x)/x^2$ is a typical wave factor representing a diffraction process. The Poisson equation then is

$$\nabla^2 V(\mathbf{r}c) = \frac{2mk_{\mathrm{F}}^2e^2}{\pi^2\hbar^2}\int d\mathbf{r}'\frac{j_1(2k_{\mathrm{F}}|\mathbf{r}-\mathbf{r}'|)}{|\mathbf{r}-\mathbf{r}'|^2}V(\mathbf{r}') \tag{A14.1.5}$$

and is to be solved subject to the conditions $V(r) \to -Ze^2/r$ for $r \to 0$ and $V(r) \to 0$ faster that r^{-1} for $r \to \infty$.

The result in Eq. (14.24) is obtained from these equations if we assume that far away from the impurity $V(r)$ varies sufficiently slowly that in Eqs. (A14.1.4) and (A14.1.5) $V(r')$ may be replaced by $V(r)$. Instead, the correct asymptotic solution is

$$\rho(r) - \rho \propto \frac{\cos(2k_{\mathrm{F}}r)}{r^3} \tag{A14.1.6}$$

and similarly for $V(r)$. The phase shift in the asymptotic formula given in Eq. (14.27) arises by a full treatment going beyond perturbation theory.

References

1. S. M. Blinder, *Advanced Physical Chemistry* (MacMillan, Toronto, 1969).
2. L. J. Milanovic, H. A. Posch and W. G. Hoover, *Molec. Phys.* **95**, 281 (1998); see also W. G. Hoover and M. Ross, *Contemp. Phys.* **12**, 339 (1971).
3. D. Tabor, *Gases, Liquids and Solids* (University Press, Cambridge, 1993).
4. T. E. Faber, *Fluid Dynamics for Physicists* (University Press, Cambridge, 1995).
5. H. Eyring, D. Henderson, B. Jones Stover and E. M. Eyring, *Statistical Mechanics and Dynamics* (Wiley, New York, 1982).
6. J. S. Rowlinson and B. Widom, *Molecular Theory of Capillarity* (Clarendon, Oxford, 1982).
7. A. H. Narten, C. G. Venkatesh and S. A. Rice, *J. Chem. Phys.* **64**, 1106 (1976).
8. J. E. Enderby and G. W. Neilson, in *Water — a Comprehensive Treatise*, Vol. 6, ed. F. Franks (Plenum, New York, 1979).
9. F. Franks, *Water* (The Royal Society of Chemistry, London, 1983).
10. V. F. Petrenko and R. W. Whitworth, *Physics of Ice* (University Press, Oxford, 1999).
11. J. D. Bernal, *Nature* **188**, 910 (1960); *Proc. R. Soc. London* **A280**, 299 (1964).
12. G. D. Scott, *Nature* **188**, 908 (1960).
13. B. J. Alder and T. E. Wainwright, *J. Chem. Phys.* **27**, 1208 (1957).
14. L. V. Woodcock, *Ann. N. Y. Acad. Sci.* **371**, 274 (1981).
15. C. A. Rogers, *Proc. London Math. Soc.* (3) **88**, 609 (1958).
16. Z. W. Salsburg and W. W. Wood, *J. Chem. Phys.* **37**, 798 (1962).
17. J. L. Finney, in *Amorphous Solids and the Liquid State*, eds. N. H. March, R. A. Street and M. Tosi (Plenum, New York, 1985).
18. M. P. Allen and D. J. Tyndesley, *Computer Simulation of Liquids* (Clarendon, Oxford, 1987).
19. W. G. Hoover and F. H. Ree, *J. Chem. Phys.* **49**, 3609 (1968).
20. See e.g. M. Baus and J. L. Colot, *Molec. Phys.* **55**, 653 (1985).
21. J. G. Kirkwood, *J. Chem. Phys.* **18**, 380 (1950); W. W. Wood, *J. Chem. Phys.* **20**, 1334 (1952).
22. M. D. Eldridge, P. A. Madden, P. N. Pusey and P. Bartlett, *Molec. Phys.* **84**, 395 (1995).

23. T. E. Faber, *Introduction to the Theory of Liquid Metals* (Cambridge University Press, Cambridge, 1972).
24. N. F. Carnahan and K. F. Starling, *J. Chem. Phys.* **51**, 635 (1969).
25. K. R. Hall, *J. Chem. Phys.* **57**, 2252 (1971).
26. B. J. Alder, W. G. Hoover and D. A. Young, *J. Chem. Phys.* **49**, 3688 (1968); B. J. Alder, D. A. Young, M. R. Mansigh and Z. W. Salsburg, *J. Comp. Phys.* **7**, 361 (1971); D. A. Young and B. J. Alder, *J. Chem. Phys.* **73**, 2430 (1980).
27. Y. Choi, T. Ree and F. H. Ree, *J. Chem. Phys.* **95**, 7548 (1991).
28. E. Velasco, L. Mederos and G. Navascues, *Molec. Phys.* **97**, 1273 (1999).
29. H. C. Longuet-Higgins and B. Widom, *Molec. Phys.* **8**, 549 (1964).
30. See for instance N. H. March and M. P. Tosi, *Atomic Dynamics in Liquids* (Dover, New York, 1991).
31. U. F. Edgal, *J. Chem. Phys.* **94**, 8179 (1991).
32. U. F. Edgal and D. L. Huber, *Phys. Rev.* **E48**, 2610 (1993).
33. S. Sastry, T. M. Truskett, P. G. Debenedetti, S. Torquato and F. H. Stillinger, *Molec. Phys.* **95**, 289 (1998).
34. D. J. Gonzales and L. E. Gonzales, *Molec. Phys.* **74**, 613 (1991).
35. R. J. Speedy, *Molec. Phys.* **80**, 1105 (1993).
36. R. J. Speedy, *J. Phys.: Condens. Matter* **9**, 8591 (1997).
37. R. J. Speedy and H. Reiss, *Molec. Phys.* **72**, 1015 (1991).
38. R. J. Speedy, *J. Chem. Soc. Faraday Trans.* II **76**, 693 (1980).
39. W. G. Hoover, W. T. Ashurst and R. Grover, *J. Chem. Phys.* **57**, 1259 (1972).
40. W. G. Hoover, N. E. Hoover and K. Hanson, *J. Chem. Phys.* **70**, 1837 (1979); K. S. Sturgeon and F. H. Stillinger, *J. Chem. Phys.* **96**, 4651 (1992).
41. K. F. Herzfeld and M. G. Mayer, *J. Chem. Phys.* **2**, 38 (1934); L. Tonks, *Phys. Rev.* **50**, 955 (1936).
42. L. van Hove, *Physica* **16**, 137 (1950).
43. See e.g. P. C. Hemmer, M. Kac and G. E. Uhlenbeck, *J. Math. Phys.* **5**, 60 (1964).
44. B. J. Alder and T. E. Wainwright, *Phys. Rev.* **127**, 359 (1962).
45. J. Vieillard-Baron, *J. Chem. Phys.* **56**, 4729 (1972).
46. J. Kolafa and I. Nezbeda, *Molec. Phys.* **84**, 421 (1995).
47. J. Warnock, D. D. Awschlom and M. W. Shafer, *Phys. Rev. Lett.* **57**, 1753 (1986).
48. J. Klafter and J. M. Drake (Eds.), *Molecular Dynamics in Restricted Geometries* (Wiley, New York, 1989).
49. D. Henderson (Ed.), *Fundamentals of Inhomogeneous Fluids* (Dekker, New York, 1992).
50. K. K. Mon and J. K. Percus, *J. Chem. Phys.* **112**, 3457 (2000).
51. See e.g. M. W. Zemansky, *Heat and Thermodynamics* (McGraw-Hill, New York, 1951).
52. See e.g. L. D. Landau and E. M. Lifshitz, *Statistical Physics*, Part 1 (Pergamon, Oxford, 1980).

53. K. S. Førland, T. Førland and S. K. Ratkje, *Irreversible Thermodynamics: Theory and Applications* (Wiley, Chichester, 1988).
54. A. B. Pippard, *Elements of Classical Thermodynamics* (Cambridge University Press, Cambridge, 1957).
55. A. R. Ubbelohde, *The Molten State of Matter* (Wiley, Chichester, 1978).
56. S. E. Babb Jr., *Rev. Mod. Phys.* **35**, 400 (1963).
57. See e.g. R. M. Wentzcovitch, R. J. Hemley, W. J. Nellis and P. Y. Yu (Eds.), *High-Pressure Materials Research* (Materials Research Society, Warrendale, 1998).
58. S. Eliezer and R. A. Ricci (Eds.), *High-Pressure Equation of State: Theory and Application* (Società Italiana di Fisica, Bologna, 1990).
59. W. J. Nellis, S. T. Weir and A. C. Mitchell, *Phys. Rev.* **A59**, 3434 (1999) and references given therein.
60. See also G. N. Angilella, R. Pucci, G. Piccitto and F. Siringo (Eds.), *Molecular and Low Dimensional Systems under Pressure*, *Physica* **B265** (1999).
61. W. J. Nellis, *Phyil. Mag.* **B79**, 655 (1999) and references given therein.
62. W. G. Hoover and M. Ross, *Contemp. Phys.* **12**, 339 (1971).
63. A. J. Greenfield, J. Wellendorf and N. Wiser, *Phys. Rev.* **A4**, 1607 (1971).
64. G. E. Bacon, *Neutron Diffraction* (Clarendon, Oxford, 1975).
65. J. E. Enderby, D. M. North and P. A. Egelstaff, *Phil. Mag.* **14**, 961 (1966).
66. L. Verlet, *Phys. Rev.* **165**, 201 (1968); J. P. Hansen and L. Verlet, *Phys. Rev.* **184**, 150 (1969).
67. A. Ferraz and N. H. March, *Solid State Commun.* **36**, 977 (1980).
68. J. G. Kirkwood and E. Monroe, *J. Chem. Phys.* **9**, 514 (1941).
69. T. V. Ramakrishnan and M. Yussouff, *Phys. Rev.* **B19**, 2775 (1979).
70. P. Hohenberg and W. Kohn, *Phys. Rev.* **136**, B864 (1964); N. D. Mermin, *Phys. Rev.* **137**, A1441 (1965); W. Kohn and L. J. Sham, *Phys. Rev.* **140**, A1133 (1965).
71. A. B. Bhatia and N. H. March, *Phys. Chem. Liquids* **13**, 313 (1984).
72. J. L. Yarnell, M. J. Katz, R. G. Wenzel and S. H. Koenig, *Phys. Rev.* **A7**, 2130 (1973).
73. See e.g. Table 2.6 in T. E. Faber, Ref. 23.
74. E. A. Guggenheim, *Thermodynamics* (North-Holland, Amsterdam, 1949); S. M. Blinder, Ref. 1.
75. L. S. Ornstein and F. Zernike, *Proc. Acad. Sci.* (*Amsterdam*) **17**, 793 (1914); *Physik. Z.* **19**, 134 (1918).
76. N. W. Ashcroft and N. H. March, *Proc. R. Soc.* (*London*) **A297**, 336 (1967).
77. M. D. Johnson and N. H. March, *Phys. Lett.* **3**, 313 (1963).
78. L. Reatto, D. Levesque and J. J. Weis, *Phys. Rev.* **A33**, 3451 (1986).
79. J. K. Percus and G. J. Yevick, *Phys. Rev.* **110**, 1 (1958).
80. E. Thiele, *J. Chem. Phys.* **39**, 474 (1963).
81. M. S. Wertheim, *Phys. Rev. Lett.* **10**, 321 (1963).
82. H. Reiss, H. L. Frisch and J. L. Lebowitz, *J. Chem. Phys.* **31**, 369 (1959).

83. J. S. Rowlinson, *Rep. Progr. Phys.* **28**, 169 (1965).
84. J. M. Bernasconi and N. H. March, *Phys. Chem. Liquids* **15**, 169 (1986).
85. H. C. Andersen, J. D. Weeks and D. Chandler, *Phys. Rev.* **A4**, 1597 (1971).
86. L. Verlet, *Phys. Rev.* **165**, 201 (1968).
87. D. Chandler and J. D. Weeks, *Phys. Rev. Lett.* **25**, 149 (1970).
88. J. E. Enderby, T. Gaskell and N. H. March, *Proc. Phys. Soc.* **85**, 217 (1965).
89. C. C. Matthai and N. H. March, *Phys. Chem. Liquids* **11**, 207 (1982).
90. N. Kumar, N. H. March and H. Wasserman, *Phys. Chem. Liquids* **11**, 271 (1982).
91. G. Senatore and N. H. March, *Phys. Chem. Liquids* **13**, 285 (1984).
92. See e.g. J. P. Hansen and I. McDonald, *Theory of Simple Liquids* (Academic, London, 1976).
93. T. Morita, *Progr. Theor. Phys.* **23**, 829 (1960).
94. Y. Rosenfeld and N. W. Ashcroft, *Phys. Rev.* **A20**, 1208 (1979).
95. R. G. Chapman and N. H. March, *Phys. Chem. Liquids* **16**, 77 (1986) and **17**, 165 (1987).
96. N. H. March and M. P. Tosi, *Phys. Chem. Liquids* **37**, 463 (1999).
97. N. H. March, F. Perrot and M. P. Tosi, *Molec. Phys.* **93**, 355 (1998).
98. See e.g. R. Winter, W. C. Pilgrim and F. Hensel, *J. Phys.: Condens. Matter* **6**, A245 (1994).
99. N. H. March and M. P. Tosi, *Phys. Chem. Liquids* **29**, 197 (1995).
100. M. E. Fisher, *Rep. Progr. Phys.* **30**, 615 (1967); C. Domb and M. S. Green (Eds.),
Phase Transitions and Critical Phenomena, Vols. 1–20 (Academic, London, 1972–2001).
101. B. Widom, *J. Chem. Phys.* **43**, 3892 (1965); L. P. Kadanoff, *Physics* **2**, 263 (1966).
102. L. Onsager, *Phys. Rev.* **65**, 117 (1944); C. N. Yang, *Phys. Rev.* **85**, 808 (1952); T. T. Wu, *Phys. Rev.* **149**, 380 (1966).
103. G. Orkoulas, M. E. Fisher and A. Z. Panagiotopoulos, *Phys. Rev.* **E63**, 051507 (2001).
104. J. S. Lin and P. W. Schmidt, *Phys. Rev.* **A10**, 2290 (1974).
105. M. Mezard, G. Parisi and M. A. Virasoro, *Spin Glass Theory and Beyond* (World Scientific, Singapore, 1987).
106. W. G. Madden and E. D. Glandt, *J. Stat. Phys.* **51**, 537 (1988); J. A. Given and G. Stell, in *Condensed Matter Theories*, eds. L. Blum and F. Bary Malik (Plenum, New York, 1993), p. 395; M. P. Tosi, *N. Cimento* **D16**, 169 (1994).
107. M. L. Rosinberg, in *New Approaches to Problems in Liquid State Theory*, eds. C. Caccamo, J.-P. Hansen and G. Stell (Kluwer, Dordrecht, 1999), p. 245.
108. F. Brochard and P. G. de Gennes, *J. Phys. Lett.* **44**, 785 (1983); P. G. de Gennes, *J. Phys. Chem.* **88**, 6469 (1984); D. Andelman and J. F. Joanny, in *Scaling Phenomena in Disordered Systems*, eds. R. Pynn and A. Skjeltrop (Plenum, New York, 1985).

109. A. J. Liu, D. J. Durian, E. Herbolzheimer and S. A. Safran, *Phys. Rev. Lett.* **65**, 1897 (1990); L. Monette, A. J. Liu and G. S. Grest, *Phys. Rev.* **A46**, 7664 (1992).
110. See e.g. L. B. Loeb, *Kinetic Theory of Gases* (McGraw-Hill, New York, 1934).
111. E. L. Cussler, *Diffusion: Mass Transfer in Fluid Systems* (Cambridge University Press, Cambridge, 1997).
112. M. v. Smoluchowski, *Phys. Z.* **17**, 557 and 585 (1916); *Z. Physik. Chem.* **92**, 129 (1917).
113. S. Chandrasekhar, *Rev. Mod. Phys.* **15**, 1 (1943).
114. D. F. Calef and J. M. Deutsch, *Amer. Rev. Phys. Chem.* **34**, 493 (1983).
115. J. K. Baird, J. S. McCaskill and N. H. March, *J. Chem. Phys.* **74**, 6812 (1981).
116. D. R. Lide (Ed.), *CRC Handbook of Chemistry and Physics* (CRC, Boca Raton, 1999).
117. E. G. Scheibel, *Ind. Eng. Chem.* **46**, 2007 (1954).
118. C. R. Wilke and P. C. Chang, *Amer. Inst. Chem. Eng. J.* **1**, 264 (1955).
119. C. A. Angell, in *Water — a Comprehensive Treatise*, Vol. 7, ed. F. Franks (Plenum, New York, 1992), p. 1.
120. L. van Hove, *Phys. Rev.* **95**, 249 (1954).
121. W. C. Marshall and S. L. Lovesey, *Thermal Neutron Scattering* (Oxford University Press, Oxford, 1973).
122. R. Kubo, *Rep. Progr. Phys.* **29**, 255 (1966).
123. See e.g. V. Arrighi and M. T. F. Telling (Eds.), *Quasi-Elastic Neutron Scattering*, *Physica* **B301**, 1 (2001).
124. See e.g. M. Giordano, D. Leporini and M. P. Tosi (Eds.), *Non-Equilibrium Phenomena in Supercooled Fluids, Glasses and Amorphous Materials*, *J. Phys.: Condens. Matter* **11**, A1 (1998).
125. J. S. Vrentas and J. L. Duda, *J. Polymer Sci.* **17**, 1085 (1979).
126. M. Tirrell, *Rubber Chem. Techn.* **57**, 523 (1984).
127. See e.g. A. Abragam, *The Principles of Magnetic Resonance* (Oxford University Press, Oxford, 1960).
128. Y. Nakamura, S. Shimokawa, F. Futamata and M. Shimoji, *J. Chem. Phys.* **77**, 3258 (1982).
129. A. J. Dianoux, in *NATO–ASI on the Physics and Chemistry of Aqueous Ionic Solutions*, eds. M.-C. Bellissent-Funel and G. W. Neilson (Reidel, Dordrecht, 1986).
130. A. Rahman and A. Paskin, *Phys. Rev. Lett.* **16**, 300 (1966).
131. R. J. Speedy, *Molec. Phys.* **62**, 509 (1987).
132. S. Chapman and T. G. Cowling, *The Mathematical Theory of Non-uniform Gases* (Cambridge University Press, Cambridge, 1970).
133. J. J. Erpenkeck and W. W. Wood, *Phys. Rev.* **A43**, 4254 (1991).
134. D. M. Heyes and J. G. Powles, *Molec. Phys.* **95**, 259 (1998).
135. B. R. A. Nijboer and A. Rahman, *Physica* **32**, 415 (1966).
136. J. R. D. Copley and S. L. Lovesey, *Rep. Progr. Phys.* **38**, 461 (1975).

137. M. H. Ernst, E. H. Hauge and J. M. J. van Leeuwen, *Phys. Rev. Lett.* **25**, 1254 (1970).
138. B. J. Alder and T. E. Wainwright, *Phys. Rev.* **A1**, 18 (1970).
139. T. Gaskell and N. H. March, *Phys. Lett.* **A33**, 460 (1970).
140. Y. Pomeau, *Phys. Rev.* **A5**, 2569 (1972); D. Levesque and W. T. Ashurst, *Phys. Rev. Lett.* **33**, 277 (1974).
141. H. B. Callen and E. A. Welton, *Phys. Rev.* **83**, 34 (1951).
142. A. Rahman, K. S. Singwi and A. Sjölander, *Phys. Rev.* **126**, 986 (1962).
143. F. H. Stillinger and T. A. Weber, *Phys. Rev.* **A25**, 978 (1982); *ibid.* **A28**, 2408 (1983).
144. R. Zwanzig, *J. Chem. Phys.* **79**, 2408 (1983).
145. K. Tankeshwar, B. Singla and K. N. Pathak, *J. Phys.: Condens. Matter* **3**, 3173 (1991).
146. R. C. Brown and N. H. March, *Phys. Chem. Liquids* **1**, 141 (1968).
147. D. C. Wallace, *Phys. Rev.* **E58**, 538 (1998).
148. D. C. Wallace, *Phys. Rev.* **E60**, 7049 (1999).
149. N. H. March and M. P. Tosi, *Phys. Rev.* **E60**, 2402 (1999).
150. E. N. da C. Andrade, *Phil. Mag.* **17**, 497 and 698 (1934).
151. P. C. Martin, O. Parodi and P. S. Pershan, *Phys. Rev.* **A6**, 2401 (1972).
152. D. Forster, *Hydrodynamic Fluctuations, Broken Symmetry, and Correlation Functions* (Benjamin, London, 1975).
153. J. Kestin and W. A. Wakeham, *Transport Properties of Fluids: Thermal Conductivity, Viscosity and Diffusion Coefficient* (Hemisphere Publishing, New York, 1988).
154. D. M. Heyes and N. H. March, *Phys. Chem. Liquids* **28**, 1 (1994).
155. J. H. Irving and J. G. Kirkwood, *J. Chem. Phys.* **18**, 817 (1950).
156. F. C. Collins and H. Raffel, *J. Chem. Phys.* **22**, 1728 (1954).
157. H. C. Longuet-Higgins and J. A. Pople, *J. Chem. Phys.* **25**, 884 (1956).
158. J. H. Ferziger and H. G. Kaper, *Mathematical Theory of Transport Processes in Gases* (North-Holland, Amsterdam, 1972).
159. J. H. Dymond and T. A. Brawn, in *Proc. 7th Symp. on Thermophysical Properties* (ASME, New York, 1997), p. 660.
160. R. C. Brown and N. H. March, in *Amorphous Materials*, eds. R. W. Douglas and B. Ellis (Wiley, London, 1972), p. 187.
161. H. S. Green, *Molecular Theory of Fluids* (North-Holland, Amsterdam, 1952).
162. W. Menz and F. Sauerwald, *Acta Met.* **14**, 1617 (1966).
163. J. Levesque, L. Verlet and J. Kurkijarvi, *Phys. Rev.* **A7**, 1690 (1973).
164. A. Rahman, *Phys. Rev.* **136**, 405 (1964).
165. D. Levesque and L. Verlet, *Molec. Phys.* **61**, 143 (1987); K. D. Hammonds and D. M. Heyes, *JCS Faraday Trans. II* **84**, 705 (1988); P. Borgelt, C. Hoheisel and G. Stell, *Phys. Rev.* **A42**, 789 (1990).
166. M. Ferrario, G. Ciccotti, B. L. Holian and J.-P. Rychaert, *Phys. Rev.* **A44**, 6936 (1991).
167. L. van Hove, *Phys. Rev.* **95**, 249 (1954).

168. S. W. Lovesey, *Theory of Neutron Scattering from Condensed Matter* (Clarendon Press, Oxford, 1984).
169. P. A. Egelstaff, *An Introduction to the Liquid State* (Clarendon Press, Oxford, 1992).
170. J. R. D. Copley and J. M. Rowe, *Phys. Rev. Lett.* **32**, 49 (1974).
171. F. R. Trouw and D. L. Price, *Annu. Rev. Phys. Chem.* **50**, 571 (1999).
172. B. J. Berne and R. Pecora, *Dynamic Light Scattering* (Wiley, New York, 1976).
173. J. P. McTague, P. A. Fleury and D. B. Dupré, *Phys. Rev.* **188**, 303 (1969).
174. F. Sette, G. Ruocco, M. Krisch, G. Masciovecchio and G. Monaco, *Science* **280**, 1550 (1998) and references given therein.
175. T. Scopigno, U. Balucani, G. Ruocco and F. Sette, cond-mat/0001190.
176. A. Rahman and F. H. Stillinger, *Phys. Rev.* **A10**, 368 (1974).
177. F. Sette, G. Ruocco, M. Krisch, U. Bergmann, C. Masciovecchio, V. Mazzacurati, G. Signorelli and R. Verbeni, *Phys. Rev. Lett.* **75**, 850 (1995); F. Sette, G. Ruocco, M. Krisch, C. Masciovecchio, R. Verbeni and U. Bergmann, *Phys. Rev. Lett.* **77**, 83 (1996).
178. V. Tozzini and M. P. Tosi, *Phys. Chem. Liq.* **33**, 191 (1996).
179. J. H. Dymond, *Physica* **75**, 100 (1974).
180. R. S. Basu and J. M. H. Levelt Sengers, *NASA Contractor Rept.* 3424 (1981).
181. Y. Rosenfeld, *Phys. Rev.* **A15**, 2545 (1977) and *Chem. Phys. Lett.* **48**, 467 (1977); W. G. Hoover, R. Glover and B. Moran, *J. Chem. Phys.* **83**, 1255 (1985); Y. Rosenfeld, *Phys. Rev.* **E62**, 7524 (2000).
182. D. M. Heyes and N. H. March, *Phys. Chem. Liq.* **33**, 65 (1996).
183. C. Hoheisel, R. Vogelsang and M. Luckas, *Molec. Phys.* **64**, 1203 (1988).
184. C. Hoheisel, *J. Chem. Phys.* **89**, 3195 (1988).
185. W. T. Ashurst and W. G. Hoover, *Phys. Rev.* **A11**, 658 (1974).
186. J. A. Given and E. Clementi, *J. Chem. Phys.* **90**, 7376 (1989).
187. D. J. Evans and G. P. Morriss, *Statistical Mechanics of Nonequilibrium Liquids* (Academic, London, 1990).
188. B. Hafskjold, F. Bresme and I. Wold, *J. Phys. Chem.* **100**, 1879 (1996).
189. D. J. Evans, *Phys. Rev.* **A91**, 457 (1982).
190. M. J. Gillan and M. Dixon, *J. Phys.* **C16**, 869 (1983).
191. G. Jacucci, G. Ciccotti and I. R. MacDonald, *J. Phys.* **C11**, L509 (1978).
192. G. P. Morriss and D. J. Evans, *Phys. Rev.* **A38**, 4142 (1988).
193. P. J. Daivis and D. J. Evans, *Phys. Rev.* **E48**, 1058 (1993).
194. N. H. March, *Liquid Metals, Concepts and Theory* (Cambridge University Press, Cambridge, 1990).
195. M. J. Rice, *Phys. Rev.* **B2**, 4800 (1970).
196. K. Tankeshwar and N. H. March, *Phys. Chem. Liq.* **31**, 39 (1996).
197. M. P. Tosi, *N. Cimento* **D14**, 559 (1992).
198. N. H. March and M. P. Tosi, *J. Plasma Phys.* **57**, 121 (1997).
199. P. M. Chaikin and T. C. Lubensky, *Principles of Condensed Matter Physics* (Cambridge University Press, Cambridge, 1995).
200. L. D. Landau and G. Placzek, *Phys. Z. Sowjetunion* **5**, 172 (1934).

201. R. D. Mountain, *Rev. Mod. Phys.* **38**, 205 (1966).
202. L. I. Komarov and I. Z. Fisher, *Sov. Phys. JETP* **16**, 1358 (1963).
203. J. Zollweg, G. Hawkins and G. B. Benedek, *Phys. Rev. Lett.* **27**, 1182 (1971).
204. R. F. Berg, M. R. Moldover and G. A. Zimmerli, *Phys. Rev. Lett.* **82**, 920 (1999) and *Phys. Rev.* **E60**, 4079 (1999).
205. L. P. Kadanoff and J. Swift, *Phys. Rev.* **166**, 89 (1968).
206. K. Kawasaki, *Phys. Rev.* **A1**, 1750 (1970).
207. D. S. Cannell and D. Sarid, *Phys. Rev.* **A10**, 2280 (1974).
208. P. C. Hohenberg and B. I. Halperin, *Rev. Mod. Phys.* **46**, 435 (1977).
209. R. Folk and G. Moser, *Phys. Rev.* **E57**, 683 and 705 (1998).
210. D. MacGowan and D. J. Evans, *Phys. Rev.* **A34**, 2133 (1986); *ibid.* **A36**, 948 (1987).
211. D. MacGowan, *Molec. Phys.* **59**, 1017 (1986).
212. D. MacGowan and D. J. Evans, *Phys. Lett.* **A117**, 414 (1986).
213. A. B. Bhatia and D. E. Thornton, *Phys. Rev.* **B2**, 3004 (1970).
214. A. B. Bhatia, D. E. Thornton and N. H. March, *Phys. Chem. Liq.* **4**, 93 (1974).
215. For a broad introduction to the subject see K. R. Atkins, *Liquid Helium* (Cambridge University Press, Cambridge, 1959). The parallelisms between superfluidity in He-II and superconductivity in metals are reviewed by D. R. Tilley and J. Tilley, *Superfluidity and Superconductivity* (Van Nostrand, New York, 1974).
216. L. Tisza, *Nature* **141**, 913 (1938).
217. E. L. Andronikashvili, *JETP* **16**, 780 (1946).
218. L. D. Landau, *J. Phys. Moscow* **5**, 71 (1941) and **11**, 91 (1947) [reprinted in Ref. 219].
219. I. M. Khalatnikov, *Introduction to the Theory of Superfluidity*, transl. P. C. Hohenberg (Benjamin, New York, 1965).
220. P. C. Hohenberg and P. C. Martin, *Ann. Phys. (NY)* **34**, 291 (1965).
221. M. L. Chiofalo, A. Minguzzi and M. P. Tosi, *Physica* **B254**, 188 (1998).
222. P. C. Hohenberg and P. M. Platzman, *Phys. Rev.* **152**, 198 (1966).
223. P. Sokol, *Bose–Einstein Condensation*, eds. A. Griffin, D. W. Snoke and S. Stringari (Cambridge University Press, Cambridge, 1995), p. 51.
224. R. A. Cowley and A. D. B. Woods, *Can. J. Phys.* **49**, 177 (1971); A. D. B. Woods and R. A. Cowley, *Rep. Progr. Phys.* **36**, 1135 (1973).
225. For a general survey of dense fluids of charged particles, see the book of N. H. March and M. P. Tosi, *Coulomb Liquids* (Academic, London, 1984).
226. For a review of the OCP see M. Baus and J. P. Hansen, *Phys. Rept.* **59**, 1 (1980).
227. P. Debye and E. Hückel, *Z. Phys.* **24**, 185 (1923).
228. G. Gouy, *J. Chim. Phys.* **29**, 145 (1903); *J. Phys.* **9**, 457 (1910).
229. D. L. Chapman, *Phil. Mag.* **25**, 475 (1913).
230. S. G. Brush, H. L. Sahlin and E. Teller, *J. Chem. Phys.* **45**, 2102 (1996).
231. M. Gillan, *J. Phys.* **C7**, L1 (1974).
232. J. P. Hansen, L. P. Pollock and I. R. McDonald, *Phys. Rev. Lett.* **32**, 277 (1974).

233. M. Rovere and M. P. Tosi, *Rep. Prog. Phys.* **49**, 1001 (1986); M. P. Tosi, D. L. Price and M.-L. Saboungi, *Annu. Rev. Phys. Chem.* **44**, 173 (1993).
234. D. G. Pettifor, *J. Phys.* **C19**, 285 (1986).
235. J. L. Tallon and W. H. Robinson, *Phys. Lett.* **A87**, 365 (1982).
236. Z. Akdeniz and M. P. Tosi, *Proc. R. Soc. London* **A437**, 85 (1992).
237. K. D. Jordan, in *Alkali Halide Vapors: Structure, Spectra and Reaction Dynamics* (Academic, New York, 1979), p. 479; M. P. Tosi and M. Doyama, *Phys. Rev.* **160**, 716 (1967); P. Brumer and M. Karplus, *J. Chem. Phys.* **58**, 3903 (1973).
238. M. E. Fisher, in *New Approaches to Problems in Liquid State Theory*, eds. C. Caccamo *et al.* (Kluwer, Dordrecht, 1999), p. 3.
239. N. F. Mott, *Metal–Insulator Transitions* (Taylor and Francis, London, 1974).
240. M. E. Fisher and D. M. Zuckerman, *J. Chem. Phys.* **109**, 7961 (1998).
241. J. E. Enderby, D. M. North and P. A. Egelstaff, *Phil. Mag.* **14**, 961 (1966).
242. D. I. Page and K. Mika, *J. Phys.* **C4**, 3034 (1971).
243. See e.g. A. Di Cicco, M. Minicucci and A. Filipponi, *Phys. Rev. Lett.* **78**, 460 (1997).
244. A. B. Bhatia and D. E. Thornton, *Phys. Rev.* **B2**, 3004 (1970).
245. S. Biggin and J. E. Enderby, *J. Phys.* **C15**, L305 (1982).
246. P. Ballone, G. Pastore and M. P. Tosi, *J. Chem. Phys.* **81**, 3174 (1984).
247. A. Waisman and J. L. Lebowitz, *J. Chem. Phys.* **56**, 3086 (1972).
248. K. R. Painter, P. Ballone, M. P. Tosi, P. J. Grout and N. H. March, *Surf. Sci.* **133**, 89 (1983).
249. F. G. Edwards, J. E. Enderby, R. A. Howe and D. I. Page, *J. Phys.* **C11**, 1053 (1978); R. L. McGreevy and E. W. J. Mitchell, *J. Phys.* **C15**, 5537 (1982).
250. B. D'Aguanno, M. Rovere, M. P. Tosi and N. H. March, *Phys. Chem. Liq.* **13**, 113 (1983); M. Rovere and M. P. Tosi, *Solid State Commun.* **55**, 1109 (1985).
251. S. Biggin and J. E. Enderby, *J. Phys.* **C14**, 3129 (1981).
252. J. A. E. Desa, A. C. Wright, J. Wong and R. N. Sinclair, *J. Non-Cryst. Solids* **51**, 57 (1982).
253. J. Mochinaga, Y. Iwadate and K. Fukushima, *Mat. Sci. Forum* **73**, 147 (1991).
254. M.-L. Saboungi, D. L. Price, C. Scamehorn and M. P. Tosi, *Europhys. Lett.* **15**, 283 (1991).
255. G. N. Papatheodorou, *J. Chem. Phys.* **66**, 2893 (1977).
256. Y. S. Badyal, M.-L. Saboungi, D. L. Price, D. R. Haeffner and S. D. Shastri, *Europhys. Lett.* **39**, 19 (1997).
257. D. L. Price, S. C. Moss, R. Reijers, M.-L. Saboungi and S. Susman, *J. Phys.: Condens. Matter* **1**, 1005 (1989).
258. M. Blander, E. Bierwagen, K. G. Calkins, L. A. Curtiss, D. L. Price and M.-L. Saboungi, *J. Chem. Phys.* **97**, 2733 (1992).
259. K. Grjotheim, C. Krohn, M. Malinovsky, K. Matiasovsky and J. Thonstad, *Aluminium Electrolysis* (Aluminium-Verlag, Dusseldorf, 1982).
260. B. Gilbert, G. Mamantov and G. M. Begun, *J. Chem. Phys.* **62**, 950 (1975); B. Gilbert and T. Materne, *Appl. Spectrosc.* **44**, 299 (1990).

261. R. Triolo and A. H. Narten, *J. Chem. Phys.* **69**, 3159 (1978); E. Johnson, A. H. Narten, W. E. Thiessen and R. Triolo, *Faraday Disc. Chem. Soc.* **1978**, 287 (1978).
262. M.-L. Saboungi, M. A. Howe and D. L. Price, *Molec. Phys.* **79**, 847 (1993).
263. J. O'M. Bockris, S. Nanis and N. E. Richards, *J. Phys. Chem.* **69**, 1627 (1965).
264. J. Trullas, A. Girò and M. Silbert, *J. Phys.: Condens. Matter* **2**, 6643 (1990).
265. K. Tankeshwar and M. P. Tosi, *J. Phys.: Condens. Matter* **3**, 7511 (1991).
266. J. Périé, M. Chemla and M. Gignoux, *Bull. Soc. Chim.* **1961**, 1249 (1961).
267. G. Ciccotti, G. Jacucci and I. R. McDonald, *Phys. Rev.* **A13**, 426 (1976).
268. C.-A. Sjöblom and A. Behn, *Z. Naturforsch.* **23a**, 495 (1968).
269. J. O. Hirschfelder, C. E. Curtiss and R. B. Bird, *Molecular Theory of Gases and Liquids* (Wiley, New York, 1964); see also E. Helfand and A. J. Rice, *J. Chem. Phys.* **32**, 1642 (1960).
270. Y. Abe and A. Nagashima, *J. Chem. Phys.* **75**, 3977 (1981).
271. A. Voronel, E. Veliyulin, T. Grande and H. A. Øye, *J. Phys.: Condens. Matter* **9**, L247 (1997).
272. A. Voronel, E. Veliyulin, V. Sh. Machvariani, A. Kisliuk and D. Quitmann, *Phys. Rev. Lett.* **80**, 2630 (1998).
273. M. H. Brooker and G. Papatheodorou, in *Adv. Molten Salt Chem.* vol. 5, ed. G. Mamantov (Elsevier, Amsterdam, 1983).
274. F. Hensel, *Adv. Phys.* **28**, 555 (1979).
275. H. Ruppersberg and H. Reiter, *J. Phys.* **F12**, 1311 (1972).
276. J. A. Meijer, W. Geertsma and W. van der Lugt, *J. Phys.* **F15**, 899 (1985).
277. W. Martin, W. Freyland, P. Lamparter and S. Steeb, *Phys. Chem. Liquids* **10**, 61 (1980).
278. M.-L. Saboungi, J. Fortner, W. S. Howells and D. L. Price, *Nature* **365**, 237 (1993).
279. R. Bini, M. Jordan, L. Ulivi and H. J. Jodl, *J. Chem. Phys.* **108**, 6849 (1998).
280. J. N. Sherwood, *The Plastically Crystalline State* (Wiley, New York, 1979).
281. V. Tozzini, N. H. March and M. P. Tosi, *Phys. Chem. Liquids* **37**, 185 (1999).
282. J. A. Pople and K. E. Karasz, *J. Phys. Chem. Solids* **18**, 28 and **20**, 1295 (1961).
283. J. E. Lennard-Jones and A. F. Devonshire, *Proc. Roy. Soc.* **A169**, 317 and **170**, 464 (1939).
284. A. Ferraz and N. H. March, *Phys. Chem. Liquids* **8**, 289 (1979).
285. J. N. Ghosli and F. H. Ree, *Phys. Rev. Lett.* **82**, 4659 (1999).
286. M. van Thiel and F. H. Ree, *Phys. Rev.* **B48**, 3591 (1993); *J. Appl. Phys.* **77**, 4805 (1995).
287. M. Togaya, *Phys. Rev. Lett.* **79**, 2474 (1997).
288. V. V. Brazhkin, S. V. Popova and R. N. Voloshin, *Physica* **B265**, 64 (1999).
289. H. Endo, K. Tamura and M. Yao, *Can. J. Phys.* **65**, 266 (1987).
290. K. Tsuji, O. Shimomura and K. Tamura, *J. Phys. Chem.* **156**, 495 (1988); K. Tsuji, *J. Non-Cryst. Sol.* **117/118**, 27 (1990).
291. B. Meyer, *Elemental Sulphur* (Interscience, New York, 1965).

292. V. V. Brazhkin, R. N. Voloshin, S. V. Popova and A. G. Umnov, *Phys. Lett.* **A154**, 413 (1991).
293. R. Naslain, *Boron and Refractory Borides*, ed. V. I. Matkovich (Springer, New York, 1977).
294. S. Krishnan, S. Ansell, J. J. Felten, K. J. Volin and D. L. Price, *Phys. Rev. Lett.* **81**, 586 (1998).
295. A. K. Soper, *J. Chem. Phys.* **101**, 6888 (1994) and *J. Mol. Liquids* **78**, 179 (1998).
296. L. Blum and A. Torruella, *J. Chem. Phys.* **56**, 303 (1972).
297. C. G. Gray and K. E. Gubbins, *Theory of Molecular Liquids*, Vol. 1 (Oxford University Press, New York, 1984).
298. A. K. Soper, F. Bruni and M. A. Ricci, *J. Chem. Phys.* **106**, 247 (1997).
299. O. Mishima and E. Stanley, *Nature* **392**, 164 (1998); *ibid.* **396**, 329 (1990); see also H. E. Stanley, S. V. Buldyrev, O. Mishima, M. R. Sadr-Lahijany, A. Scala and F. W. Starr, *J. Phys.: Condens. Matter* **12**, A403 (2000).
300. M. Pichal and O. Sifner, *Properties and Water and Steam* (Hemisphere, New York, 1989).
301. Y. S. Badyal, M.-L. Saboungi, D. L. Price, S. D. Shastri, D. R. Haeffner and A. K. Soper, *J. Chem. Phys.* **112**, 9206 (2000).
302. P. G. de Gennes, *Scaling Concepts in Polymer Physics* (Cornell University Press, Ithaca, 1979); M. Doi and S. F. Edwards, *The Theory of Polymer Dynamics* (Clarendon, Oxford, 1986); J. des Cloiseaux and G. Jannink, *Polymers in Solution: their Modelling and Structure* (Clarendon, Oxford, 1990); M. Doi, *Introduction to Polymer Physics* (Clarendon, Oxford, 1996).
303. P. J. Flory, *Principles of Polymer Chemistry* (Cornell University Press, Ithaca, 1953).
304. A. Keller, *Rep. Progr. Phys.* **31**, 623 (1968); A. Keller, in *Polymers, Liquid Crystals and Low-Dimensional Solids*, eds. N. H. March and M. P. Tosi (Plenum, New York, 1984).
305. F. C. Frank and M. P. Tosi, *Proc. Roy. Soc.* **A263**, 323 (1961).
306. P. G. de Gennes, *The Physics of Liquid Crystals* (Oxford University Press, Oxford, 1974); S. Chandrasekhar, *Liquid Crystals* (Cambridge University Press, Cambridge, 1977); P. S. Pershan, *Structure of Liquid Crystal Phases* (World Scientific, Singapore, 1988); I.-C. Khoo, *Liquid Crystals* (Wiley, New York, 1995).
307. For this and other details on the statistical mechanics of liquid crystals, the reader may refer to D. Frenkel, in *Liquids, Freezing and Glass Transition*, eds. J.-P. Hansen, D. Levesque and J. Zinn-Justin (North-Holland, Amsterdam, 1991), p. 689.
308. L. D. Landau and E. M. Lifshitz, *Statistical Physics* (Pergamon, London, 1980).
309. R. Eppenga and D. Frenkel, *Mol. Phys.* **52**, 1303 (1984).
310. A. Matsuyama and T. Kato, *Phys. Rev.* **E59**, 763 (1999).

311. A. ten Bosch, P. Maissa and P. Sixou, *J. Chem. Phys.* **79**, 3462 (1983); M. Warner, J. M. F. Gunn and A. Baumgartner, *J. Phys.* **A18**, 3007 (1985); X. J. Wang and M. Warner, *J. Phys.* **A19**, 2215 (1986); P. Maissa and P. Sixou, *Liq. Cryst.* **5**, 1861 (1989).
312. W. Maier and A. Saupe, *Z. Naturforsch.* **13a**, 564 (1958).
313. T. J. Sluckin and P. Shukla, *J. Phys.* **A16**, 1539 (1983).
314. R. B. Meyer, in *Molecular Fluids: Les Houches 1973* (Gordon and Breach, New York, 1973) p. 273.
315. P. G. de Gennes, *Solid State Commun.* **10**, 753 (1972); *Mol. Cryst. Liq. Cryst.* **21**, 49 (1973).
316. K. K. Kobayashi, *J. Phys. Soc. Jpn.* **29**, 101 (1970); *Mol. Cryst. Liq. Cryst.* **13**, 137 (1971); W. L. McMillan, *Phys. Rev.* **A4**, 1238 (1971).
317. R. B. Meyer and T. G. Lubensky, *Phys. Rev.* **A14**, 2307 (1976).
318. S. Hess, *Physica* **A267**, 58 (1999).
319. S. Hess and B. Su, *Z. Naturforsch.* **54a**, 559 (1999).
320. L. Onsager, *Phys. Rev.* **62**, 558 (1942) and *Ann. N. Y. Acad. Sci.* **51**, 627 (1949).
321. M. A. Colter, *J. Chem. Phys.* **67**, 4268 (1976).
322. T. Gruhn and M. Schoen, *Phys. Rev.* **E55**, 2861 (1997); *Mol. Phys.* **93**, 681 (1998); *J. Chem. Phys.* **108**, 9124 (1998).
323. S. Hess, M. Kröger, W. Loose, C. Pereira Bergmaier, R. Schramek, H. Voigt and T. Weider, in *Monte Carlo and Molecular Dynamics of Condensed Matter Systems*, eds. K. Binder and G. Ciccotti (IPS, Bologna, 1996), p. 825.
324. M. A. Osipov and S. Hess, *J. Chem. Phys.* **99**, 4181 (1993).
325. See e.g. A. E. Owen, in *Amorphous Solids and the Liquid State*, eds. N. H. March, R. A. Street and M. Tosi (Plenum, New York, 1985), p. 395.
326. C. A. Angell, *Science* **267**, 1924 (1995).
327. H. Vogel, *J. Phys.* **22**, 645 (1921); G. S. Fulcher, *J. Am. Ceram. Soc.* **8**, 339 (1925); G. Tammann and W. Z. Hesse, *Anorg. Allgem. Chem.* **156**, 245 (1928).
328. C. A. Angell, *J. Phys. Chem. Solids* **49**, 863 (1988).
329. W. Kauzmann, *Chem. Rev.* **43**, 219 (1948).
330. J. Jäckle, *Rep. Progr. Phys.* **49**, 171 (1986).
331. D. Turnbull and J. C. Fisher, *J. Chem. Phys.* **17**, 71 (1949).
332. F. Franks, in *Water — a Comprehensive Treatise*, Vol. 7, ed. F. Franks (Plenum, New York, 1982), p. 215.
333. D. Turnbull, in *Reviews of Solid State Science*, Vol. 3 (World Scientific, Singapore, 1989), p. 291.
334. P. R. ten Wolde and D. Frankel, *Science* **277**, 1975 (1997).
335. P. T. Sarjeant and R. Roy, *Mater. Res. Bull.* **3**, 265 (1968).
336. F. F. Abraham, *Homogeneous Nucleation Theory* (Academic, New York, 1974).
337. D. L. Price, *Current Opinion in Solid State and Material Science* **1**, 572 (1996).
338. D. L. Price and M.-L. Saboungi, in *Local Structure from Diffraction*, eds. S. J. L. Billinge and M. F. Thorpe (Plenum, New York, 1998), p. 23.

339. W. H. Zachariasen, *J. Am. Chem. Soc.* **54**, 3841 (1932).
340. S. C. Moss and D. L. Price, in *Physics of Disordered Materials*, eds. D. Alder, H. Fritzsche and S. R. Ovshinsky (Plenum, New York, 1985), p. 77; P. S. Salmon, *Proc. R. Soc. London* **A445**, 351 (1994).
341. S. Susman, K. J. Volin, D. L. Price, M. Grimsditch, J. P. Rino, R. K. Kalia and P. Vashishta, *Phys. Rev.* **B43**, 1994 (1991).
342. V. N. Novikov and A. P. Sokolov, *Solid State Commun.* **77**, 243 (1991).
343. R. J. Nemanich, *Phys. Rev.* **B16**, 1674 (1977).
344. U. Buchenau, M. Prager, N. Nücker, A. J. Dianoux, N. Ahmad and W. A. Phillips, *Phys. Rev.* **B34**, 5665 (1986).
345. A. C. Wright, A. G. Clare, B. Bachra, R. N. Sinclair, A. C. Hannon and B. Vessal, *Trans. Am. Cryst. Assoc.* **27**, 239 (1991).
346. Z. Badirkhan, M. Rovere and M. P. Tosi, *Phil. Mag.* **B65**, 921 (1992).
347. B. Stenhouse, P. J. Grout, N. H. March and J. Wenzel, *Phil. Mag.* **36**, 129 (1977).
348. I. Prigogine and R. Defay, *Chemical Thermodynamics* (Longman, London, 1954).
349. T. M. Nieuwenhuizen, *J. Phys.: Condens. Matter* **12**, 6543 (2000).
350. J. H. Gibbs and E. A. DiMarzio, *J. Chem. Phys.* **28**, 373 (1958).
351. A. B. Pippard, *The Elements of Classical Thermodynamics* (Cambridge University Press, Cambridge, 1957).
352. R. O. Davies and G. O. Jones, *Adv. Phys.* **2**, 370 (1953).
353. M. Goldstein, *J. Chem. Phys.* **39**, 3369 (1963); *ibid.* **51**, 3728 (1969); *ibid.* **67**, 2246 (1977).
354. P. W. Anderson, *Ill-Condensed Matter*, eds. R. Balian, R. Maynard and G. Toulouse (North-Holland, Amsterdam, 1979), p. 161.
355. F. H. Stillinger, *Science* **267**, 1935 (1995).
356. W. Götze and L. Sjögren, *Rep. Progr. Phys.* **55**, 241 (1992).
357. W. Götze, *J. Phys.: Condens. Matter* **11**, A1 (1999); H. Z. Cummins, *J. Phys.: Condens. Matter* **11**, A95 (1999).
358. See e.g. D. Perera and P. Harrowell, *Phys. Rev. Lett.* **81**, 120 (1988); M. D. Ediger, C. A. Angell and S. R. Nagel, *J. Phys. Chem.* **100**, 13200 (1996).
359. R. Yamamoto and A. Onuki, *Phys. Rev. Lett.* **81**, 4915 (1998).
360. C. T. Moynihan, S. N. Crichton and S. M. Opalka, *J. Non-Cryst. Solids* **131–133**, 420 (1991).
361. I. M. Hodge, *Science* **267**, 1945 (1995).
362. S. Torquato, *Phys. Rev.* **E51**, 3170 (1995).
363. M. D. Rintoul and S. Torquato, *Phys. Rev.* **E58**, 532 (1998).
364. A. Lindsay Greer, *Science* **267**, 1947 (1995).
365. C. A. Angell, *Solid State Ionics* **105**, 15 (1998).
366. A. Bunde, K. Funke and M. D. Ingram, *Solid State Ionics* **105**, 1 (1998).
367. B. Roling, A. Happe, K. Funke and M. D. Ingram, *Phys. Rev. Lett.* **78**, 2160 (1997).

368. B. Frick and D. Richter, *Science* **267**, 1939 (1995).
369. S. Ichimaru, *Basic Principles of Plasma Physics* (Benjamin, Reading, 1973).
370. See e.g. P. G. de Gennes, *The Physics of Liquid Crystals* (Clarenden, Oxford, 1974).
371. P. G. de Gennes, *Scaling Concepts in Polymer Physics* (Cornell University Press, Ithaca, 1979).
372. M. Doi and S. F. Edwards, *The Theory of Polymer Dynamics* (Oxford University Press, New York, 1986).
373. R. F. Loring, *J. Chem. Phys.* **88**, 6631 (1988).
374. See e.g. G. Marrucci, in *Polymers, Liquid Crystals, and Low-Dimensional Solids*, eds. N. H. March and M. P. Tosi (Plenum, New York, 1984), p. 143.
375. P. E. Rouse, *J. Chem. Phys.* **21**, 1272 (1953); B. H. Zimm, *J. Chem. Phys.* **24**, 269 (1956).
376. See e.g. S. Chandrasekhar, in *Polymers, Liquid Crystals, and Low-Dimensional Solids*, eds. N. H. March and M. P. Tosi (Plenum, New York, 1984), p. 189.
377. F. C. Frank, *Disc. Faraday Soc.* **25**, 19 (1958).
378. I. Jànossy, *Optical Effects in Liquid Crystals* (Kluwer, Dordrecht, 1991).
379. G. Durand, in *Polymers, Liquid Crystals, and Low-Dimensional Solids*, eds. N. H. March and M. P. Tosi (Plenum, New York, 1984), p. 239.
380. J.-P. Hansen and P. N. Pusey, *Europhys. News* **30**, 81 (1999).
381. P. N. Pusey, in *Liquids, Freezing and Glass Transition*, eds. J.-P. Hansen, D. Levesque and J. Zinn-Justin (North-Holland, Amsterdam, 1991), p. 763.
382. G. N. Choi and I. M. Krieger, *J. Coll. Interface Sci.* **113**, 94 (1986).
383. R. E. Rosenweig, *Ferrohydrodynamics* (Cambridge University Press, Cambridge, 1985).
384. E. Blums, A. Cebers and M. Maiorov, *Magnetic Fluids* (de Gruyter, Berlin, 1997).
385. J. M. Rubì and J. M. G. Vilar, *J. Phys.: Condens. Matter* **12**, A75 (2000).
386. R. E. Rosenweig, *Science* **271**, 614 (1996).
387. F. Gazeau, C. Baravian, J. C. Bacri, R. Perzynsky and M. I. Shliomis, *Phys. Rev.* **E56**, 614 (1997).
388. See e.g. G. L. Gaines, *Insoluble Monolayers at Liquid–Gas Interface* (Wiley, New York, 1966).
389. G. A. Hawkins and G. B. Benedek, *Phys. Rev. Lett.* **32**, 524 (1974); M. W. Kim and D. S. Cannell, *Phys. Rev. Lett.* **35**, 889 (1975) and *Phys. Rev.* **A13**, 411 (1976).
390. W. M. Gelbart, A. Ben-Shaul and D. Roux, *Micelles, Membranes, Microemulsions and Monolayers* (Springer, New York, 1994).
391. R. Lipowski, *Vesicles and Biomembranes* in *Encyclopedia of Applied Physics* Vol. **23** (VCH Publishers, New York, 1998), p. 199.
392. D. Langevin, *Europhys. News* **30**, 72 (1999).
393. See e.g. W. Helfrich, *J. Phys.: Condens. Matter* **6**, A79 (1994).
394. P. Bradshaw, *An Introduction to Turbulence and its Measurement* (Pergamon, Oxford, 1971).

395. H. Tennekes and J. L. Lumley, *A First Course in Turbulence* (MIT, Cambridge, 1972).
396. D. Ruelle and F. Takens, *Commun. Math. Phys.* **20**, 167 and **23**, 343 (1971).
397. P. G. Drazin and W. H. Reid, *Hydrodynamic Stability* (Cambridge University Press, 1981).
398. M. Lesieur, *Turbulence in Fluids* (Kluwer, Dordrecht, 1997).
399. E. Guyon, J. Hulin and L. Petit, *Hydrodynamique Physique* (Inter-Editions, Paris, 1981).
400. G. K. Batchelor, *An Introduction to Fluid Dynamics* (Cambridge University Press, 1967).
401. S. A. Thorpe, *J. Fluid Mech.* **32**, 693 (1968).
402. A. Libchaber, C. Laroche and S. Fauve, *J. Phys. Lett.* **43**, L-211 (1982); A. Libchaber and J. Maurer, in *Nonlinear Phenomena at Phase Transitions and Instabilities*, ed. T. Riste (Plenum, New York, 1982), p. 259.
403. M. J. Feigenbaum, *Los Alamos Science* **1**, 4 (1980).
404. M. Giglio, S. Musazzi and U. Perini, *Phys. Rev. Lett.* **47**, 243 (1981).
405. P. Bergé, M. Dubois, P. Manneville and Y. Pomeau, *J. Phys. Lett.* **41**, L-341 (1980).
406. F. Heslot, B. Castaing and A. Libchaber, *Phys. Rev.* **A36**, 5870 (1987); B. Castaing, G. Gunaratne, F. Heslot, L. Kadanoff, A. Libchaber, S. Thomae, X.-Z. Wu, S. Zaleski and G. Zanetti, *J. Fluid Mech.* **204**, 1 (1989).
407. S. B. Pope, *Turbulent Flows* (Cambridge University Press, Cambridge, 2000).
408. W. D. McComb, *The Physics of Fluid Turbulence* (Clarendon, Oxford, 1990); T. Bohr, M. H. Jensen, G. Paladin and A. Vulpiani, *Dynamical Systems Approach to Turbulence* (Cambridge University Press, Cambridge, 1998).
409. A. N. Kolmogorov, *Dokl. Akad. Nauk SSSR* **30**, 301 and **31**, 538 (1941); *J. Fluid Mech.* **12**, 82 (1962).
410. H. L. Grant, R. W. Stewart and A. Moilliet, *J. Fluid Mech.* **12**, 241 (1962).
411. J.-D. Fournier and U. Frisch, *Phys. Rev.* **A28**, 1000 (1983).
412. D. Forster, D. R. Nelson and M. J. Stephen, *Phys. Rev.* **A16**, 732 (1977).
413. J.-D. Fournier and U. Frisch, *Phys. Rev.* **A17**, 747 (1978).
414. R. H. Kraichman, *J. Fluid Mech.* **5**, 497 (1959).
415. E. Vanden Eijnden, *Phys. Rev.* **E58**, 5229 (1998).
416. R. H. Kraichman, *J. Math. Phys.* **2**, 124 (1961).
417. G. K. Batchelor, *Proc. Camb. Phil. Soc.* **48**, 345 (1952); see also L. F. Richardson, *Proc. Roy. Soc.* **A110**, 709 (1926) and G. I. Taylor, *Proc. Roy. Soc.* **A164**, 15 (1938).
418. See e.g. M. Mareschal (Ed.), *The Microscopic Approach to Complexity in Non-Equilibrium Molecular Simulations*, *Physica* **A240**, 1 (1997).
419. U. Frisch, B. Hasslacher and Y. Pomeau, *Phys. Rev. Lett.* **56**, 1505 (1986); D. d'Humières, P. Lallemand and U. Frisch, *Europhys. Lett.* **2**, 291 (1986).
420. For reviews see R. Benzi, S. Succi and M. Vergassola, *Phys. Rep.* **222**, 145 (1992); S. Succi, G. Amati and R. Benzi, *J. Stat. Phys.* **81**, 5 (1995); Y. H. Qian, S. Succi and S. Orszag, *Ann. Rev. Comput. Phys.* **3**, 195 (1995).

421. F. Toschi, G. Amati, S. Succi, R. Benzi and R. Piva, *Phys. Rev. Lett.* **82**, 5044 (1999).
422. J. E. Moyal, *Proc. Camb. Phil. Soc.* **48**, 329 (1952).
423. P. Manneville, *Dissipative Structures and Weak Turbulence* (Academic, Boston, 1990).
424. For a more specialized review on surface thermodynamics and three-phase equilibria, see B. Widom, in *Liquids, Freezing, and the Glass Transition*, eds. J. P. Hansen, D. Levesque and J. Zinn-Just (North-Holland, Amsterdam, 1991), p. 507.
425. G. R. Freeman and N. H. March, *J. Chem. Phys.* **109**, 10521 (1998).
426. P. A. Egelstaff and B. Widom, *J. Chem. Phys.* **53**, 2667 (1970).
427. See also R. C. Brown and N. H. March, *Phys. Rep.* **24C**, 77 (1976).
428. J. W. Gibbs, *Collected Works* **1**, 219 (Longman, Green and Co., New York, 1928); see also C. Herring, in *Metal Interfaces* (Am. Soc. Metals, Cleveland, 1952).
429. J. Frenkel, *Kinetic Theory of Liquids* (Dover, New York, 1955).
430. F. H. Stillinger and A. Ben-Naim, *J. Chem. Phys.* **47**, 4431 (1967); see also C. A. Croxton, *Statistical Mechanics of the Liquid Surface* (Wiley, Chichester, 1980).
431. J. W. Cahn and J. E. Hilliard, *J. Chem. Phys.* **28**, 258 (1958).
432. B. Widom, *J. Chem. Phys.* **43**, 3892 (1965).
433. A. B. Bhatia and N. H. March, *J. Chem. Phys.* **68**, 4651 (1978).
434. D. G. Triezenberg and R. Zwanzig, *Phys. Rev. Lett.* **28**, 1183 (1972).
435. A. J. M. Yang, P. D. Fleming and J. H. Gibbs, *J. Chem. Phys.* **67**, 74 (1977).
436. N. H. March and M. P. Tosi, *J. Chem. Phys.* **111**, 1786 (1999).
437. A. B. Bhatia and N. H. March, *J. Chem. Phys.* **68**, 1999 (1978).
438. S. Fisk and B. Widom, *J. Chem. Phys.* **50**, 3219 (1969); see also B. U. Felderhof, *Physica* **48**, 451 (1970).
439. B. Widom, *J. Chem. Phys.* **43**, 3892 (1965).
440. D. W. Oxtoby and R. Evans, *J. Chem. Phys.* **89**, 7521 (1988); X. C. Zeng and D. W. Oxtoby, *J. Chem. Phys.* **94**, 4472 (1991); R. Nyquist, V. Talanquer and D. W. Oxtoby, *J. Chem. Phys.* **103**, 1175 (1995).
441. R. Evans, *Adv. Phys.* **28**, 143 (1979).
442. L. Gránásy, Z. Jurek and D. W. Oxtoby, *Phys. Rev.* **E62**, 7486 (2000).
443. M. Iwamatsu and K. Horii, *Aerosol. Sci. Technol.* **27**, 563 (1997).
444. J. G. Kirkwood and F. Buff, *J. Chem. Phys.* **17**, 338 (1949).
445. I. Egry, *Scr. Metall. Mat.* **26**, 1349 (1992).
446. M. Born and H. S. Green, *Proc. Roy. Soc. London* **A190**, 455 (1947).
447. N. H. March and J. A. Alonso, *Molec. Phys.* **95**, 353 (1998); J. A. Alonso and N. H. March, *Phys. Rev.* **E60**, 4125 (1999).
448. J. A. Barker, *Molec. Phys.* **80**, 815 (1993).
449. See e.g. M. Born and E. Wolf, *Principles of Optics* (Pergamon, Oxford, 1975).
450. G. H. Gilmer, W. Gilmore, J. Huang and W. W. Webb, *Phys. Rev. Lett.* **14**, 491 (1965).

451. R. A. L. Jones and R. W. Richards, *Polymers at Surfaces and Interfaces* (Cambridge University Press, Cambridge, 1999).
452. D. A. Edwards, H. Brenner and D. T. Wasan, *Interfacial Transport Processes and Rheology* (Butterworth-Heinemann, Boston, 1991).
453. R. Dorshow and L. Turkevich, *J. Chem. Phys.* **98**, 5762 (1993).
454. J. U. Brackbill, D. B. Kothe and C. Zemach, *J. Comput. Phys.* **100**, 335 (1992).
455. Y. C. Chang, T. Y. Hou, B. Merriman and S. Osher, *J. Comput. Phys.* **124**, 449 (1996).
456. B. T. Nadiga and G. Zaleski, *Eur. J. Mech.* **B15**, 885 (1996).
457. Q. Zou and X. He, *Phys. Rev.* **E59**, 1253 (1999).
458. A. Lerda, *Anyons* (Springer, Berlin, 1992).
459. N. Bogoliubov, *J. Phys. USSR* **11**, 23 (1947).
460. R. P. Feynman, *Phys. Rev.* **94**, 264 (1954); R. P. Feynman, in *Progress in Low Temperature Physics*, Vol. 1, eds. C. J. Gorter and D. F. Brewer (North-Holland, Amsterdam, 1955), p. 17.
461. See e.g. A. J. Leggett, *Rev. Mod. Phys.* **71**, S318 (1999).
462. J. Gavoret and P. Noziéres, *Ann. Phys.* **28**, 349 (1964).
463. F. Y. Wu and E. Feenberg, *Phys. Rev.* **122**, 739 (1961).
464. L. Reatto and G. V. Chester, *Phys. Rev.* **155**, 88 (1967).
465. E. Feenberg, *Theory of Quantum Fluids* (Academic, New York, 1969).
466. R. P. Feynman and M. Cohen, *Phys. Rev.* **102**, 1189 (1956).
467. D. Pines and P. Noziéres, *Theory of Quantum Fluids* (Benjamin, New York, 1966).
468. H. W. Jackson and E. Feenberg, *Rev. Mod. Phys.* **34**, 686 (1962).
469. N. R. Werthamer, *Phys. Rev. Lett.* **28**, 1102 (1972); H. Horner, *Phys. Rev. Lett.* **29**, 556 (1972); H. R. Glyde, *Can. J. Phys.* **52**, 2281 (1974); V. Tozzini and M. P. Tosi, *Europhys. Lett.* **32**, 67 (1995).
470. A. Griffin, *Excitations in a Bose-Condensed Liquid* (Cambridge University Press, Cambridge, 1993).
471. D. E. Galli, E. Cecchetti and L. Reatto, *Phys. Rev. Lett.* **77**, 5401 (1996).
472. H. E. Hall, *Adv. Phys.* **9**, 89 (1960); G. W. Rayfield and F. Reif, *Phys. Rev.* **136**, A1194 (1964).
473. G. V. Chester, R. Metz and L. Reatto, *Phys. Rev.* **175**, 275 (1968).
474. A. L. Fetter and A. A. Svidzinsky, *J. Phys.: Condens. Matter* **13**, R135 (2001).
475. E. A. Cornell, J. R. Ensher and C. E. Wieman, in *Bose–Einstein Condensation in Atomic Gases* (IOS, Amsterdam, 1999), p. 15.
476. W. Ketterle, D. S. Durfee and D. M. Stamper-Kurn, in *Bose–Einstein Condensation in Atomic Gases* (IOS, Amsterdam, 1999), p. 67.
477. D. Kleppner, T. J. Greytak, T. C. Killian, D. G. Fried, L. Willmann, D. Landhuis and S. C. Moss, in *Bose–Einstein Condensation in Atomic Gases* (IOS, Amsterdam, 1999), p. 177.
478. W. Ketterle and E. A. Cornell, *Coherent Matter Waves*, eds. R. Kaiser, C. Westbrook and F. David (EDP Sciences, Paris, 2001).
479. C. C. Bradley, C. A. Scakett and R. G. Hulet, *Phys. Rev. Lett.* **78**, 985 (1997).

480. F. Pereira Dos Santos, J. Léonard, J. Wang, C. J. Barrelet, F. Perales, E. Rasel, C. S. Unnikrishnan, M. Leduc and C. Cohen-Tannoudji, *Phys. Rev. Lett.* **86**, 3459 (2001).
481. S. R. de Groot, G. J. Hooyman and C. A. ten Seldam, *Proc. R. Soc. London A* **203**, 266 (1950); V. Bagnato, D. A. Pritchard and D. Kleppner, *Phys. Rev.* **A35**, 4354 (1987). We also remark that in a rigorous viewpoint, the critical behaviour known as Bose–Einstein condensation may only arise when we take the limit $N \to \infty$, since all thermodynamic quantities are analytical in a finite system. This implies some smoothing of the phase transition around the N-dependent critical temperature estimated in the text.
482. G. Baym and C. J. Pethick, *Phys. Rev. Lett.* **76**, 6 (1996).
483. D. S. Jin, J. R. Ensher, M. R. Matthews, C. E. Wieman and E. A. Cornell, *Phys. Rev. Lett.* **77**, 420 (1996); M.-O. Mewes, M. R. Andrews, N. J. van Druten, D. M. Kurn, D. S. Durfee, C. G. Townsend and W. Ketterle, *Phys. Rev. Lett.* **77**, 988 (1996); S. Stringari, *Phys. Rev. Lett.* **77**, 2360 (1996); M. Edwards, P. A. Ruprecht, K. Burnett and C. W. Clark, *Phys. Rev.* **A54**, 4178 (1996).
484. J. R. Ensher, D. S. Jin, M. R. Matthews, C. E. Wieman and E. A. Cornell, *Phys. Rev. Lett.* **77**, 4984 (1996); A. Minguzzi, S. Conti and M. P. Tosi, *J. Phys.: Condens. Matter* **9**, L33 (1997).
485. M. R. Andrews, C. G. Townsend, H. J. Miesner, D. S. Durfee, D. M. Kurn and W. Ketterle, *Science* **275**, 637 (1997).
486. B. P. Anderson and M. A. Kasevich, *Science* **281**, 1686 (1998); I. Bloch, T. W. Hänsch and T. Esslinger, *Phys. Rev. Lett.* **88**, 3008 (1999); E. W. Hagley, L. Deng, M. Kozuma, J. Wen, K. Helmerson, S. L. Rolston and W. D. Phillips, *Science* **283**, 1076 (1999).
487. S. Burger, F. S. Cataliotti, C. Fort, F. Minardi, M. Inguscio, M. L. Chiofalo and M. P. Tosi, *Phys. Rev. Lett.* **86**, 4447 (2001).
488. K. W. Madison, F. Chevy, W. Wohlleben and J. Dalibard, *Phys. Rev. Lett.* **84**, 806 (2000); F. Chevy, K. W. Madison and J. Dalibard, *Phys. Rev. Lett.* **85**, 2223 (2000); K. W. Madison, F. Chevy, V. Bretin and J. Dalibard, *Phys. Rev. Lett.* **86**, 4443 (2001); J. R. Abo-Shaeer, C. Raman, J. M. Vogels and W. Ketterle, *Science* **292**, 476 (2001); S. Inouye, S. Gupta, T. Rosenband, A. P. Chikkatur, A. Görlitz, T. L. Gustavson, A. E. Leanhardt, D. E. Pritchard and W. Ketterle, *Phys. Rev. Lett.* **87**, 080402 (2001).
489. L. D. Landau, *Sov. Phys. JETP* **3**, 920 (1957).
490. W. R. Abel, A. C. Anderson and J. C. Wheatley, *Phys. Rev. Lett.* **17**, 74 (1966).
491. See e.g. M. P. Tosi, in *Electron Correlations in the Solid State*, ed. N. H. March (Imperial College Press, London, 1999), p. 1.
492. J. Friedel, *N. Cimento Suppl.* **7**, 287 (1958).
493. J. Langer and S. H. Vosko, *J. Phys. Chem. Solids* **12**, 196 (1960); N. H. March and A. M. Murray, *Phys. Rev.* **120**, 830 (1960).
494. I. Langmuir, *Proc. Nat. Acad. Sci.* **14**, 627 (1928).

495. P. C. Gibbons, S. E. Schnatterly, J. J. Ritsko and J. R. Fields, *Phys. Rev.* **B13**, 2451 (1976).
496. E. P. Wigner, *Phys. Rev.* **46**, 1002 (1934); *Trans. Faraday Soc.* **34**, 678 (1938).
497. D. M. Ceperley and B. J. Alder, *Phys. Rev. Lett.* **45**, 566 (1980); G. Ortiz, M. Harris and P. Ballone, *Phys. Rev. Lett.* **82**, 5317 (1999).
498. B. Tanatar and D. M. Ceperley, *Phys. Rev.* **B39**, 5005 (1989); D. Varsano, S. Moroni and G. Senatore, *Europhys. Lett.* **53**, 348 (2001).
499. C. C. Grimes and G. Adams, *Phys. Rev. Lett.* **42**, 795 (1979).
500. H. Buhmann, W. Joss, K. von Klitzing, I. V. Kukushkin, A. S. Plaut, G. Martinez, K. Ploog and V. B. Timofeev, *Phys. Rev. Lett.* **66**, 926 (1991).
501. S. V. Kravchenko, G. V. Kravchenko, J. E. Furneaux, V. M. Pudalov and M. D'Iorio, *Phys. Rev.* **B50**, 8039 (1994); S. V. Kravchenko, D. Simonian, M. P. Sarachik, W. Mason and J. E. Furneaux, *Phys. Rev. Lett.* **77**, 4938 (1996).
502. J. Yoon, C. C. Li, D. Shahar, D. C. Tsui and M. Shayegan, *Phys. Rev. Lett.* **82**, 1744 (1999).
503. D. R. Tilley and J. Tilley, *Superfluidity and Superconductivity* (Van Nostrand, New York, 1974).
504. J. Bardeen, L. N. Cooper and J. R. Schrieffer, *Phys. Rev.* **106**, 162; *ibid.* **108**, 1175 (1957); J. R. Schrieffer, *Theory of Superconductivity* (Benjamin, Reading, 1964).
505. L. N. Cooper, *Phys. Rev.* **104**, 1189 (1956).
506. G. Baym, C. J. Pethick and D. Pines, *Nature* **224**, 673 (1969).
507. P. G. de Gennes, *Superconductivity of Metals and Alloys* (Benjamin, New York, 1966).
508. B. D. Josephson, *Phys. Rev. Lett.* **1**, 251 (1962).
509. See e.g. D. Vollhardt and P. Wolfle, *The Superfluid Phases of Helium 3* (Taylor and Francis, London, 1990).
510. See e.g. A. J. Leggett, in *Quantum Liquids*, eds. J. Ruvalds and T. Regge (North-Holland, Amsterdam, 1978), p. 167.
511. N. H. March, *Current Science* **75**, 1246 (1998); see also N. H. March and M. P. Tosi, *Laser Part. Beams* **16**, 71 (1998).
512. N. H. March, *Liquid Metals — Concepts and Theory* (Cambridge University Press, Cambridge, 1990).
513. L. Reatto, *Phil. Mag.* **95**, 353 (1998).
514. See e.g. R. Winter, W. C. Pilgrim and F. Hensel, *J. Phys.: Condens. Matter* **6**, A245 (1994).
515. P. A. Egelstaff, N. H. March and N. C. McGill, *Can. J. Phys.* **52**, 1651 (1974).
516. G. A. de Wijs, G. Pastore, A. Selloni and W. van der Lugt, *Phys. Rev. Lett.* **75**, 4480 (1995).
517. G. Baym, *Phys. Rev.* **135**, A1691 (1964).
518. J. M. Ziman, *Phil. Mag.* **6**, 1013 (1961).
519. E. N. da C. Andrade, *Phil. Mag.* **17**, 497 (1934); see also R. C. Brown and N. H. March, *Phil. Chem. Liq.* **1**, 141 (1966) and T. E. Faber, Ref. 23.

520. See e.g. N. W. Ashcroft, *J. Phys.: Condens. Matter* **12**, A129 (2000).
521. S. T. Weir, A. C. Mitchell and W. J. Nellis, *Phys. Rev. Lett.* **76**, 1860 (1996); G. W. Collins, L. B. DaSilva, P. Celliers, D. M. Gold, M. E. Foord, R. J. Wallace, A. Ng, S. V. Weber, K. S. Budil and R. Cauble, *Science* **281**, 1178 (1998).
522. D. Saumon and G. Chabrier, *Phys. Rev.* **A46**, 2084 (1992).
523. W. R. Magro, D. M. Ceperley, C. Pierleoni and B. Bernu, *Phys. Rev. Lett.* **76**, 1240 (1996).
524. M. Ross, *Rep. Progr. Phys.* **48**, 1 (1985).
525. P. A. Egelstaff, D. I. Page and C. R. T. Heard, *J. Phys.* **C4**, 1453 (1971).
526. J. G. Kirkwood, *J. Chem. Phys.* **3**, 300 (1935).
527. M. Born and H. S. Green, *Proc. R. Soc. (London)* **A188**, 10 (1946).
528. K. I. Golden, N. H. March and A. K. Ray, *Molec. Phys.* **80**, 915 (1993).
529. See e.g. X. Y. Liu, P. van Hoof and P. Bennema, *Phys. Rev. Lett.* **71**, 109 (1993).
530. E. B. Sirota, *J. Chem. Phys.* **112**, 492 (2000); see also E. B. Sirota and A. B. Herhold, *Science* **283**, 529 (1999).
531. W. Ostwald, *Z. Physik. Chem.* **22**, 286 (1897).
532. A. R. Gerson, K. J. Roberts, J. N. Sherwood, A. M. Taggart and G. Jackson, *J. Cryst. Growth* **128**, 1176 (1993).
533. E. B. Sirota and M. D. Singer, *J. Chem. Phys.* **101**, 10873 (1994); E. B. Sirota, *Langmuir* **14**, 3133 (1998).
534. E. B. Sirota, *J. Chem. Phys.* **112**, 402 (2000).
535. H. Marmanis, *Phys. Fluids* **10**, 1428 and 3031 (1998).
536. G. Russakoff, *Amer. J. Phys.* **38**, 1188 (1970).
537. H. W. Wyld, *Ann Phys. (NY)* **14**, 143 (1961).
538. L. L. Lee, *Ann Phys. (NY)* **32**, 292 (1965).
539. R. Phythian, *J. Phys.* **A2**, 181 (1969); see also S. F. Edwards, *J. Fluid Mech.* **18**, 239 (1964) and J. R. Herring, *Phys. Fluids* **8**, 2219 (1965).
540. R. Kraichnan, *J. Fluid Mech.* **5**, 497 (1958); *Phys. Fluids* **7**, 1723 (1964).
541. J. M. Burgers, *Proc. Roy. Neth. Acad. Soc.* **43**, 2 (1940).
542. M. Vergassola, B. Dubrulle, U. Frisch and A. Noullez, *Astron. Astrophys.* **289**, 325 (1994).

Index